MW00761155

Multichip Module Design, Fabrication, and Testing

Electronic Packaging and Interconnection Series
Charles M. Harper, Series Advisor

Related Books of Interest

To order or receive additional information on these or any other McGraw-Hill titles, in the United States please call 1-800-822-8158. In other countries, contact your local McGraw-Hill representative.

BC14BCZ

Multichip Module Design, Fabrication, and Testing

James J. Licari

McGraw-Hill, Inc.

New York   San Francisco   Washington, D.C.   Auckland   Bogotá
Caracas   Lisbon   London   Madrid   Mexico City   Milan
Montreal   New Delhi   San Juan   Singapore
Sydney   Tokyo   Toronto

Library of Congress Cataloging-in-Publication Data

Licari, James J., date.
  Multichip module design, fabrication, and testing / James J.
Licari.
        p.   cm.
  Includes index.
  ISBN 0-07-037715-4
    1. Electronic packaging.   2. Multichip modules (Microelectronics)—
Design and construction.   3. Multichip modules (Microelectronics)—
Testing.   I. Title.
TK7870.15L53   1995
621.381'046—dc20   94-27455
                                                                  CIP

1 2 3 4 5 6 7 8 9 0   DOC/DOC   9 0 9 8 7 6 5 4

ISBN 0-07-037715-4

*The sponsoring editor for this book was Stephen S. Chapman, the editing
supervisor was David E. Fogarty, and the production supervisor was
Pamela A. Pelton. It was set in Palatino by McGraw-Hill's Professional Book
Group composition unit.*

Printed and bound by R. R. Donnelley & Sons Company.

Contents

Contributors

Magdy S. Abadir High Value Electronics Division, Microelectronics and Computer Technology Corporation (CHAPTER 2)

Howard Davidson Distinguished Engineer, Sun Microsystems (CHAPTER 8)

Hugh L. Garvin Head of Thin Film Processes and Integrated Circuit Packaging Department, Exploratory Studies Laboratory of Hughes Research Laboratory, Malibu, Calif. (CHAPTER 7)

Charles P. Minning Senior Scientist, Hughes Aircraft, Electro-Optical Systems Group, Aerospace and Defense Sector (CHAPTER 5)

Peter A. Sandborn Multichip Systems Design Advisor Project and High Value Electronics Division, Microelectronics and Computer Technology Corporation (CHAPTER 2)

Foreword

It has now become clearly established that the packaging and interconnecting of groups of bare semiconductor chips offer the greatest advantage in both packaging density and electronic product performance.
Further, it has recently become clear that the multichip module (MCM)
version of microelectronic packaging technology offers at least an order-
of-magnitude improvement over earlier multichip packaging and interconnection technologies. The challenge, however, is to achieve reliable
design and fabrication, materials, processes, and technologies to reproducibly manufacture the small physical features which are the heart of
MCMs. Earlier forms of multichip microelectronic packages are still
widely used under the generic term *hybrid microcircuits*. These hybrid circuits usually consist of a group of less complex bare semiconductor chips
and several other miniature components mounted and interconnected on
a substrate formed by depositing thick- or thin-film circuitry. MCM technology is an advanced extension of bare-chip hybrid technolgy. The two
major distinctions between hybrid technology and MCM technology are
(1) the chip-to-substrate area ratios that can now be achieved (>30 percent) with MCMs due largely to the fine-line patterning that has been
developed (<100-μm pitch, for example, 25-μm-wide lines and 75-μm
spacings for thin films) and (2) the ability to multilayer thin-film interconnect substrates using low-dielectric-constant dielectrics such as polyimides, benzocyclobutene (BCB), or silicon dioxide. Three forms of substrates and deposited circuits lead to three categories of MCMs, namely,
MCM-D for deposited thin-film circuitry (as in integrated circuits),
MCM-C for fine-line thick-film circuitry (as in hybrids), and MCM-L for
fine-line circuitry on plastic laminates (as in printed-wiring boards). The
fine features of all these MCM forms can only be achieved reliably

through precise design and fabrication techniques. This book, then, is dedicated to providing design and fabrication information and guidelines for best achieving reliable, high-performance MCMs—so important for emerging and future electronic systems.

Regarding subjects covered, this book addresses all the critical elements for producing high-performance MCMs, from design through fabrication and finally testing and repair. The book begins with two excellent chapters covering an introduction to multichip modules and fundamentals for multichip module design. This is followed by a set of chapters dealing with the very important considerations for materials and processes for interconnect substrates. Next comes a chapter on thermal management for these high-heat-density modules, a chapter on the assembly manufacturing process, and a chapter on the leading-edge three-dimensional packaging concepts, including microwave packaging. Finally, two excellent chapters are devoted sequentially to inspection and test, and rework and repair. The sum of all this is an outstanding presentation of MCM design and fabrication from front end to product completion.

CHARLES A. HARPER
President, Technology Seminars, Inc.
Lutherville, Maryland

Preface

What are multichip modules (MCMs)? Why all the fuss about them? Why didn't we have them 20 years ago?

Well, as a matter of fact, we've had them all the while, since the invention of the integrated circuit, but they were called *hybrid microcircuits*. A hybrid circuit is an interconnection and packaging approach in which several bare chips are interconnected to form a functional circuit. What are called MCMs today are but extensions of hybrid circuits which provide at least an order-of-magnitude increase in density and/or performance. For the past 20 years we have been too busy designing and developing more complex, higher-functionality dice going from SSI to LSI to VHSIC and VLSIC, ever increasing the number of transistors, gates, bits, and clock speeds per square centimeter. So fascinated have we been with the advancements in ICs that we have neglected to address the packaging of these chips. We are now faced with a situation analogous to buying a high-speed high-performance car—a Lamborghini—and trying to drive it on a Los Angeles freeway. One needs a clear, unobstructed racetrack to get the maximum performance from such a car or a VHSIC chip. Hybrid circuits gave us an order-of-magnitude reduction in weight and volume and an increase in reliability over printed wiring assemblies. By interconnecting bare dice instead of prepackaged dice, a whole level of interconnections was eliminated together with all the associated packages. This resulted in improved reliability and reduced costs as well as increased density. With hybrid microcircuits, the semiconductor dice were small, approximately 50 to 100 mils square and were well spaced on the interconnect substrate. The silicon-to-substrate area ratio seldom exceeded 5 percent and more often was only 1 to 2 percent. Clock speeds for digital circuits seldom exceeded 20 MHz. Now with the development of large ASIC, memory, and high-speed chips, there is a need for denser packaging and packaging that will elicit the full performance of the chip. MCMs are the new generation of hybrid microcircuits that fullfill these needs. Current MCMs with their fine-line interconnections are achieving 30 to 80 percent silicon-to-substrate area ratios and are even exceeding 100 percent for three-dimensional structures. Also MCMs are capable of operating at clock speeds greater than 100 MHz.

This book combines a mix of mature technologies that have been used for many years for hybrid microcircuits and are the basics for MCMs with leading-edge unique technologies that are specific to multichip modules. The latter technologies are rapidly changing. Hence the book should be considered a snapshot in time. Although several other books have been written on MCMs, this book presents the perspective of the author whose career in microelectronics has spanned 35 years, beginning in the early 1960s with the application of the newly invented IC, through printed-wiring boards, hybrid microcircuits, and now MCMs. Subjects which normally are briefly touched upon have been elevated to chapter status because of their greater importance in advanced electronics. Among these are rework and three-dimensional modules.

In writing this book, I was faced with choosing the sequence for the chapters. Should *design* be treated first, or should *materials and processes*? Can *materials* be treated separately from *processes*? It is difficult to discuss design without an understanding of materials, processes, and *testing*. Further, where does *thermal management* come in? The problem with advanced electronics is that it has become increasingly difficult to treat the many technologies involved sequentially as separate elements. Rather, the technologies should be treated as a totality because of their interdependence. In practice, however, first electronic modules are designed, then materials and processes are selected, and the substrate is fabricated. This is followed by assembly, testing, and rework, if any. I have opted for this more traditional sequence, starting with design. Therefore, there will be some overlap among the chapters which should reinforce the individual texts.

General Notes

- The terms *die, chip, device, IC,* and *component* are often used interchangeably. Dice is used as the plural of *die*.

- Although some effort was made to standardize on units of measurement, this was not always possible. To assist the reader, tables of conversion factors are provided in Appendix A.

- Numerous abbreviations and acronyms are common in the electronics industry. A list of those used in this book with meanings is given in Appendix B.

Acknowledgments

Multichip module technology is much more multidisciplinary than previous electronic packaging approaches. Electrical performance and thermal performance are now an integral part of the packaging, and an understanding of the interrelationships between electrical properties and materials properties, even at the molecular and atomic levels, assumes importance. A comprehensive book on MCMs therefore cannot be written completely by one person since so many specialties are involved. Recognizing this, I have relied on several experts to write chapters that complement mine. I am very grateful to Dr. Peter Sandborn and Magdy Abadir of MCC for writing Chap. 2 on MCM design, to Dr. Charles Minning of Hughes Aircraft for Chap. 5 on thermal management, to Dr. Howard Davidson of Sun Microsystems for Chap. 8 on inspection and testing, to Dr. Hugh Garvin of Hughes Research Laboratory for Chap. 7 on three-dimensional packaging, to Raymond Brown for his contributions to the LTCC portions of Chaps. 3 and 4, and to Helen Congleton for reviewing some of the material.

J. J. LICARI

1

Introduction

1.1 Hybrid Microcircuits: The Basic Technology for Multichip Modules

Hybrid microcircuit thin- and thick-film technologies, known and used for over 30 years, are the basics for *multichip modules* (MCMs). In fact, in spite of all the definitions proposed for MCMs to distinguish them from hybrid circuits and in spite of their wide publicity, MCMs are essentially an extension of hybrid circuits. Basically, MCMs may be considered the new generation of hybrid circuits that provides at least an order-of-magnitude higher electrical performance and orders-of-magnitude reductions in weight and volume. Most of the fabrication and assembly processes used for MCMs are the same as those used for hybrids, except that for MCMs these processes have been extended to produce finer conductor lines, spacings, and vias and to incorporate low-dielectric-constant dielectrics that reduce crosstalk and increase signal speeds. For years, bare chips have been interconnected on thick-film multilayer ceramic substrates, thick-film cofired ceramic multilayer substrates, or single-layer thin-film ceramic substrates. However, two distinctions between hybrid circuits and MCMs may be drawn: (1) The high silicon-to-substrate area ratios that can now be achieved (>30 percent) with MCMs are due largely to the fine-line patterning that has been developed (<100-μm pitch, for example, 25-μm-wide lines and 75-μm spacings for thin films). (2) Multilayer thin-film interconnect substrates can be created by using low-dielectric-constant dielectrics such as polyimides, benzocyclobutene (BCB), or silicon dioxide. Multichip modules in which the interconnect substrate is fabricated by thin-film multilayer processes are referred to as MCM-D (the D is for *deposited*).

Advances in thick-film fine-line and in plastic laminate fine-line interconnect substrates are now paralleling those of thin-film, and multichip modules produced by these technologies are referred to as MCM-C (the C is for *ceramic*) and MCM-L (*laminate*), respectively. There is a trend toward using combinations of the three basic technologies. For example, because cofired ceramic thick film is limited in the degree of line resolution that can be produced (generally 5-mil lines and spacings), the top several layers can be transitioned to thin film to accommodate IC chips having I/O pitches of 4 mils. IBM's thermal conduction module falls in this category with its 63 conductor layers cofired with a glass ceramic and its top layers fabricated by using the thin-film polyimide copper process. Furthermore, the three technologies are merging, and compromises are being made to reduce costs yet maintain the performance and density benefits. Hence, a design engineer should be familiar with the basic processes and capabilities of thick-film, thin-film, and printed-wiring board processes in order to select the lowest-cost process commensurate with the electrical, thermal, mechanical, and environmental requirements of the system. These processes are dealt with extensively in several books and will be treated here only to the extent that they apply directly to the fabrication of MCMs.[1-6]

1.2 Multichip Module Drivers

There are two main motivations for the use of MCMs: high electrical performance to match that of high-speed integrated circuits (ICs) and increased circuit density. Improved performance derives from the choice of materials, design, and layout of the interconnect substrate to match the impedance of the ICs and to reduce signal paths and interconnections. Hence performance is a direct function of the close packing of ICs and short interconnect paths. However, for many commercial and/or consumer applications, density alone is the driver.

With the development and realization of high-performance, very high-density integrated circuits such as very large-scale integrated circuits (VLSICs) and very high-speed integrated circuits (VHSICs), and GaAs integrated circuits, the classical interconnect and packaging approaches became limiting factors. Dice sizes have approached 1 inch square—the size that an entire hybrid substrate used to be (Fig. 1-1). I/Os per die of 200 to 500 are now common and are projected to surpass 1000 in the next several years (Fig. 1-2). I/O pitches have also been reduced to 4 mils, and IC dice having millions of gates and tran-

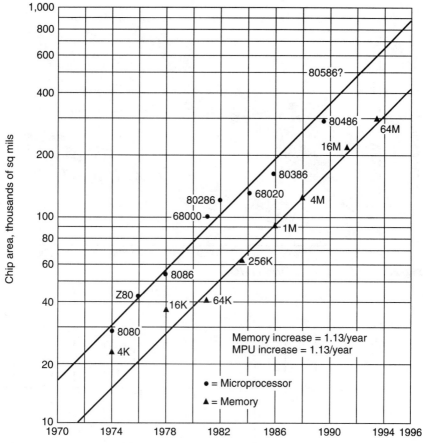

Figure 1-1. Size trends for ICs. (*Source: Intel/ICE.*)

sistors capable of storing megabits of memory are now commonplace (Fig.1-3). Clock speeds have increased beyond 100 MHz (Fig. 1-4). Because of these advances at the die level, the electrical, thermal, and density characteristics of the interconnect substrates and of the packages have had to be readdressed. Besides their traditional roles of providing mechanical support and environmental protection, substrates and packages must now also assume a role in the electrical functioning of the circuit. Capacitance, impedance, reflection losses, and inductance are key to achieving the high-speed performance of the IC dice and of the circuit as a whole. Further, conductor line and via pitches of 10 mils or greater, typical of printed-wiring boards and thick-film hybrid circuits, are too gross for interconnecting the new generation of high-performance IC chips with their micron-size I/O pads.

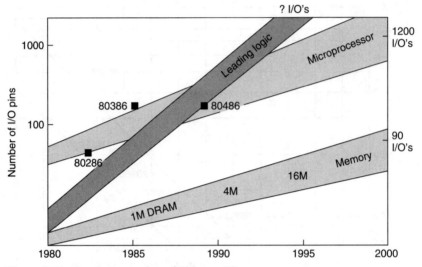

Figure 1-2. Trends in number of I/Os per IC.

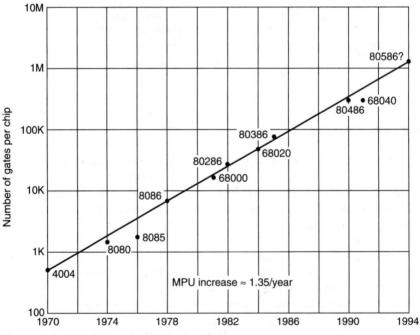

Figure 1-3. Trends in gate density per chip.

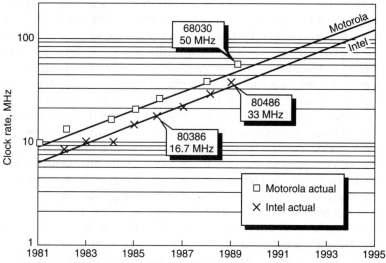

Figure 1-4. Microprocessor clock rates. (*Source: Dataquest.*)

To meet these demands, multichip modules were developed. MCMs are, in reality, advanced hybrid microcircuits designed and constructed to maximize die performance and achieve a higher level of miniaturization and integration. MCMs fit the definition of a hybrid circuit in that the bare dice of various functional types are interconnected and integrated on an interconnect substrate. The main difference is that the MCM substrate is designed and fabricated to minimize parasitics, crosstalk, signal reflections, and signal paths. The main characteristics of MCMs may thus be summarized as follows:

- MCMs contain several bare chip devices, generally application-specific integrated circuits (ASICs) or VLSICs.

- Chips are mounted on high-density interconnect substrates having fine conductor lines and spacings, generally with 4-mil (100-μm) pitch or less.

- There is a high ratio of silicon die area to substrate area, greater than 30 percent and as high as 80 to 90 percent; it is over 100 percent for three-dimensional stacked modules (see Chap. 7).

- There are controlled dimensions for controlled impedance.

- Interlayer dielectric has a low dielectric constant, preferably lower than 4.

- High-density multilayered circuits have short interconnect paths and high routing densities. Line densities range from 250 to over 500 per linear inch.

Thin-film multilayer interconnect substrates in which the capacitance and crosstalk of the dielectric are minimized through the use of low-*k* materials and controlled geometries meet these requirements. Early attempts at fabricating large-area multilayer thin-film substrates in the 1960s were abandoned because of the inability to deposit large-area defect-free and stress-free films, either organic or inorganic. The driving force at that time was primarily high density more than performance since VLSICs had not yet been developed. Fortunately, thick-film multilayer substrates met most of the requirements for the ICs of that period. Today, with advances in plasma chemical vapor deposition (CVD), plasma etching, thermally and dimensionally stable polymers having low dielectric constants, and clean-room improvements, the stage was set for the emergence of the high-density MCM-D modules. Line widths of 25 μm and spacings of 75 μm are now fairly routine and are being reduced even further to 15-μm lines and 35-μm spacings. These dimensions cannot be achieved yet with other processes such as thick-film or printed wiring boards. The reduction in size, volume, weight, and number of modules per system achieved by using MCMs is indeed the greatest ever achieved in the history of electronic packaging. Figure 1-5 shows an analysis of how 28 conventional printed-circuit assemblies for a data processor can be replaced by five MCMs and result in increased performance. Figure 1-6 shows a 12-to-1 reduction in a radar signal processing system from 12 boards to one 6 × 9 in *standard avionics module* (SAM) containing only six MCMs (three

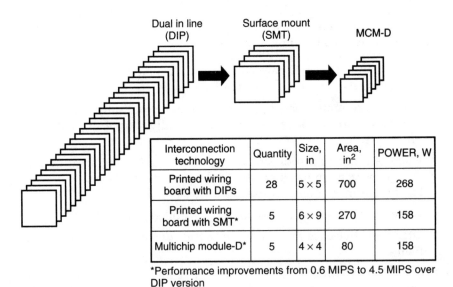

Interconnection technology	Quantity	Size, in	Area, in^2	POWER, W
Printed wiring board with DIPs	28	5 × 5	700	268
Printed wiring board with SMT*	5	6 × 9	270	158
Multichip module-D*	5	4 × 4	80	158

*Performance improvements from 0.6 MIPS to 4.5 MIPS over DIP version

Figure 1-5. MCM-D insertion effect for data processor.

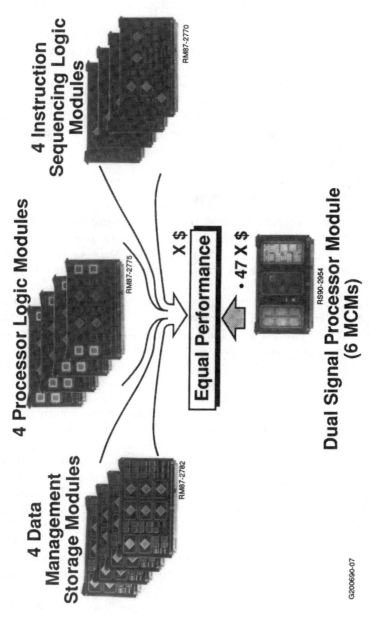

Figure 1-6. Size and cost reduction for radar systems modules (12-to-1 size reduction at half cost). (*Courtesy Hughes Aircraft Co.*)

on each side of the SAM). Each MCM contains a 2 × 4 in interconnect substrate constructed of five conductor layers (ground plane, power plane, two signal levels, and a top metallization for die attachment and interconnection).

Other MCMs are less complex consisting of only a few IC chips interconnected and packaged in small packages and treated and sold as a single component. These "few chip" MCMs may be packaged in cofired ceramic, in transfer-molded plastic, or in metal packages and are generally mass-produced and treated as throwaways, if they fail. The cost of these few-chip MCMs is not so much dependent on the substrate technology as on the cost and testing of devices and the labor in assembling and packaging.

1.3 Historical Background

It is hard to say when MCMs came into being because their evolution from hybrid circuits is rather fuzzy. Some attribute the first serious work in very high-density fine-line circuits (MCM-D) to Honeywell, Kyocera, and NEC in the early to mid-1980s.[7-9] Others consider the multilayer cofired ceramic modules of the early 1970s produced by IBM and several Japanese firms for their mainframe computers as the first multichip modules. These early modules were fabricated from high-temperature cofired alumina and had as many as 73 layers of refractory metal conductors (tungsten). The thick films were transitioned to fine-line thin-film circuitry for the top layers in order to accommodate fine-pitch closely spaced IC chips. More recently, IBM is producing *thermal conduction modules* (TCMs) from a low-temperature cofired glass-ceramic tape that is compatible with copper conductors.[10]

Thin-film multilayer interconnect substrates, although attempted over 30 years ago, did not materialize until recently for several reasons. First, the IC dice of the 1960s and 1970s were not high-density or high-speed and did not demand advances in packaging beyond what was then available. Second, the multilayer thick-film and the single-layer thin-film processes available at the time satisfied most of the high-density digital and analog circuit requirements and were quite inexpensive. Third, the technologies for fabricating large-area thin-film multilayer substrates were not developed. For example, the early polymeric and inorganic dielectrics could not be deposited as thin films over large areas without developing pinholes, other imperfections, and stresses. In the early 1960s, attempts were made to form thin-film multilayer interconnect substrates by depositing inorganic dielectric insulation such as silicon dioxide on silicon wafers by chemical vapor deposition. Attempts were also made to form an oxide dielectric on aluminum by anodizing. In both approaches the oxide dielectric was too porous to provide a reliable separation between the conductor layers. The anodized aluminum oxide was also hygroscopic, absorbing water and changing in capacitance and electrical insulating properties. A successful precursor of MCM-D might be considered the air-gap microbridge interconnect substrates developed by Bell Laboratories in 1968.[11,1] Although these structures consisted of only two levels of conductors (actually crossovers), the separation by air provided high performance, and the hundreds to thousands of microbridges provided a large increase in density. Later, structures were made in which the microbridges were supported with polyimide, and thin-film resistors were integrated.[12]

One of the first successful attempts at developing an MCM-D, in the sense that we understand it today, was made by Licari and Meinhard in 1971 under a NASA–Marshall Space Flight Center contract[13] but was not reported in the open literature. In that program, both Vespel SP (a molded polyimide from Du Pont) and aluminum were used as the substrates. Du Pont's polyimide ML varnish coating was then spin-coated and baked to remove solvents. Next, vias were wet-etched in the polyimide with an ammoniacal alkaline solution through a via-patterned photoresist as a conformal mask. On removing the photoresist, the remaining polyimide precursor was fully imidized by heating at 400°C. Aluminum or copper was then sputter-deposited and photoetched to metallize the vias and create the next circuit pattern. It is interesting that this process for wet-etching a polyimide precursor has recently been resurrected in the drive to reduce the cost of processing MCM-Ds.[14,15]

1.4 Classifications of Multichip Modules

So pervasive has been the interest in MCMs that every technical journal and society dealing with electronics is addressing and featuring multichip modules; indeed, each appears to be claiming MCMs as its own. The definition of MCM has thus expanded from the initial multilayer thin-film processes to thick-film ceramic processes and even to fine-line printed-circuit board processes. Societies and journals have even changed their names to reflect the broader electronic packaging concept that MCMs embrace. The interconnect substrate, of course, is the key to high performance and high density, but it must not be forgotten that an MCM is the entire assembled module and includes the processes for die attachment, die bonding, and sealing, even though most of the attention has focused on the substrate. Further, substrates are being integrated with packages or are formed as part of the package as in cofired ceramic, and modules are being integrated with the next level of assembly. Hence the distinctions and boundaries between substrates, package, and next-level motherboard assembly are slowly disappearing

The IPC Institute for Interconnecting and Packaging Electronic Circuits, previously known as the Institute for Printed Circuits, took the initiative in categorizing and defining MCMs.[16] The IPC categorized multichip modules in three groups, primarily according to the dielectric and interconnect substrate used in fabricating the multilayer circuitry. The modules are classified according to the materials and processes used to fabricate the interconnect substrate rather than according to any subsequent assembly or packaging methods. The three categories are

MCM-D (deposited)

MCM-C (ceramic)

MCM-L (laminated)

1.4.1 Multichip Module Deposited

MCM-Ds are constructed by using unreinforced thin-film deposited materials as interlayer dielectrics and metallization. Sequential photolithographic processes, commonly used in the semiconductor industry, are used to apply and pattern the dielectric and metallization. The thin-film metallizations comprising ground planes, voltage planes, signal layers, and die bonding pads are formed by traditional sputtering or vacuum evaporation methods followed by photolithographic patterning. Polyimide coatings are almost universally used as the dielectric to separate the metal layers and to provide the low dielectric constant (less than 5 but typically 3) needed to reduce capacitance. The multilayer circuits are formed on dimensionally stable bases such as alumina ceramic, silicon, aluminum, or aluminum nitride. A variation of MCM-D involves producing the multilayer structure on an expendable mandrel, peeling it from the mandrel as a decal, then reattaching it to a permanent substrate. Parallel processing can also be used in which each layer is individually processed, peeled off, laminated or otherwise bonded with adhesive, and mounted on a permanent substrate. Drilling of through-holes and electroplating are necessary to interconnect the layers after lamination. The key advantage of parallel processing is the ability to inspect and test each layer individually, thus improving the final yield and avoiding the cost of scrapping the module at the last stages of production where most of the value has already been accrued.

1.4.2 Multichip Module Ceramic

MCM-Cs are constructed from ceramic functioning both as the inter-layer dielectric and as the supporting substrate. The ceramic may be in the form of low-temperature cofired ceramic (LTCC) (see Chap. 3), high-temperature cofired ceramic (HTCC), or sequential screen-printed and fired thick film. Widely used ceramic compositions include alumina, alumina-glass, mullite, cordierite, and aluminum nitride. The dielectric constants of these materials generally range from 5 to 10, although some low-temperature cofired ceramics having a dielectric constant of approximately 4 are also available. Thick-film conductor pastes are used with MCM-Cs but must be proved compatible with the ceramic processing temperatures and conditions. For example, only refractory metal conductor pastes can be used with HTCC because the sintering temperatures are very high (1200 to 1500°C) and even so must be fired in a reducing atmosphere. The advantage of LTCC and glass ceramics is that they are compatible with the highly electrically conductive pastes such as copper, silver, and gold because of their relatively low firing temperatures (850 to 950°C). And z-direction vias are formed in the dielectric paste simultaneously with the screen printing process. The vias are subsequently filled with conductor paste, also by screen printing, and then fired. Vias may also be formed by laser ablation, punching, or drilling during the green tape processing. MCM-C involves parallel processing; thus each layer can be inspected and tested prior to lamination and cofiring.

Many companies that started with MCM-D designs later found that there is a limit to how much the designs can reduce the MCM-D costs, and hence companies have redesigned to ceramic MCM-C versions or even to plastic MCM-L versions. Reducing costs and preserving high performance have been a challenge. Honeywell, e.g., designed, built, and compared its five-chip *advanced spaceborne computer module* (ASCM) both in thin-film copper/polyimide and in cofired ceramic (Fig. 1-7). The MCM-C module is 2.1 inches square, is constructed of 14 layers of high-temperature cofired ceramic, has six signal layers plus power and ground planes, and has 200 external leads on 35-mil pitch. In this case, even though the ceramic version requires more layers and has grosser line pitch than its MCM-D counterpart (12 versus 4 mils), the cost is lower and performance was still adequate for the application.

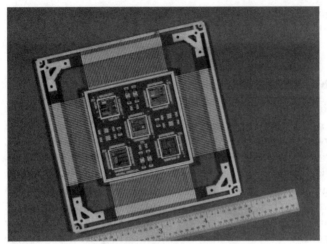

Figure 1-7. MCM-C fabricated from high-temperature cofired ceramic, 5-mil lines on 12-mil pitch. (*Courtesy Honeywell Solid State Electronics Center.*)

1.4.3 Multichip Module Laminate

MCM-Ls are produced by the conventional printed-circuit board processes except that finer conductor lines, spacings, and vias must be formed to achieve higher densities. The dielectric materials may consist of reinforced or unreinforced epoxies, polyimides, or cyanate esters. Copper foil laminates are photoetched to form the conductor patterns; vias are drilled and then electroplated. The layers are processed individually (parallel processing), then stacked and laminated under pressure and temperature, resulting in a monolithic multilayer structure. For the lowest-cost consumer or commercial applications, MCM-L designs should be considered first.

1.4.4 Variations

A trend is emerging in which combinations of the three basic approaches are being used as a compromise among cost, performance, and density. The lower-cost cofired ceramics and plastic laminates can be used for most of the layers provided that a transition can be made to thin-film fine lines for the top layers where fine-pitch ICs are to be closely interconnected. Thus variations such as combinations of MCM-C with D on top and MCM-L with D on top are being produced.[17] Figure 1-8 shows a cross section of a fine-line copper/polyimide multilayer processed on cofired ceramic. Thermal modules for IBM's and

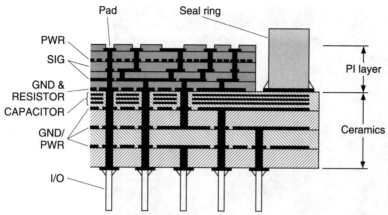

Figure 1-8. Cross section of MCM-D on MCM-C, copper/polyimide multilayer on cofired ceramic substrate. (*Courtesy NTK Technical Ceramics.*)

Fugitsu's mainframe computers are examples of the MCM-D/C combination. Thin films may also be processed for the top bonding layers of plastic laminate multilayers. IBM in Japan has developed this combination of MCM-D/L, which they have named the *surface laminar circuit* (SLC).

Although most MCMs are assembled with bare chips to achieve a high silicon-to-substrate area ratio, the definition of multichip modules has even been expanded to include mixtures of bare dice and prepackaged dice assembled on the same substrate, also closely packed to achieve high density. These combinations are being referred to as MCM-H (multichip module hybrid).

1.5 Applications

Applications for MCMs cover a wide range, from the very low-cost "low-end" consumer products to high-performance "high-end" electronics used in military, space, and medical systems.

1.5.1 High-End MCMs

The MCMs referred to as high-end include the more complex, higher-performance MCMs which generally also cost more. Examples in this category are MCMs for mainframe computers, super-high-speed computers, military radar, communications and navigation systems, spacecraft, satellites, and medical implants. For these applications, MCM-D, MCM-C, or combinations of these two are used. MCM-Ds are pre-

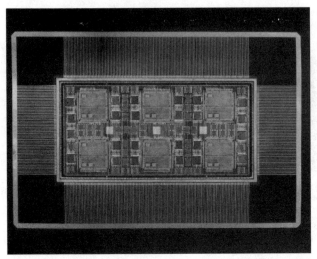

Figure 1-9. Data memory structure MCM-D (1.8 × 3.8 in,
five-conductor-layer aluminum/polyimide substrate).
(*Courtesy Hughes Aircraft Co.*)

ferred where the highest density, finest lines, and highest clock speed
performance are required. High-speed signal and data processors used
in military radar systems are being produced by using MCM-D technologies.[18] An example is Hughes Aircraft's data memory structure
module designed for a radar application (Fig. 1-9). It contains six large
ASIC dice (570 mils-square) assembled with other memory chips and
decoupling capacitors on a 1.8 × 3.8 in, five-conductor-layer alu-
minum/polyimide substrate. By implementing MCMs such as the data
memory structure, Hughes has demonstrated significant improvements in both miniaturization and performance for its F-22 and F-18
radar systems.

In the *signal processor packaging design* (SPPD) program, Rockwell
International developed a 75-g, 400-MFLOP digital signal processor for
Eglin Air Force Base. The program successfuly demonstrated the use of
MCM-Ds based on processing thin-film aluminum and polyimide on
silicon wafers. The original product was designed and built on four
printed-circuit boards and had a volume of 4233 cm³ whereas the MCM
version reduced this volume to 65 cm³. The module (Fig. 1-10) consists
of two identical substrates interconnected by a flex cable and assembled in a single aluminum nitride package. The dice are bonded by
using flip-TAB which provides a zero-fan-out connection between the
dice and the substrate for highest packing density. Rockwell has also
developed an enhanced version of the SPPD (Fig. 1-11). This MCM
packs six digital signal processors and 1.25 Mbytes of static RAM on a

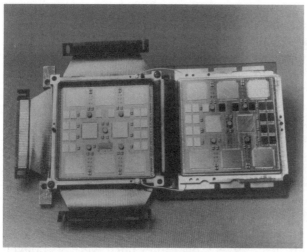

Figure 1-10. A 400-MFLOP digital signal processor fabricated by using high-density MCM-D processes. (*Courtesy Rockwell International.*)

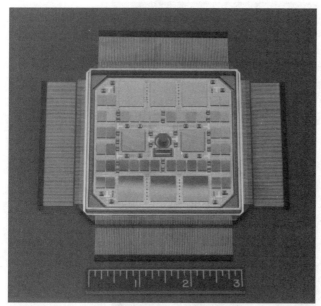

Figure 1-11. Enhanced signal processor packaging design, 49 ICs flip-TAB bonded on a 3 × 3 in, two-signal-layer interconnect substrate. (*Courtesy Rockwell International.*)

single interconnect substrate approximately 3×3 in. The high density was achieved by using flip-TAB and thin-film multilayer interconnect substrates where the signal traces were 25 μm on 50-μm pitch. All 49 ICs were electrically pretested and burned in, resulting in almost 100 percent first-pass assembly yield. Only two signal layers were required to route and interconnect the dice.[19]

The MCM-D technology was pushed to new heights in a Navy-funded program where 3.8×3.8 in, seven-conductor-layer substrates were designed and fabricated from 6-in-diameter silicon wafers. The demonstation modules assembled from these interconnect substrates contained fourteen 32K gate array ICs, some of which functioned as shift registers while others were ripple counters (Fig. 1-12). The ripple counters, some of which were TAB-bonded and others wire-bonded, were tested at clock speeds up to 350 MHz through the substrate and package pins.[20] Aluminum nitride packages 2×4 in and 4×4 in having 452 and 612 leads, respectively, on 25-mil pitch were also developed for this program.[21,22] The aluminum nitride packages[23] provided a 50 percent improvement over alumina packages in heat dissipation using chips that generated 5 W.

Figure 1-12. Navy demonstration MCM-D, 4×4 in, seven-conductor-layer polyimide/aluminum system. (*Courtesy Hughes Aircraft Co.*)

Due to the high costs of custom devices, interconnect substrates, hermetic packages, and the labor involved in assembly and test, these high-density MCM-Ds cannot be considered throwaways. Reliable rework procedures must be followed to correct any malfunctioning parts (see Chap. 9). Replacement of defective dice can be minimized by pretesting and burning in the dice prior to assembly, or what is known as using *known good dice* (see Chap. 8).

1.5.2 Low-End MCMs

Low-end MCMs are used for the very low-cost, high-volume commercial and/or consumer products. This market includes all the small computing systems such as desktop, laptop, palmtop, and notebook computers; portable phones; personal organizers; smart cards; automotive applications; appliances; and other consumer products yet to be invented. These MCMs must be produced cheaply and cannot even bear the cost of a hermetic enclosure. They are generally MCM-L types produced by low-cost printed-circuit board processes from epoxy FR-4 multilayer laminates (Fig. 1-13) or MCM-C types produced from high-temperature or low-temperature cofired alumina ceramics. After the devices are assembled, they are encapsulated, coated, or glob-topped

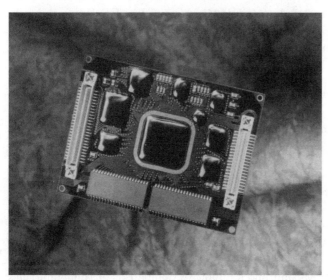

Figure 1-13. Chip-on-board packaging for low-cost laptop computers showing epoxy glob-topped dice on MCM-L. (*Courtesy S-MOS Systems and Electronic Packaging & Production Magazine.*)

with inexpensive epoxies or sealed in commercially available plastic packages. The main constraint in using plastic packaging is poor thermal dissipation. Normal plastic packaging is limited to 1 to 2 W of heat dissipation. With copper leadframes and heat spreaders, this can be increased to approximately 5 W. Plastic-encapsulated modules, if they fail, should be treated as throwaways, since the removal of fully cured plastics, especially thermosetting plastics such as epoxies, is difficult, costly, and almost impossible without damaging the rest of the module.

From an implementation standpoint, consumer electronics should focus on pushing as much of the functionality and cost to the silicon die level as possible. Packaging at the component level would then add only a small extra cost. Current MCM-D technology (large number of closely packed high-density dice) is therefore not suited to consumer electronics since packaging becomes the dominant cost of the end product. Consumer modules should therefore consist of a few ICs, typically a microprocessor chip, one or two ASICs, several memory dice, and some peripheral chips (capacitors).[24] However, this approach may change as the cost of interconnection and packaging, especially in high volumes, decreases.

References

1. J. J. Licari and L. R. Enlow, *Hybrid Microcircuit Technology Handbook,* Noyes Publications, Park Ridge, NJ, 1988.

2. L. I. Maissel and R. Glang, *Handbook of Thin Film Technology,* McGraw-Hill, New York, 1970.

3. J. J. Licari, "Hybrid Microelectronic Packaging," chap. 7 in C. Harper, ed., *Electronic Packaging and Interconnection Handbook,* McGraw-Hill, New York, 1991.

4. J. J. Licari, "Thin and Thick Films," chap. 8 in C. Harper and R. Sampson, eds., *Electronic Materials and Processes Handbook,* 2d Ed., McGraw-Hill, New York, 1993.

5. R. F. Bunshah, ed., *Handbook of Deposition Technologies for Films and Coatings,* Noyes Publications, Park Ridge, NJ, 1994.

6. D. W. Hamer and J. V. Biggers, *Thick Film Hybrid Microcircuit Technology,* Wiley, New York, 1972.

7. R. J. Jenson and H. Vora, *IEEE Transactions of Components, Hybrids, and Manufacturing Technology,* **CHMT-7**(4): 384, December 1984.

8. M. Terasawa, S. Minami, and J. Rubin, "A Comparison of Thin Film, Thick Film, and Cofired High Density Ceramic Multilayer with the Combined Technology," *International Journal of Hybrid Microelectronics,* **6**(1): 607, 1983.

9. T. Watari and H. Murano, *IEEE Transactions of Components, Hybrids, and Manufacturing Technology,* **CHMT-8**(4): 462, December 1985.

10. R. R. Tummala and J. Knickerbocker, "Advanced Cofired Multichip Technology at IBM," *Proceedings of the International Electronics and Packaging Society,* San Diego, CA, September 1991.

11. M. P. Lepselter, "Air Insulated Beam Lead Crossover for Integrated Circuits," *Bell System Technical Journal,* **48,** February 1968.

12. J. J. Licari, J. E. Varga, and W. A. Bailey, "Polyimide Supported Microramps for High Density Circuit Interconnections," *Solid State Technology,* vol. 19, no. 7, July 1976, pp. 41–44.

13. J. E. Meinhard and J. J. Licari, "Investigation of Multilayer Printed Circuit Board Materials," NASA contract NAS8-21477, Final Report, June 1971.

14. H. J. Neuhaus, "A High Resolution Anisotropic Wet Patterning Process Technology for MCM Production," *Proceedings of the First International Conference on Multichip Modules,* Denver, 1992.

15. J. Summers et al., "Wet Etching Polyimides for Multichip Module Applications," ibid.

16. *Guidelines for Multichip Module Technology Utilization,* IPC-MC-790, Institute for Interconnecting and Packaging Electronic Circuits, Lincolnwood, IL, August 1992.

17. E. Myszka et al., "The Development of a Multilayer Thin Film Copper/Polyimide Process for MCM-D Substrates," *Proceedings of the International Conference on Multichip Modules,* Denver, 1992.

18. D. A. Doane and P. D. Franzon, eds., *Multichip Module Technologies and Alternatives,* Van Nostrand Reinhold, New York, 1993.

19. R. K. Scannell and J. D. Spear, "Developing Multichip Modules for a Rapidly Changing Technology Market," *2d International Conference on MCMs,* Denver, April 1993.

20. P. Sullivan and J. Chernicky, "Design Considerations, High Speed Test and Reliability of High Density Multichip Modules," ibid.

21. E. Luh, J. Enloe, J. Lau, and L. Dolhert, "Aluminum Nitride for Advanced Multichip Modules," *Proceedings of the International Electronics and Packaging Society,* San Diego, CA, September 1991.

22. L. Dolhert et al. "Hot Pressed AlN Packaging: Thermal Performance and Mil-Std-883 Reliability Testing," *21st Joint Hybrid Microelectronics Symp. of International Society of Hybrid Microelectronics,* Cherry Hill, NJ, May 1992.

23. A. Kovacs and P. H. Reaves, "Thermal Performance of Multichip Modules," *ISHM, Proceedings of International Society of Hybrid Microelectronics,* Orlando, FL, October 1991.

24. J. D. LaVere, "Low-Cost Multichip Packaging for Consumer Electronics Applications," *Printed Circuit Design,* August 1992.

2

Multichip Module Design

P. A. Sandborn and M. Abadir

2.1 Introduction

Design is the process of translating ideas, concepts, and constraints into specifications, data, and drawings from which a working system can be constructed. The process of designing many types of complex systems has been highly refined in recent years. Among these systems are integrated circuits (ICs). IC design is characterized by a comparatively well-constrained set of design options, materials, and technologies. In addition, many well-developed computer-aided design (CAD) tools for managing behavioral descriptions, synthesis, simulation, and physical design of ICs exist. The design of multichip systems is characterized by a greater diversity of system and subsystem components (and assembly approaches) than IC design, requiring an interdisciplinary analysis, highly dependent upon physical and functional partitioning.

2.1.1 The Differences between Designing ICs, PCBs, and MCMs

At first glance, the design of multichip modules (MCMs) may appear to be a simple extension of techniques developed and refined for application-specific integrated circuits (ASICs) and printed-circuit boards (PCBs). However, MCMs use new and different materials and assembly approaches, and they introduce new thermal, electrical, test, and manufacturing problems. These problems are compounded by the concurrency of traditional IC and PCB design problems. For example, PCB design has traditionally not been concerned with on-chip drivers or the layout of IC bond pads, since the systems under design are low-performance and the packaging technology is low-density. In MCMs, however, many IC properties have to be designed concurrently with the MCM for successful high-performance system operation. For example, a system comprising off-the-shelf chips can be easily implemented on a large-area PCB, but may be impossible to route on a high-density MCM because the chip I/Os are not matched to support easy top-layer routing. Routing on internal layers requires a large number of vias which may or may not be available. In addition to the concurrency of IC design and module design, MCMs, in general, require a high degree of concurrency among design tools. While traditional packaging design is approached by using "divide and conquer" techniques (i.e., the use of many unintegrated point design tools), the higher density of MCMs rarely allows design decisions to be made based on a single-point design view.

New design issues also populate the MCM landscape. Traditional PCB designs do not include provisions for testability since PCB assembly techniques have a very high yield and the ICs are individually packaged in single-chip packages which can be readily tested and burned in prior to assembly. In MCMs, however, the ICs are bonded directly to the interconnect substrate (leaving out the single-chip package). Therefore, test and burn-in of ICs prior to assembly are not guaranteed and may not be economically practical. MCM designers must therefore pay extra attention to designing modules that may trade performance for cost, so that the modules can be readily tested and repaired. An additional complication is that physical design tools derived from PCB packages are often unlikely to have the fine database resolution and flexibility necessary for MCM design. In general, physical design tools like routers need to be "rules-driven" in order to ensure that manufacturable designs are created. The finer design rules associated with MCMs and the higher performance that usually accompanies their use require new routers which understand the rules and constraints required by MCM design.

In addition to the general design issues expressed above, MCMs pose unique electrical and thermal design problems. Small-line geometries and spacings in MCMs translate to higher ohmic losses and the potential for greater crosstalk between lines and layers. Because of these issues, MCMs are more dependent on three-dimensional electromagnetic simulation. Because dice are connected directly to an interconnect in MCMs, many of the traditional methods of managing excess heat, such as heat sinks, may prove impossible or unwieldy to use. In addition, the close proximity of the dice to the interconnect requires that the thermal expansion behavior of materials in MCMs be closely matched to that of the dice.

2.1.2 The Packaging Design Environment

Design processes can be viewed as a sequence of transformations on various design representations, at various levels of abstraction. In general, each transformation increases the information about the system. The design activities required for PCBs and MCMs can be broadly divided into the four areas shown in Fig. 2-1: specification, synthesis, physical design, and simulation. *Specification* is the formulation of an executable system specification from ideas and constraints provided by the user. *Synthesis* is the generation of a netlist, pin definitions, and test vectors from the executable specifications. *Physical design* is the conversion of netlists and other specification information into module and system layouts. *Simulation* is the detailed analysis of designed structures in support of all the other design activities. The combination of synthesis with elements of specification (supported by simulation) is sometimes referred to as a *package compiler* or, if chip design is included, a *system compiler.*

Design of ICs follows a path similar to the one shown in Fig. 2-1; however, specification, synthesis, analysis, and verification activities are performed for various levels of design representation prior to physical design. The design representations which must be considered include functional, logical, and circuit. PCB and MCM system design with predefined component sets does not require as detailed a breakdown of behavioral and structural design synthesis activities; however, it generally consists of a more complex physical and technology-based specification and synthesis.

Traditionally, a serial methodology (Fig. 2-2) has been used to perform the activities discussed above. In an ideal serial methodology, each phase of the design is completed before the next phase is initiated. In practice, however, design iterations are always necessary, causing delays in the time to market while consuming valuable resources. Traditionally, serial design methodologies relegate issues such as system testability and manufacturability too late in the design process (often after prototypes have been constructed) whereas their treatment early in the design cycle could lead to substantial savings in many cases. Serial design methodologies have evolved from a "correct by verification" sequential design approach, which tends to push designers into a depth-first style that favors synthesis of a design before verifying that it satisfies all the initial constraints.

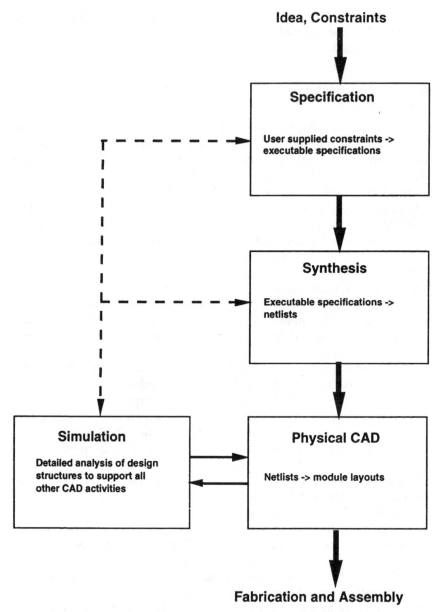

Figure 2-1. The design process activities for an electronic module.

Modern design philosophy advocates the use of concurrent design methodologies, which encourage pursuing all facets of design and development in parallel. Concurrent design philosophies are built upon a correct-by-design hypothesis. One representation of a concur-

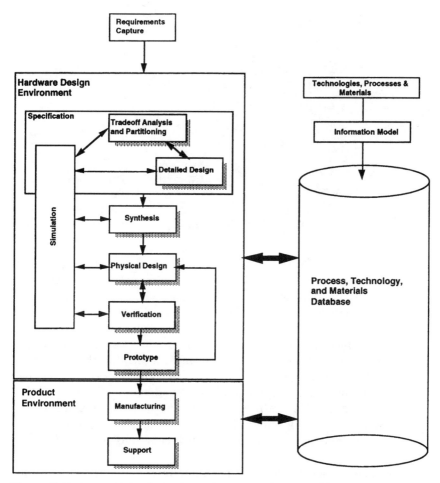

Figure 2-2. An example of a traditional serial methodology for complex electronic systems design.[1] Ideally each activity is completed before the next is initiated.

rent design approach for multichip systems is shown in the design life-cycle vision in Fig. 2-3.[2] This model is referred to as a *data-centric* model; i.e., the designer begins at the center with very little data or information about the system (just ideas), and the increasingly larger concentric rings represent more detailed design progression and increased system knowledge. The rings represent various *levels* of activity in the design, production, and life cycles. The specification level formulates an executable system specification from the requirements and is the focus of the conceptual design activities discussed in Sec. 2.2. The concentric rings in the model are divided into sections representing various *views* which must be considered in the design life cycle. The topics included on the figure are important, but do not rep-

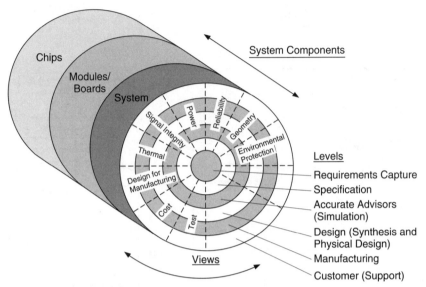

Figure 2-3. Data-centric model for the concurrent design and life cycle of complex electronic systems.[2] (©1993, IEEE.)

resent all the factors which must be considered for a successful design. The pie-shaped elements cover all the rings, indicating that each subject should be considered at all levels of design and production (i.e., design for manufacturability should be considered at the specification level in addition to its traditional treatment at the conclusion of the first serial design cycle). Integration (information sharing) is necessary between design views, between design levels, and between system components.

A concurrent methodology which contains the analysis capabilities discussed above infers a "breadth-first" approach. This approach provides the opportunity to explore many possible implementations prior to the synthesis of a design, thereby reducing the risk of extensive redesign later in the life cycle.

Independent of whether a serial or concurrent design methodology is used, computer-aided engineering (CAE) tools need to be linked in a seamless manner to CAD layout and routing tools and with computer-integrated manufacturing (CIM) tools which drive automated fabrication, assembly, and test equipment. The term *seamless* implies that no human intervention is necessary between steps using the various tools. These links are essential for achieving competitive design or revision cycle times. Table 2-1 summarizes some of the essential elements in a seamless CAD/CAE/CIM system. Not all the tools listed in Table 2-1 exist today, and many which exist do not yet work together

Table 2-1. Desired Features of Seamless CAD/CAE/CIM Tools for MCM Design

Design features	Desired features
Information capture and representation	HDL (hardware description language) interpretation/ simulation Standards checking Schematic capture Netlist generation Component list generation Capture of component operation vectors Library of component data
Interconnect design	Automatic assignment of pad patterns for components Automatic parts placement Autorouting Automatic generation of power and ground grids Design rule checking Z-axis checking
Electrical simulation	Calculation of line resistance Calculation of relevant capacitance and inductance matrices Circuit simulation Clock distribution and timing analysis Power and ground distribution analysis Calculation of decoupling capacitance required Calculation of termination resistance required Bus design Analysis of electromagnetic interference
Thermal simulation	Calculation of internal thermal resistances associated with components Calculation of effectiveness of external cooling environment Acoustic noise analysis
Producibility (manufacturability)	Generation of mask-making data in appropriate format Generation of test data for bare substrates Generation of CIM data for automated assembly Generation of test vectors for assembled modules Design-for-testability check Design-for-manufacturability check Design for repairability Design for maintainability Design for upgradeability Cost analysis (recurring and nonrecurring) Learning curve
Reliability	Design for reliability Vibration and shock analysis Chemical and corrosion check Automatic prediction of mean time between failure (MTBF)

SOURCE: After Ref. 3.

in a seamless manner. However, there is considerable industry effort under way to develop such systems.[4]

The following sections discuss the design activities introduced in Fig. 2-1 and various design analysis topics related to MCMs, including size and routing, electrical analysis, reliability, cost analysis, and design-for-test approaches. Since any one of these topics could easily be the topic of a complete chapter or an entire book, the level of discussion in this chapter is apologetically brief; however, extensive references are provided for the interested reader.

2.2 Conceptual Design (Specification)

Conceptual design refers to module and system design activities that are necessary to create the specification for a module or system (i.e., information modeling, partitioning, and tradeoff analysis). System synthesis would include these activities plus system architectural design, software specification, the development of netlists, and other activities necessary before physical design is begun. The conceptual design of multichip modules and systems has already been described in detail.[1]

One of the serious shortcomings in the present CAD tool set for packaging and interconnect system-level design is in the area of conceptual design support and specifically formal tradeoff analysis and partitioning. The need for higher system performance has brought about the requirement for more advanced packaging and interconnection approaches which do not compromise the higher performance of advanced chips. Traditionally, packaging and interconnect CAD tools have evolved from more mature tools used in IC design. This scenario is changing as designers realize that IC and board-level tools cannot evolve to support the unique problems posed in the multichip system's design space. Tradeoff analysis for these systems requires expertise in electronics, thermodynamics, mechanical engineering, manufacturing (including cost analysis), testing, and reliability. The presence of a large set of technologies, materials, and process alternatives in system packaging and interconnect design only complicates the process.

For most system designers, the goal is to find a design which balances performance with ease of manufacture and support, while minimizing cost. Early in the design cycle, numerous tradeoff decisions have to be made intelligently and quickly before major investments are committed or detailed design work begins. Besides time and budget constraints, system packaging tradeoff decisions are hampered by the lack of adequate and accessible alternative-technology data. These critical decisions, made within the first 10 percent of the total design cycle time, will ultimately define up to 80 percent of the final product cost. Therefore, making the most appropriate choices early in the design cycle will have a significant impact throughout the design, production, and life cycles.

The lack of a complete set of technological information is an especially serious problem in the packaging and interconnect field, since the number of technologies, processes, materials, and approaches is substantial and selecting optimums is arduous and nontrivial if one truly wants a balance in cost and performance. Alternative technologies and materials include interconnects (printed-circuit boards, ceramic, thin-film, etc.), packaging materials, bonding techniques (wire bond, TAB, flip chip), test techniques, and manufacturing methods. The designer may not be aware of all the technological choices that exist, and few designers can comprehend all the interdependencies and ramifications that the technologies and materials chosen may have on a particular design's cost and manufacturability. Furthermore, the refinement of the tradeoff analysis must continue throughout the design cycle, so that as more accurate information is obtained, there is a way to reconfirm that the selected technologies, processes, and materials are, in fact, still providing the "best" system solution.

The selection of an optimum combination of technologies depends on three items: (1) the characteristics of the components to be integrated, (2) the application for which the system will be used (i.e., performance requirements, cost, operating environment, and support requirements), and (3) the availability of previously designed structures and their design history. In addition, the problem is not confined to just the selection of packaging technologies for a module or just the selection of a chip technology. Complete tradeoff analysis must include chip or component through whole system design and hence must occur during all phases of the design process.

Tradeoff analysis activities could be carried out by using detailed point-solution simulation tools. However, because not enough information about the system is available during the conceptual design phase, the usefulness of detailed simulators during this portion of the design process is limited. In addition, the lack of a well-defined theory relating behavior and structure at the system level often makes approaches based on detailed computations less favorable than knowledge-base-supported estimation methods for conceptual system design and synthesis. Relying on detailed simulations to provide the information necessary to make system-level implementation decisions can be a dangerous, resource-intensive undertaking. Providing intelligent assistance to the designer in the specification process often proves to be more useful than simulation in the early stages of design. Estimation-based conceptual tradeoff techniques are intended not to replace detailed simulation, but to provide support for early (conceptual) specification design when simulation may not be practical.

The two approaches which have appeared to help close this gap are methodology managers for detailed simulation tools and automated model building for detailed simulation tools. The first approach consists of tools which provide guidance and automatic data manipulation for the use of detailed simulation tools. The integration of various modeling tools under a single user interface does not, by itself, make detailed point solutions cost-effective or practical at the conceptual level. However, adding elements of simulation or methodology management can make the use of modeling tools more practical during conceptual design. An existing tool which operates as a simulation manager is the *packaging design support environment* (PDSE)[5] developed at the University of Arizona. The PDSE integrates stand-alone parameter extraction modeling tools for simulating transmission lines.

Besides the management or guidance of point-solution tools, the required tool inputs need to be made compatible with the information available at the conceptual level of design. The second approach is to automatically build models for use in detailed simulators and post-process their results back into the specification design space. Several examples of software tools which construct models for more detailed simulators have appeared.[6,7]

Conceptual design includes three fundamental activities: *information modeling, tradeoff analysis,* and *partitioning.*

2.2.1 Information Modeling

Information models are used to capture the content and meaning of the elements of a design problem and record them as a high-level description. The elements which need to be represented include design requirements, constraints and budgets, technology and material descriptions, and process details. Information models are formed by identifying and classifying the objects corresponding to technology, materials, and processes necessary for performing tradeoff analysis and design. Once objects are identified, they can be converted to standard representations like EXPRESS[8] and compiled into databases for use in the design process. Since packaging technology involves a close relationship between design, processes, and materials, it is not always sufficient to characterize materials or technologies in terms of only their measurable properties in the absence of design knowledge and process understanding. The three interact in a nontrivial fashion which is critical to defining complex packaging systems. Thus the key issue in setting up a database is to determine how to capture, represent, and manipulate different bodies of knowledge and their interactions.

Information management may be subdivided as follows:

- Information representation
- Information content
- Information dynamics
- Information availability and retrieval
- Information version control

Information *representation* ensures that the information obtained is in a form that is understandable to the designers and their CAD tools. Information *content* refers to the meaning of the information. Information *dynamic* refers to the evolution of information content and the capture of the underlying design decisions. Information *availability and retrieval* refer to the ease with which the necessary information can be found and obtained. Information *version control* ensures that the information obtained is the most up-to-date version that exists.

In many instances the representation of information is a less restrictive design activity than the simple act of obtaining the appropriate information with which to design MCMs. Data availability is always on a design project's critical path and can often be a significant bottleneck. Design data on ICs that are assembled into MCMs can be an especially serious problem for MCMs.[9] MCM designers require detailed information on die yield, testability, geometry (die size and bond pad placements), power dissipation, driver and receiver construction, and power supply requirements. Since bare-die sales are not routine for most IC manufacturers, the information listed above often is not readily available yet is absolutely necessary for successful MCM design.

2.2.2 Tradeoff Analysis

Tradeoff analysis is the process of comparing performance gains in one region of the design space with associated performance losses in another. When simple enough systems are considered or when the design constraints are strong enough, tradeoff analysis may lead to design optimization. However, in general, tradeoff analysis is performed at a "what if" level. Studies which provide comparisons of various packaging technologies contain vital information, but attempting to draw general conclusions about the applicability of one technology over another from these studies can be risky. Simple technology comparison studies don't provide formal methodologies to facilitate trade-off analysis or partitioning, and therefore their conclusions are hard to apply to specific designs.

Tradeoff analysis can be addressed in either advisory or estimation-based predictive ways. Design advice explains to the designer how options compare and tracks compatibility among selected choices. Design advice is usually implemented by using knowledge-based approaches. Besides obtaining qualitative design advice, the designer can benefit from quantitative feedback on the implications of design choices. Unfortunately, detailed simulation and physical design activities are usually too complex and time-consuming to be a viable part of a broad tradeoff performance evaluation, so simplified estimation metrics are used to help make detailed tradeoff practical.

Figure 2-4 shows the interdependency between the various categories of metrics which must be considered during tradeoff analysis. It is obvious from Fig. 2-4 that the problem cannot be solved by simple superposition of prediction results; rather it is a problem of simultaneity. For example, adding wiring layers to an interconnect (substrate) might result in a smaller (area) module for an application. The smaller-area module may, however, be more difficult to cool because the compo-

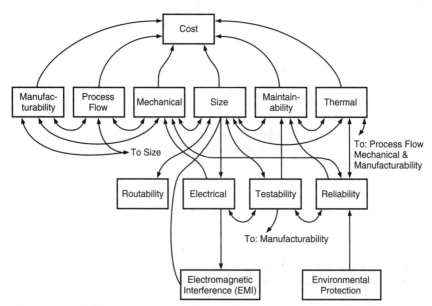

Figure 2-4. Estimation metrics necessary for the analysis of module-level packaging and interconnect problems.[1] The links describe the interdependency of the estimation metrics.

nents are closer together or because the cooling path through the additional layers has a greater thermal resistance associated with it, requiring the addition of thermal vias in the interconnect. The thermal vias, in turn, block wiring tracks in the interconnect, reducing the wiring capacity of the module sufficiently that the area advantage gained through the addition of extra layers is negated. Alternatively, suppose the additional thermal vias were not required in this example and the designer obtained the desired module shrinkage, but the module reliability is reduced as a result of the increased power dissipation density. The point of this example is that simple design changes sometimes do not produce the desired results or produce the intended result at the expense of reduced performance in another region of the tradeoff space.

2.2.3 Partitioning

A complex system can be partitioned in a variety of ways. Partitions are needed to accommodate varied design constraints, take advantage of existing components and subsystems, facilitate parallel development activities, provide access, environmentally protect, provide for testability and maintainability, and achieve reconfigurability. System partitioning is usually accomplished by making highly subjective judgments.

CAD tools and methodologies for partitioning functionality into a variable number of chips have existed for some time. Unfortunately these tools generally ignore the packaging and interconnect issues associated with connecting the multiple chips. For example, the optimum area-delay combination of chips for a given behavior may be of little value if the packaging and interconnect environment required to realize the optimum is an order of magnitude more expensive than an environment required for a poorer area-delay combination. The next level in the hierarchy is the partitioning of components (chips and passive components) onto modules or boards. Module partitioning is essentially analogous to its chip-level counterpart, though requiring more interdisciplinary "objective" functions. The variety of possible partitions for a multichip system can be categorized into three modes.[10]

Functional partitioning has the goal of decreasing the coupling between the substructures so that interface constraints are minimized. This type of partitioning results in a set of substructures optimized to be as stand-alone as possible. The coupling to be minimized may be any application-dependent combination of electrical, mechanical, thermal, and informational parameters.

Modular partitioning has the goal of maximizing the reuse of sub-structures. This type of partitioning sacrifices optimal separability, capitalizing on previous design experience by repeated application of a single design solution within a system.

Hierarchical partitioning acknowledges that the creation of a system generally proceeds through several stages. The highest levels consist of relatively stable substructures (stable in a product lifetime sense), while the lower levels consist of substructures which are more rapidly changing. Hierarchical partitioning optimizes the system reconfigurability.

The couplings between the components in a multichip system may be mechanical, chemical, electrical, and/or thermal. Partitioning a system is usually based on the minimization of one or more of these couplings.

System partitioning is not just a matter of determining which chips go on which modules. Partitioning also entails the decisions surrounding software development; i.e., wise physical partitioning choices can significantly ease parallel software development requirements. Hardware and software partitioning is discussed in detail in Ref. 11.

2.3 Synthesis

Design synthesis is the process of creating new design representations or providing refinement to existing design representations. In general, synthesis produces an artifact that satisfies some high-level behavioral and structural specification. A stricter definition of synthesis is the translation from a behavioral (functional) description to a structural description and the translation of a structural description to a physical description,[12] as shown in the Y chart in Fig. 2-5. The axes of the Y chart represent three design perspectives: (1) functional (or behavioral)—the input/output of the system; (2) structural—the interconnection of components to realize the required functionality; and (3) physical—the physical manifestation of the components and interconnections used to realize the required structure. Along each axis, multiple levels of abstraction, or detail, are represented.

In general, synthesis activities map perspectives from one axis to another in a counterclockwise direction. Synthesis activities near the center of the Y chart are generally better defined and more mature than those farther from the center. The translation from the functional description to the structural description generates structures that are not bound in physical space. The translation from the structural defini-

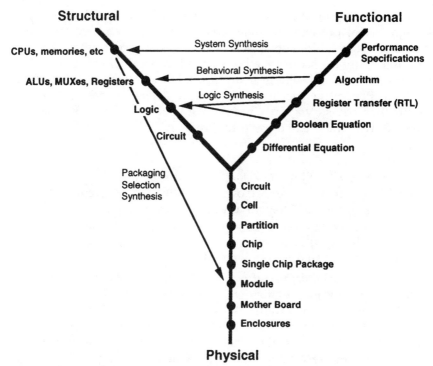

Figure 2-5. Synthesis tasks represented on a Y chart. Common synthesis activities are plotted as mappings on the chart.

tion to the physical definition adds the physical information necessary to produce a working version of the element.

System-level synthesis is not as concisely defined as lower-level synthesis activities. There are no well-developed languages analogous to a hardware description language or Boolean equations to define and represent system-level functionality. The lack of a well-defined theory relating behavior and structure at the system level means that approaches used in logic and behavioral design cannot be directly extended to system-level design.[13] Efforts are under way to develop system design languages.[14] Reference 15 contains a detailed discussion of chip and system synthesis activities.

2.4 Design Analysis and Simulation

In this section we briefly review the hierarchy of modeling and analysis approaches that must be used in the design of MCMs. Design analysis can be approached in either adviser or predictive ways. Design advice presents the qualitative advantages and/or disadvantages that expert designers typically consider in accepting or rejecting a design option. Advisory methods are adequate during the specification portion of the design in many cases, but never sufficient for final designs. Predictive methods yield more accurate, design-specific results than advisory approaches do, but usually require considerably more effort to initiate and perform. Predictive methods can be categorized as

Figures of merit. Simple packaging figures of merit are easily computable. These estimates are often technology-dependent but application-independent. They are useful for comparing the general merits of one interconnect or bonding technology over another but do not provide any application-specific performance detail, e.g., the calculation of saturated near-end crosstalk or the power-delay product.

Closed-form design-dependent models. These models attempt to capture simple design-specific details such as the effect of the number and size of loads on a transmission line. Analysis in this case is limited to the formulation of closed-form expressions. This type of modeling is practical at the conceptual design level; however, it usually results in performance measures which are acceptable for relative comparison of one technology and architecture combination to another but not acceptable on an absolute-accuracy basis. An example is the estimation of a critical net delay by adding closed-form expressions for the RC charging delay and time-of-flight delays. Such a summation is useful for comparison purposes but does not model real nets accurately since delays on most real-world nets also depend on reflections, crosstalk, and switching noise.

Two types of closed-form models can be used. Heuristic models use general guidelines or empirical studies to model a system. A general guideline is a rule of thumb. The risk of basing prediction solely on general guidelines is that the guidelines can become outdated as technologies and materials change. Heuristic models also make use of empirical study results. Quantitative empirical study results usually take the form of simple expressions obtained from fitting experimental or detailed simulation data for a specific situation (versus deriving an expression from basic principles). An example of an empirical result is Rent's rule[16] which is used to relate the number of I/Os and gates or chips in a system. Heuristic models do not possess a "deep" understanding of the design process or the fundamental engineering principles. Their validity is not necessarily known under circumstances other than the specific cases from which they were formulated. The advantage of heuristic models is that they require very little computational overhead and yield quick answers.

Analytical models use closed-form formulations derived from basic principles or probabilistic techniques to predict the performance of a system. Closed-form analytical expressions for predicting performance metrics can be derived from basic principles under a set of well-understood assumptions. Because the assumptions are part of the model development, the applicability (and validity) of a model to a particular situation can be easily determined. Probabilistic models involve capturing the characteristics of a design activity in a set of random variables with an associated set of probability density functions. An example of a probability-based model is the wiring model developed by Sutherland and Oestreicher.[17] Analytical models tend to be more general than heuristic models; however, probabilistic models often lead to overly pessimistic results.

Simulation models. Simulation is the detailed analysis of a critical portion of a design usually consisting of the solution of a large number of simultaneous equations. Accurate results can be obtained from closed-form expressions only under very special circumstances which may not represent realistic situations. Simulation generally results in accurate solutions but often at the cost of large amounts of computation time and/or model setup time. As stated in Sec. 2.2, it is often difficult to formulate a system-level view large enough to aid in decision making without performing a very large number of simulations.

In the remainder of this section, we briefly review the analyses which must be performed in MCM design. None of the discussions that follow are complete, and the reader is encouraged to seek additional information in the references cited.

2.4.1 Size and Routing Analysis

Almost all measures of performance and manufacturability will be affected in some way by the size of the module. The size of the module determines the length of critical nets, which may be an important factor in determining system cycle times and other electrical performance issues. How closely the chips are packed will determine the power dissipation density of the module, which may determine how effectively or economically the module can be cooled. The size of the module will also contribute to its reliability and yield which affect the cost. The size of the module depends on not only the size and number of components that must be placed on or in it, but also the requirements placed upon the module by the internal connection of the components and the external connection of the module.

Module size is determined by a collection of effects which include wiring capacity, routability, bond pad pitch, escape routing, power dissipation density, off-module connections, and placement, in addition to the actual physical sizes of the components.

The prediction of MCM size is usually accomplished by computing a set of footprints for each component (active and passive) in an MCM. The limiting footprint of a particular component is determined by the largest size limitation imposed by the factors

$$F_{p(\text{interconnect capacity})} \qquad F_{p(\text{escape routing})}$$
$$F_{p(\text{via density})} \qquad F_{p(\text{die})}$$
$$F_{p(\text{bond pad density})} \qquad F_{p(\text{power dissipation density})}$$

where F_p designates a particular footprint. The interconnect capacity footprint is the size limitation based on the amount of wiring required to connect a component within the module. It depends on the wiring capacity of the interconnect, the quality of the routing, and the placement of the components. Interconnect capacity limitations are discussed in detail in Refs. 1 and 18. The via density footprint accounts for the number of vias available to connect component I/Os to wiring layers. For example, Ref. 19 shows that the maximum number of vias required per net in order to ensure routability (assuming enough layers can be added) is four. The number of vias required in MCMs is determined by the type of application and the degree to which the I/Os on adjacent chips are matched to allow top-layer routing. The bond pad density footprint accounts for the minimum distribution of bond pads on the surface of the interconnecting module.[1] The escape routing footprint treats difficulties encountered in routing component I/Os out from under the dice to wiring tracks on the surface of interconnects or to vias which connect to other wiring layers. Escape routing is a significant concern for modules that contain area array flipchip bonded components. The surface laminar circuit (SLC) MCM interconnect technology, developed by IBM,[20] was specifically designed to accommodate escape-route needs. A model for predicting the number of layers required to "escape route" a component is contained in Ref. 1. The die footprint represents the physical size of the die and its surrounding bonds and minimum spacing to adjacent components. The power dissipation density footprint depends on module cooling constraints.[21]

To obtain the module area, the footprints representing each component are appropriately accumulated; i.e., if all the components in the module are identical and are attached to only one side of the interconnect, then (ignoring placement constraints) the minimum module area is the number of components multiplied by the maximum of the terms above. For modules which contain chips of many different sizes and I/O counts, the formulation of the module area has to take into account that a chip whose footprint is limited by $F_{p(die)}$ can be placed within the footprint of an adjacent chip whose footprint is limited by $F_{p \text{ (interconnect capacity)}}$. The possible placement of components on both sides of the module must also be considered.

In addition to the chip footprints, the module size may be limited by placement concerns and connection limitations to the next level of packaging.

2.4.2 Thermal Analysis

Supporting high heat fluxes while maintaining the chips at relatively low temperatures is a major challenge facing today's electronic system designers. Acceptable operational temperature levels are dictated by performance constraints, reliability demands, and the requirement that complex systems operate over a wide variety of environmental conditions (see Chap. 5).

Power is removed from an electronic system by heat transfer. In order for an electronic system to operate reliably, the temperature of transistor junctions must be maintained below specified critical levels by conducting the heat dissipated by the device out of the system. Heat must be conducted from small hot areas to large cooler areas, where the heat can be removed by convection using either a gas or a liquid.

The thermal management in an MCM is a victim of its own success.[22] The tight chip spacing (high packaging density) and increased operating speeds cause MCMs to dissipate more power in a smaller area (higher power dissipation density) than traditional packaging systems.

The analysis of the thermal performance of MCMs involves the calculation of internal thermal resistances from critical component junctions to heat sinks or surfaces where the heat can be removed. Detailed thermal analysis is the subject of Chap. 5 and is treated further in several other books.[23,24]

2.4.3 Electrical Analysis

Communicating information from one chip to another within an MCM requires the signal level on a transmission line to change. The transition to a new signal level does not occur instantaneously due to propagation delay and the nonzero transition times of drivers on chips. Decreasing the signal transition times facilitates better performance (i.e., faster computers), but also magnifies a variety of phenomena that are undesirable and often detrimental to a module performing its intended task, namely, noise.

Three electrical parameters—resistance, capacitance, and inductance—are always present in packaging and interconnect systems. Resistances cause dc drops on signal lines and contribute to *RC* charging delays, both of which are unwanted effects. Resistances, however, can help damp unwanted noise in systems. Resistance in signal lines increases as line geometries shrink and can be several ohms per centimeter in thin-film MCM interconnects. For very fast systems, resistance is also a function of frequency due to skin effects. Capacitance is the primary cause of delays in packaging and interconnect systems. Capacitive loading on signal lines is decreased by using bare dice instead of dice in single-chip packages and by reducing net lengths. Capacitive coupling between power and ground in power distribution systems is beneficial for managing switching noise. Capacitance and resistance can be either lumped (discrete load or termination) or distributed (such as the resistance and capacitance associated with a transmission line). Inductance contributes to switching noise and time-of-flight delays in packaging systems. The inductance can be reduced by shortening the length of electrical connections in modules (such as bond wires or connector traces) or by replacing bus structures with planes. The effective inductance in modules is also reduced by increasing the circuit return-path mutual inductance, which can be accomplished by decreasing the pitch between conductors carrying current in opposite directions. Capacitive and inductive coupling between adjacent signal lines causes crosstalk.

The basic elements of interest in electrical performance are delays and noise. These elements are characterized through the measurement of crosstalk, switching noise, dc drops, attenuations, and critical path delays. An excellent treatment of the electrical analysis of packaging and interconnect structures is contained in Ref. 25.

Delay. Probably the most important area in electrical performance for potential comparison between different packaging combinations is delay. Packaging delays, along with the delays associated with gates, receivers, and memory access times, will determine the maximum operating frequency of multichip systems. The packaging delay depends on not only the length of nets, but also the distribution of loads on the nets, the size of the loads, and the characteristics of the circuits driving the nets.

The total delay in a circuit path can be approximately decomposed into the following elements: RC charging delays, time-of-flight delays ($\sqrt{LC}$ delays), delays associated with noise (crosstalk, switching noise, reflections), and delays associated with on-chip circuitry.

A simple RC charging delay is formulated in Eq. (2-1).

$$t_{charging} = \alpha_1 R_{on} (C_1 + C_2 + C_3 + C_4 + C_5) + \alpha_2 R_1 C_2 \qquad (2\text{-}1)$$
$$+ \alpha_1 R_1 (C_3 + C_4 + C_5) + \alpha_2 R_2 C_4 + \alpha_1 R_2 C_5$$

Equation (2-1) corresponds to the interconnect modeled in Fig. 2-6 where C_1, C_3, and C_5 are lumped loads on the transmission line, C_2 and C_4 are distributed loads associated with the capacitance of the transmission line, R_{on} is the on resistance of the output stage, and α_1 and α_2 are coefficients which depend on the signal level to which one wishes to measure the delay. In (2-1), α_1 is used for a lumped RC network, and α_2 is used when a uniformly distributed RC network is required. For a step input (signal line rise time $t_r = 0$) with the delay measured at the 90 percent level of the signal, $\alpha_1 = 2.3$ and $\alpha_2 = 1.0$. At the 50 percent signal level, $\alpha_1 = 0.7$ and $\alpha_2 = 0.4$.

Equation (2-1) models the entire packaging delay (neglecting noise) on a transmission line when $R \gg \omega L$ over the entire frequency spectrum of the signal (ω is the frequency in radians). This condition is true for most on-chip interconnects, since they are highly lossy, and for

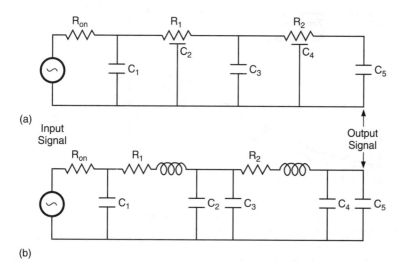

(a)

(b)

Figure 2-6. Simple equivalent circuit representations of loaded nets. (a) Distributed RC circuits. (b) Lossy transmission lines (characterized by characteristic capacitance and inductance); R_1 and R_2 represent the loss in the two transmission line segments.

many low-speed interconnect systems. For transmission lines in which $R \leq \omega L$ (i.e., high-speed or approximately lossless transmission structures), the packaging delay is usually estimated by using a time-of-flight (tof) approximation

$$t_{tof} = \text{length} \sqrt{\epsilon_0 \epsilon_{eff} \mu_0} \qquad (2\text{-}2)$$

where ϵ_0 and μ_0 are the permittivity and permeability of free space, respectively, and ϵ_{eff} is the effective dielectric constant of the material embedding the line. The effective dielectric constant is not only a function of the dielectric material; it can also be a function of the line geometry and construction. Examples of ϵ_{eff} calculations may be found in the literature.[26]

The total delay of a packaging and interconnect system is some combination of the RC charging delay [Eq. (2-1)] and the time-of-flight delays [Eq. (2-2)] plus delays associated with noise. Figure 2-7 shows the relative sizes of the RC charging delay and the time-of-flight delays for a point-to-point net in a fine-line printed-circuit board (MCM-L). For this simple comparison, a load capacitance of 3 pF and a driver with an on resistance of 25 Ω were chosen. A 50-Ω copper stripline was assumed to be in the line, and the delay was measured to the 50 percent level.

The RC charging and time-of-flight delays can be computed and

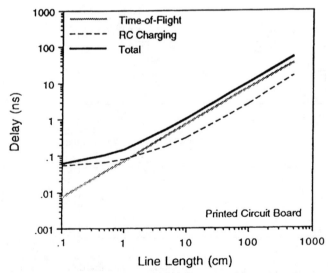

Figure 2-7. PCB delay components. An FR-4 board was assumed ($\epsilon = 4.5$) with a signal line width of 8 mils and thickness of 1.4 mils.

used as a simple comparison of MCM technologies and design alternatives; however, accurate prediction of actual delays on nets requires the use of detailed electrical simulation. Simulation of transmission structures present in MCMs usually requires several steps: (1) the prediction of inductance L and capacitance C matrices for the multiconductor system, (2) synthesis of equivalent circuit models for the transmission lines from the L and C matrices, (3) attachment of driver and receiver models, and (4) circuit simulation of the resulting network. The L and C matrices can be formulated by using several numerical techniques including finite element modeling (FEM) and full-wave analysis.

Controlled Impedance. When a signal line is placed above a voltage reference plane and the dimensions of the line are held constant, the performance of the signal transmission structure can be accurately controlled and predicted. Maintaining a constant (i.e., controlled) characteristic impedance along a signal transmission structure requires that the inductance L and capacitance C remain constant along the length of the line. The characteristic impedance of a signal transmission structure is given by

$$Z_0 = \sqrt{\frac{L}{C}} \qquad (2\text{-}3)$$

where L and C are the inductance and capacitance per unit length of the transmission line. The most common approaches to constructing controlled impedance interconnections is to use stripline, microstrip, or coplanar lines. Stripline and microstrip structures require reference planes, and coplanar structures require reference lines (Fig. 2-8).

Traditional, low-performance packaging systems do not contain controlled impedance interconnections; i.e., reference planes are not used, and reference lines do not surround critical signal connections. Therefore, the characteristic impedance of signal transmission structures is not constant over the length of a line. As electrical performance requirements become more stringent, it becomes increasingly important that the performance of critical signal transmission paths be tightly controlled. MCMs readily provide the signal transmission structures shown in Fig. 2-8 and allow controlled impedance constraints to be considered during physical design (see Sec. 2.5).

Maintaining controlled impedance transmission structures over meshed reference planes is a design complication encountered in many high-density MCMs. In thin-film MCMs, for example, reference planes are meshed in order to provide holes for outgassing of organic dielectrics. The design of controlled impedance lines over meshed reference planes has been reported.[27,28]

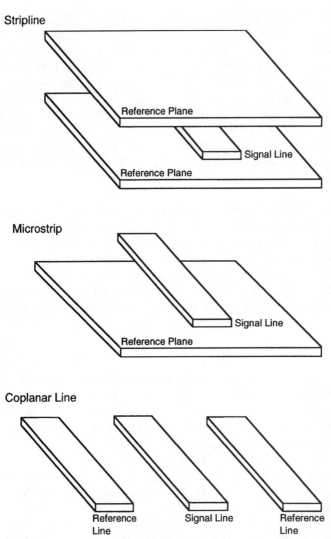

Figure 2-8. Examples of controlled impedance transmission structures.

Noise. In high-performance systems, the management of noise is a major design issue. Noise can limit the achievable system performance by degrading edge rates, increasing delays, reducing noise margins, and causing false switching of logic gates. Noise magnitude and noise frequency are closely tied to the approach used for packaging and interconnection. The effects of the noise depend on logic rise and fall times, gate currents, system size, system timing, net and bond lengths and pitches, interconnect electrical design and power, and ground distribution on the module.

Crosstalk, sometimes called *coupled noise,* is a result of mutual inductance and capacitance between neighboring transmission lines; i.e., electric and magnetic fields associated with lines which are close together interact, and the lines communicate information to each other. Crosstalk increases as the distance between lines (spacing) decreases, the distance between the lines and their circuit return path (ground plane, e.g.) increases, and the length of their close proximity (coupled length) increases.

A crosstalk figure of merit which can be computed and used to gain some limited insight on the comparison of technologies is saturated near-end crosstalk. When the coupled length of matched transmission lines becomes long enough to make the propagation delay greater than half the rise time of the signal, then the near-end crosstalk reaches its maximum amplitude (saturates). Under saturated conditions the line length, signal rise time, and dielectric constant of the embedding material do not affect the magnitude of the backward crosstalk; i.e., near-end crosstalk is solely a function of the signal-line geometry and the signal swing. The saturated crosstalk on a quiet line due to one excited line is given by Catt[29] as

$$\text{Saturated crosstalk (dB)} = 20 \log_{10} \left| \frac{Z_{oe} - Z_{oo}}{Z_{oe} + Z_{oo}} \right| \qquad (2\text{-}4)$$

where Z_{oe} and Z_{oo} are the even and odd mode impedances of the coupled transmission lines, given by

$$Z_{oe} = \sqrt{\frac{L_o + L_m}{C_o - C_m}} \quad \text{and} \quad Z_{oo} = \sqrt{\frac{L_o - L_m}{C_o + C_m}}$$

where L_o and L_m are the self-inductance and mutual inductance per unit length and C_o and C_m are the self-capacitance and mutual capacitance per unit length of the coupled lines. Figure 2-9 shows the saturated crosstalk between two lines for three different 50-Ω stripline designs as predicted by Eq. (2-4). The ratio of the number of excited to quiet lines also affects crosstalk and saturated crosstalk. If the number of identical excited lines next to the quiet line considered above and in Eq. (2-4) were doubled, the crosstalk due to each excited line would add by superposition and the total crosstalk on the quiet line would approximately double.

As with delay calculations, accurate crosstalk magnitudes can be determined only by simulation of coupled-line systems.

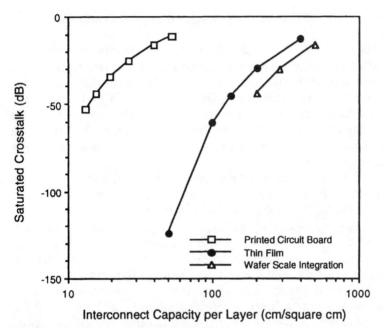

Figure 2-9. Saturated crosstalk as a function of interconnect capacity per layer for 50-Ω coupled stripline configurations.[1] The printed-circuit board had a conductor thickness = 0.7 mil, conductor width = 5 mils, ground plane separation = 15 mils, and a dielectric constant = 4.5. The thin-film interconnect had a conductor thickness = 0.2 mil, conductor width = 0.6 mil, ground plane separation = 1.7 mils, and a dielectric constant = 3.4. The wafer-scale integration example had a conductor thickness = 2 μm, conductor width = 10 μm, ground plane separation = 30 μm, and dielectric constant = 3.9.

A potentially serious problem faced by high-performance systems is switching or ΔI noise. This type of noise originates when drivers switch current drawn from a power distribution system.

Switching noise is defined by the expression for the voltage drop across an inductor

$$V_{\text{noise}} = nL_{\text{eff}} \frac{di}{dt} \tag{2-5}$$

where V_{noise} = switching noise magnitude
$\quad\quad n$ = number of simultaneously switching drivers
$\quad L_{\text{eff}}$ = effective inductance of path through which di/dt flows
$\quad di/dt$ = current slew rate of one driver

Like crosstalk, switching noise is a quantity dependent on the system design. The contributions to switching noise at the module level come from the inductance of bonding leads, single-chip packages, on-module power and ground delivery system (i.e., planes, buses), and power and ground delivery to the module.

Bonds usually represent the most immediate source of inductance to a switching gate. It is important to decrease the inductance of the bonding leads as much as possible. Two ways of reducing inductance are to make the bonds short and to increase the number of power and ground leads (decreasing the signal-to-ground ratio) by placing the leads which comprise the circuit return path (usually ground leads) as close to the leads carrying the switching current as possible, thus increasing the mutual-inductance component associated with the bonds. The effects of bond pitch and lead distribution among signal, ground, and power are discussed in detail by Rainal.[30] As an example, the effective inductance associated with a ground lead can be estimated by considering the expression for the effective inductance of a ground lead, which can be written as[31]

$$L_{eff} = \frac{L_{self} I_{ground} - (\Sigma L_{mutual} I_{signal})}{I_{ground}} \tag{2-6}$$

where L_{self} is the self-inductance of a ground lead, ΣL_{mutual} is the mutual inductance of the ground lead due to I_{signal} flowing in the surrounding signal lines, and I_{ground} is the current flowing in the ground lead. Note that the ratio of I_{signal} to I_{ground} is the signal-to-ground lead ratio. Figure 2-10 shows the effective inductance plotted as a function of lead length for various bonding approaches.

The potential for lower switching noise when flip-chip (solder-bump) attachment approaches are used is obvious from Fig. 2-10. Increasing the signal-to-ground ratio has the effect of decreasing the mutual inductance and thereby increasing L_{eff} and V_{noise}, and the overall result is to shift the curves in Fig. 2-10 upward. Increasing the pitch of the leads decreases the mutual inductances and will also shift the curves in Fig. 2-10 upward.

To formulate the overall L_{eff} for a module requires that estimates of the inductance associated with the bonding technology, chip packages, and interconnect power distribution network be obtained. Techniques for estimating L_{eff} for module designs are outlined in Refs. 33 and 34. Analytical closed-form approaches to the modeling of switching noise are discussed and reviewed in Ref. 35.

The management of switching noise can be approached in several ways. The most direct is to attack the problem at the chip level by using complementary or balanced driver logic implementations. In principle, a pair of complementary outputs should contribute equal

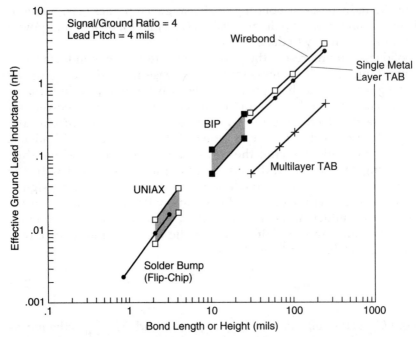

Figure 2-10. Effective inductance as a function of bond length for various bonding technologies.[1] The signal-to-ground ratio and lead pitch were held constant in all cases. The following specific dimensions were assumed: Wire bond: diameter = 1 mil; single-metal-layer TAB: width = 2 mils and thickness = 1 mil; multilayer TAB: separation between traces and ground ring = 2 mils, BIP[32] (bonded interconnect pin): diameter = 17 to 75 μm; UNIAX, Raychem Limited: diameter = 25 to 75 μm, solder bump: diameter = 75 μm.

but opposite noise voltages to the line to which the outputs are terminated, and the opposing noise voltages are expected to cancel. The effectiveness of using complementary outputs for the management of switching noise in emitter-coupled logic (ECL) systems is discussed in Ref. 36. Active regulation schemes, usually taking the form of a current source, and passive bypass capacitors implemented on the chip can also be used to regulate noise. The effectiveness of on-chip bypassing is discussed in Ref. 25. Numerous options also exist for managing switching noise by using off-chip methods, the most popular of which is the use of bypass capacitors. Bypass capacitors ideally should be placed to minimize the amount of inductance between the capacitor and the switching event. One approach is to place bypass capacitors on the die connected to the inner-lead bond pads.[37] Bypass capacitors may be placed in or on a single-chip package or on the interconnect or board. However, in this case, the effective inductance of the bonding leads will not be bypassed. If bypass capacitors are placed on the interconnect, the use of low-impedance interconnects has the advantage that their proximity to the switching event is not critical as long as they are located close enough that the electrical delay between the gate and the bypass is much less than the signal rise and fall times. Thin- and thick-film interconnects have the added advantage that bypass capacitors can be built into the interconnect, and the use of closely coupled power and ground planes can often supply much of the needed bypass capacitance.

Reference 38 contains a detailed discussion of modeling switching noise.

2.4.4 Reliability Analysis

Reliability is the statistical art of determining the length of time over which a device or system will function correctly. The reliability of electronic systems depends on both the hardware, such as integrated circuits, single-chip packages, bonding techniques, interconnects, cooling schemes, and power supplies; and on the software, such as operating systems and application programs. The cost of repairing failures has been rising almost as quickly as electronic system costs have dropped, so reliability may well be the most important design goal for many systems, including inexpensive consumer electronics applications.

Reliability is a system characteristic which must be built into the design: it can not be added after product development. Therefore, aggressive design for reliability can result in designs which have orders-of-magnitude improvement in system reliability through an early attention to reliability concurrent with other critical performance views of a system.

The reliability function $R(t)$ of a component or system represents the conditional probability that it will operate correctly at time t, given that it was operational at time zero. In a system without repair (after leaving the factory), the most widely used reliability function is the exponential function

$$R_c(t) = e^{-\lambda_c t} \tag{2-7}$$

where λ_c is the failure rate expressed in failures per time, t represents the time period, and $R_c(t)$ is the probability of correct operation at time t.

System reliability is an integration of the reliabilities of the lower-level subsystems or components which make up the system. Nonredundant systems must have all their components operational for the system to function properly; therefore this type of system reliability is given by

$$R_s(t) = \prod_{i=1}^{n} R_{ci}(t) = e^{-\lambda_s t} \tag{2-8}$$

where $R_s(t)$ is the probability that the system will be operational at time t and $R_{ci}(t)$ represents the reliability of the ith component (out of a total of n). If the reliability function is exponential, the system failure rate is given by

$$\lambda_s = \sum_{i=1}^{n} \lambda_{ci} \tag{2-9}$$

The *mean time to failure* (MTTF) is the average period of time for which a system will operate before failing relative to time zero. The MTTF for a system is given by

$$\text{MTTF} = \frac{1}{\lambda_s} \tag{2-10}$$

The functions presented by Eqs. (2-8) to (2-10) are a model of the operational life of a system.

The analysis of reliability concentrates on determining failure rates λ. There are several methods for determining failure rates of individual components and packaging elements. Models developed to predict the effect of temperature on chemical reaction rates have been widely applied to electronic components. The two most prevalent models are the Arrhenius and Eyring models. These models use exponential failure distributions (and therefore constant failure rates), and they generally make the assumption that the dominant component failure mechanisms depend on the steady-state temperature. This assumption is used to derive activation energies or thermal acceleration factors for the component based on a weighted-average methodology. The failure rates in these cases are characterized as an exponential function of the steady-state temperature.

The best-known reliability prediction methods are published in Mil-Hdbk-217, where failure rate predictions embed Arrhenius models.[39] Mil-Hdbk-217 models have been shown to be a conservative estimator of absolute failure rate (i.e., they predict higher failure rates than those actually observed), but if used correctly, they can provide a method of comparing the reliability of design alternatives. A discussion of the shortcomings of Mil-Hdbk-217 may be found in Ref. 40.

An alternative to the statistically based failure rate predictions in Mil-Hdbk-217 are physically based models. Physics-of-failure methods[41] model actual failure mechanisms to determine failure rates and do not depend on the existence of field data to provide predictions.

A topic upon which MCM reliability depends and which is of particular interest in MCM design is chip protection. In traditional packaging approaches, all chips are contained within their own single-chip package which inherently provides protection from their immediate environment as well as protection from handling during assembly. Methods of encapsulating bare dice assembled on MCM interconnects have been developed (see Chap. 6). The cost penalty for sealing an assembled MCM in a hermetic package can be substantial. If the MCM is small enough, it can be placed in the cavity of a standard single-chip package; otherwise it must be placed in a ceramic or metal case which, if large, may require support posts or reinforced lids. The lid seal must be rugged and hermetic and must survive temperatures and procedures used for assembly to the next level of system packaging (in addition to possible repair and replacement procedures). It is also desirable that the lids be removable without damaging the enclosed MCM.

2.4.5 Cost Analysis

The performance advantages of multichip modules over conventional printed-circuit board packaging have been demonstrated many times. While multichip modules do not perform significantly better for every application, they almost never underperform conventional approaches which use single-chip packages. If the performance of multichip modules is potentially superior, why are so few applications committed to volume production using multichip modules? The reason has little to do with size, electrical performance, or thermal performance but is a result of manufacturing issues including cost, lack of infrastructure, and the reluctance of manufacturers to change to new technologies. Manufacturing issues at the module level include cost; testability; time to build, repair, and rework; manufacturability; learning curves; and infrastructure.

Cost is the most important metric in many packaging and interconnect systems. The relative costs of bonding techniques and interconnect technologies are available from numerous sources, and these comparisons provide important tradeoff information. However, it is the completed system cost including rework and testing which is the ultimate determining factor in cost analysis.

To completely assess the cost of a system, the following factors should be considered:

- Nonrecurring design and development costs
- Learning curves
- Capital costs

- Tooling costs
- Recurring fabrication and assembly costs of components and packaging
- Repair and rework costs
- Testing costs
- Support and maintenance

Interconnects are defined as the substrate or board on which multiple chips are interconnected. Interconnect costs have been compared by using several different measures, few of which, in themselves, allow an "apples to apples" comparison of technologies (i.e., no simple application-dependent comparisons are available).

A prevalent method is to compute the cost per unit area of interconnect. This measure can be misleading since it does not take into account that an application packaged by using a high-wiring-density technology (thin film, e.g.) may require a smaller interconnect area than if it were packaged by using a low-density technology. Cost per area can be used to describe an interconnect only if that figure is accompanied by data on the resulting module size. Cost per area can also be deceiving for technologies which require the drilling of holes, since their cost may be highly dependent on the number of holes which must be drilled, their diameters, and spacings. Figure 2-11 shows this type of interconnect cost relation. Versions of Fig. 2-11 that contain more design detail, such as the number of layers and whether the board is single-sided or double-sided, have been generated. [43] These relations are not the result of a detailed derivation, but rather a summary of quotes provided by interconnect suppliers.

A second measure of interconnect cost is the cost per wiring line length. This metric suffers from approximately the same problem as the previous one: It does not account for the fact that, because of smaller module area requirements, applications which make use of high-density interconnects usually require a shorter total length of wiring. This metric is also affected by the quality of the placement and routing techniques used.

A more realistic metric is the cost per interconnect bond pad.[44] This metric is more useful for a stand-alone comparison of interconnects because it is the common top surface function that all interconnection technologies deliver. Cost per interconnect pad is unambiguous, can be readily calculated, and does not rely on rule-of-thumb estimates or other subjective interpretations. Cost per chip is another similar metric that has been suggested.[45]

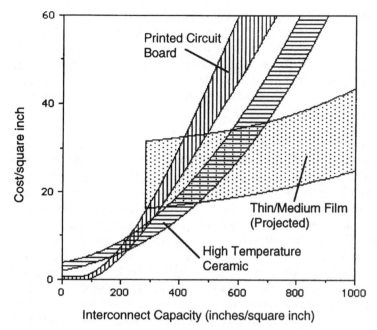

Figure 2-11. Cost per square inch of major interconnect technologies.[42] (©1992 IEEE.)

None of the above metrics is inaccurate, if qualified with information about the application being packaged. However, they do not represent fair comparisons when taken out of context.

More accurate computation methods for interconnect costs have been developed and can be used. The computation of interconnect costs requires the characterization of detailed process steps. A review of detailed computational methods is provided in Ref. 46. Detailed cost analysis is broken down into three categories: traditional cost analysis, activity-based cost analysis, and manufacturing simulation approaches. Traditional cost analysis computes the cost of a product based on the labor content and materials required. Burden or overhead is added as a percentage of direct labor. This approach works well as long as the labor cost is the most significant cost driver. Activity-based cost analysis determines the cost of products based on the activities performed to create the products. Activity-based analysis does not depend directly on the labor content of an activity but rather depends on the number of times the activity is required and the number of changeovers in those activities. Unlike the first two approaches, simulation-based approaches are not accounting-based. Simulation approaches provide the best cost estimations at the expense of increased detail and complexity. Two approaches have been used. The basic approach uses factory simulation in which each process step is modeled and the process steps are then executed in the appropriate sequence simulating the actual manufacturing flow. An alternate approach employs technical cost modeling (or process-based cost modeling). Technical cost modeling is an a priori cost model based on manufacturing simulation.

Known Good Die. One of the significant challenges which confronts the manufacturing of multichip modules is the test and burn-in of the unpackaged semiconductor chips. In traditional packaging approaches, all chips are packaged individually in separate enclosures and thus can be readily tested and burned in. Full functional tests can be performed by socketing the single-chip packages and running them at full operating frequency and over the full temperature range. For dice in single-chip packages, the as-received chip yield can be 0.99999 or higher. Unpackaged dice do not have the same high probability of being good since similar test and burn-in capabilities are not widely available. Unpackaged dice receive some testing at the wafer level (before sawing); however, these tests usually check only a limited number of parameters and are not performed at speed. Limited wafer testing usually results in as-received yields of 0.5 to 0.99 which are several orders of magnitude lower than those for packaged dice.[3] The relatively large expense of multichip module packaging is further increased when extensive rework and repair to replace defective chips are required because the chips could not be fully tested prior to module assembly.

The relationship between chip yield and first-pass module yield is a simple function of the number of devices on the module. Figure 2-12 suggests that in modules which contain relatively large numbers of devices with low yields, it may be cost-effective to devote extra resources to the test and burn-in of the dice before assembly. Alternatively, it may be economical to consider using TAB bonding or single-chip packages over wire bonding or flip-chip because of their test and burn-in advantages. There is a tradeoff between the cost of chip testing before assembly and the cost of rework after assembly. Depending on the application, it may be less expensive to repair the module than to perform extra testing of the chips (or add extensive built-in test to the chips). In other applications the cost of rework may be high enough to make extensive chip testing and burn-in economical. For example, chip-first overlay interconnect approaches[47] are extremely sensitive to chip yield since repairing a completed module requires complete removal of the interconnect layers, thus making extensive chip testing prior to assembly economical for many applications.

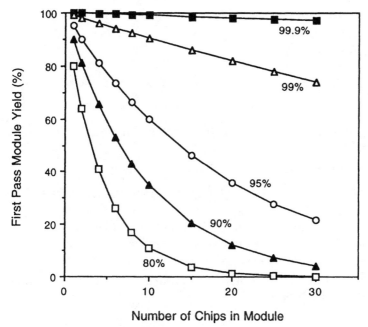

Figure 2-12. First-pass module yield as a function of the number of chips in the module and the average chip yield at assembly (chip yields of 80 percent to 99.9 percent are shown in the figure).[1] The module yield is given by $Yield_{module} = (Yield_{chip})^N$ where N is the number of chips in the module.

In Sec. 2.6 we discuss in greater detail the impact of various test and rework strategies on the cost and quality of MCMs.

2.5 Physical Design

Physical design is the process that begins with a module specification and synthesis information (component list, netlist, technology and material descriptions, and design constraints) and adds geometric information that can be used in the fabrication of the system. The information added which constitutes the physical design includes component locations, connection (etch) dimensions and locations, and hole and/or via dimensions and location. Traditionally physical design represents four activities:

- Placement and floor planning
- Routing

- Verification (design-rule check and extraction)
- Transfer of physical design information to the prototyping and manufacturing process

The following illustrates a typical process flow for an MCM physical design:

1. Create new pad stacks and new die symbols (or import them from existing libraries).
2. Import or create, and format, a schematic or a netlist.
3. Create all possible vias between layers (according to the manufacturing design rules).
4. Place the dice (symbols) on the substrate (either manually or automatically), and check the thermal and electrical performance predictions. Iterate if necessary.
5. Run routers. Interactive routers and autorouters are used to complete the routing process.
6. Route powers and grounds.
7. Conduct design rule checking (DRC) and correction.
8. Analyze the electrical and thermal characteristics of the design. Refine the placement or routing if necessary.
9. Transfer physical design data to the MCM foundry in the appropriate format.

To minimize the number of iterations, both placement and routing tools must be driven by the performance constraints of the target design, e.g., by a performance-driven layout system.

2.5.1 Process- and Technology-Driven Physical Design

There are many different MCM technologies and foundries from which the designer can choose. Each has its own unique manufacturing design rules. Hence, the physical design process must be customized to create an interconnect layout that is consistent with the rules for the chosen technology and MCM foundry. Thus the MCM physical design tools must accept process and technology rules on which libraries, design rule checking, placement, routing, and computer-aided manufacturing (CAM) outputs are based.

Technology Rules. MCM technology rules cover line widths and clearances, via sizes, holes and metal pads, via geometrics (such as staggered, staircased, or stacked), power and ground plane construction, layer rules and build order, thermal structures, and chip assembly rules (for TAB, flip TAB, chips first, flip-chip solder bump, wire bond, etc.). In addition to this list of absolute specification rules, there are often many design and yield enhancement rules, such as using specific types of vias or providing greater clearances between metal features.

Table 2-2 summarizes some typical technology rule values for MCM-C, -L, and -D. The technology and process rules are used during the various steps of the physical design process to ensure the creation of a correct manufacturable design layout. For example, library components are checked for conformity with the technology rules, the placement of the chips is optimized based on knowledge of the assembly technology and its rules, and the manufacturing and testing processes influence autorouting and design rule checking based on chip proximity.

Interconnects (substrates) are defined according to a sequence of processing steps resulting in the creation of pad layers, vias, interconnect layers, and metal plane layers. Metal plane layers are an important structure in the MCM layout. Process rules describe specific dimensions of allowable patterns. High-speed digital circuits require power and ground planes; however, solid metal areas are difficult to process. Metal areas and plane layers have to accommodate staggered via patterns that traverse the plane and facilitate the removal of outgassing from organic dielectrics during their cure.

A via is a structure that connects two layers. Using a single via type to connect traces on different layers (e.g., a stacked via) can introduce

Table 2-2. Typical MCM Technology and Process Rule Values

	MCM-D	MCM-C	MCM-L
Line widths*	10–25	100–250	25–100
Line/line spacing*	10–25	100–250	25–100
Via hole diameter*	10–50	100–250	25–150
Process layers	2–6	2–100	2–10
Component spacing*	400–800	800–2000	800–2000
CAM outputs	GDSII, DXF, HDICON	Gerber, GDSII, DXF	GDSII, Gerber, DXF

*All values in micrometers.

thermal mismatches between the various layers. Yet the complexity of MCMs necessitates routing approaches that connect devices by using the least amount of interconnect. To help connect complicated traces, MCMs typically accommodate a wide variety of via structures. These include blind, buried, staggered, staircased, and spiraled structures, as shown in Fig. 2-13.[4]

Staggered vias can be placed directly underneath each other—as long as they are separated by at least one dielectric layer. Staircased vias, however, do not generally fall underneath each other and are laid out across the substrate in a stairstep fashion. Spiral vias, like staggered vias, can be placed directly underneath each other, but must be separated by at least two dielectric layers.

MCM designers are often restricted to using via structures that are compatible with their existing manufacturing processes. Technology and processing rules define whether vias may be stacked to create a through-hole structure. Alternatively, vias may be staggered with minimum and maximum via-to-via distances on a single net. The most difficult rules involve vias which are partially stacked or allow for vias that may align vertically on complex staggered patterns. For example, in low-temperature cofired ceramic processes, two or three layers of vias can be stacked. Beyond one layer of staggered vias, via stacking rules become very sensitive to the processing technology. Different rules may even be needed for different layers. For example, wider trace clearances on initial metal layers of high-temperature cofired ceramic MCM-C substrates improve the yield.

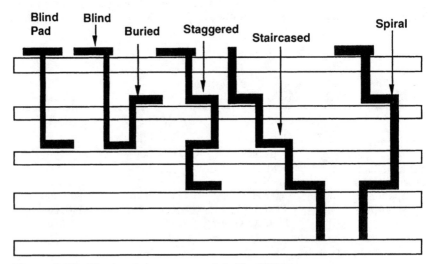

Figure 2-13. Via structures accommodated by MCMs.

2.5.2 Placement and Floor Planning

Placement and floor planning are closely associated with routing and represent a critical first step in physical design. Placement is the process of concisely fixing the relative positions of all components in the module. Poor placement can result in a difficult or impossible routing problem, whereas good placement can substantially ease potential routing problems. Since placement directly affects the length of interconnects between pins, the performance of critical nets is highly dependent on placement. The operation of placing the components is aided by a "rat's nest" (Fig. 2-14). The rat's nest displays all nets that must be used later and can be used to manually facilitate routing by visually reducing potential net congestion.

Floor planning and assignment are related to placement, and many of the same techniques can be used. In both floor planning and assignment, however, an extra degree of freedom exists—flexibility in the interfaces of the components that comprise the design. In designing ICs this results in flexibility to assign pad locations on the die. This extra degree of freedom significantly expands the size of the design space that must be considered in component placement. The assignment process addresses pin and gate assignment issues within chips. In a design environment such as that shown in Fig. 2-3, assignment issues would be treated concurrently with packaging-level placement and floor planning; however, assignment is usually not a degree of freedom that the packaging CAD user is allowed. Floor planning issues related to packaging CAD are generally confined to treating the order of the I/Os, leaving the MCM as a variable when the overall system design is considered.

Automatic Placement. MCM layout systems give designers access to a combination of manual and automatic placement. The designer may place the critical components manually, then allow the CAD tool to automatically place the remaining components. Automatic placement stems from the necessity to honor goals such as minimizing layout area, minimizing crosstalk, distributing heat uniformly, maximizing electrical performance and noise margins, and reducing cost. To automate placement, these goals are cast into an *objective function*. The objective function computes a "score" which measures the quality of a candidate placement. Traditionally the metrics which make up the objective function are divided into two classes: those that assume that the nets do not interact with other nets and those that account for net-to-net interaction. Noninteracting net metrics usually consist of some measure of the total wire length required to complete the problem. Net-to-net interaction

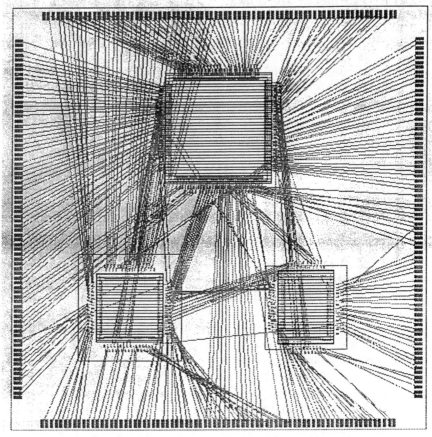

Figure 2-14. An example of a rat's-nest drawing for a three-chip MCM. The rat's-nest drawing shows the minimum spanning trees of the signal nets and is a good visual indicator of the quality of a placement from a wiring standpoint.

metrics focus on congestion. Congestion can be observed by using a rat's nest drawing like the one shown in Fig. 2-14. Numerous metrics which measure congestion have been used. These include counting the number of nets that cross a cut line or the track density within a routing channel.

Since many real design goals are difficult to cast into objective functions that can be efficiently evaluated by a computer, more restrictive objective functions must be substituted. The hope is that a placement which results in a "good" score with respect to a restrictive objective function is in fact also as good with respect to the actual goals. Unfortunately, the wire-length- and congestion-oriented objective functions, while necessary, are not sufficient for component placement in MCMs. Placement of components in MCMs requires attention to manufacturability, cost, test, and performance issues in addition to wiring. The authors are unaware of any automatic placement or partitioning efforts which address MCM manufacturability, cost, or testing.

Once an objective function is determined, the placement (or similarly partitioning) problem represents a multivariable optimization exercise; and many of the mathematical methods used for the solution are based on more generic optimization approaches which are also used outside the area of microelectronics. Placement (and partitioning) algorithms are divided into two categories: constructive and iterative improvement methods. An introduction to basic placement algorithms is contained in Ref. 48.

2.5.3 MCM Routing

Difficulties in routing conducting traces on MCM interconnects generally stem from the scale of the problem and the need for high performance (controlled impedance) at low cost. An MCM design may contain hundreds or thousands of nets. The sheer number and the potential configurations of MCM vias further complicate the problem. Because most MCMs have a fine pitch and several layout layers, the complexity is much higher than that of a typical printed-circuit board. These challenging complexities necessitated the relaxation of normal routing-geometry rules. Some of the requirements unique to MCMs are the relaxation of one-layer, one-direction rules; arbitrary-angle routing; variable-width traces; spacing on a per-net basis; and evenly distributed wiring. A brief review of MCM routing issues is given in Ref. 49.

Also important is the ability to support an incremental-design style to accommodate the numerous design revisions that typically occur during the design cycle. Thus, changes must be easy to make and must have a minimal effect on the rest of the design.

Evolution of MCM Autorouting Algorithms. Until the mid 1980s, all autorouting algorithms were grid-based. Some of the standard algorithms included maze routing, shove-aside, squeeze-through, and rip-up and reroute. However, many grid-type systems did not adapt well to the increased density and variety of pin spacings needed for MCMs. Most people assume that a grid is an array of orthogonal lines. But none of the standard grid-based algorithms demand an orthogonal grid. They only demand that the relative position of each grid maintain an orderly pattern. Grid-based routing algorithms are reviewed in Ref. 50.

Accordingly, a new class of routing algorithms was developed which distort, or warp, an existing orthogonal grid to optimize available free channels and to meet off-grid feature requirements.[51] Manufacturing and clearance design rules drive the grid warping technique. Hence, warp-based routing qualifies as a general method for increasing routing usage percentages and improving manufacturability. The Prance-XL autorouter, which is part of Cadence Design Systems Inc.'s Allegro system, is a good example of warp technology autorouters.

Recently, a gridless routing approach, called *shape-based routing*, has been pioneered commercially by an autorouter called Spectra from Cooper & Chyan Technology (CCT) Inc. Autorouters that use grids decompose the information into an array of points and retain information about each point, or grid cell, in memory. As the grids become finer and more dense, they occupy a larger amount of memory and an increasing amount of the computational effort is spent on grid-cell management rather than on solutions of the autorouting problem. On the other hand, shape-based autorouters use geometric shapes to provide more freedom in creating dense routing and simplify processing while conserving memory. Design rules and electrical information are attached to each shape, hence memory use grows only linearly with the number of shapes associated with the design.

Another approach for routing MCMs, called *rubber-band routing,* has been proposed. A tool named SURF, developed at the University of California at Santa Cruz,[52] models routing with a canonical form of topological wiring known as a *multilayer rubber-band sketch*. Each layer of the sketch is composed of a set of planar wire branches. Each branch is modeled as an elastic band that contracts to the shortest possible length while maintaining the same topology. The wires are not allowed to cross over one another in each plane. The endpoints of the rubber bands may represent terminals, vias, or branch points. If an endpoint moves, the wires automatically stretch (as would an infinitely elastic band). The rubber-band sketch can be transformed easily into rectilinear or octilinear wiring. It can also naturally support arbitrary or all-angle wiring patterns. Because this method is not grid-based, not all wires need to be the same width. In fact, the user can make the wire width a function of the wire length in order to control reflections.

Most commercially available autorouting software programs use a variety of algorithms to route high speed MCMs. These algorithms include grid-based methods as well as shape- and pattern-based ones. Designers have found that using autorouters can dramatically reduce total layout time. However, no autorouter can guarantee a 100 percent solution to all routing problems. As products stretch technological limits, designers need interactive routing to preroute critical high-speed signals and to fine-tune their designs, especially to meet timing and other system-level requirements.

Process- and Performance-Driven Routing. Physical MCM layout systems let designers specify rules and constraints for complex layer and via structures as well as manufacturing ground rules. There are two basic methods to ensure that technology- or foundry-specific constraints are met. First, use an iterative design-then-analyze approach. In this approach, the design is completed and then analyzed for thermal and signal integrity as well as for design and manufacturing rule compliance. Second, a correct-by-design approach guarantees that the produced layout satisfies electrical, physical, or thermal considerations required to ensure signal integrity, reliability, testability, and overall manufacturability. Most commercially available physical MCM design systems are undergoing a gradual evolution to support the correct-by-design approach. This evolution is apparent in the tight coupling between the design and analysis tools, as well as the incorporation of more and more tradeoff analysis and prediction tools early in the design cycle (see Sec. 2.2).

2.5.4 MCM Design Kits

The industry's need for low-volume access to high-volume MCM foundries and the general need for reduced cycle times and costs have encouraged teaming up among MCM foundries and electrical design automation (EDA) tool vendors to develop and standardize MCM design guidelines and kits. These design kits have a purpose similar to ASIC design kits. They provide the means for customizing a specific MCM design to fit the technology and manufacturing parameters of a specific MCM foundry. The typical contents of a design kit include the pad and symbol libraries, technology files, MCM substrate design template, and a sample design. The design kit for a particular foundry contains all the information necessary to manufacture that design at that foundry. Efforts are under way to eventually ensure that the design information passed between designers and foundries is standardized so that designers know what information will be provided by a foundry and how to interpret it. The Application-Specific Electronic Module (ASEM) CAD Alliance sponsored by the Advanced Research Projects Agency (ARPA) is an example of such an information identification, modeling, and standardization effort.

2.6 Design for Test

Testing MCMs is a difficult task because they contain multiple high-pin-count chips connected together into one circuit with high-density interconnections. MCMs (more than any other packaging approach) stress the limits of available test solutions at the chip, substrate, and module levels.

2.6.1 Selecting an Optimal Test Strategy

Why are conventional test methods inadequate for MCMs? Traditionally, bed-of-nails testers have been used to test printed-wiring board assemblies. However, because of the high chip density and the small interconnect line dimensions of emerging MCM and other packaging technologies, this approach cannot be used to gain access to internal lines. Even if it were possible, this type of tester is extremely costly. Also, external testers are rapidly approaching another physical limitation—the circuits under test are faster than the delay along the tester probe circuitry.

High-quality, cost-effective multichip modules must be designed with test and fault diagnosis as critical design requirements. However, deciding on where and when to test—and whether to apply design for test (DFT) and built-in self-test (BIST) at the IC, multichip module, or board level—requires considerable study and evaluation to determine the economics of the various solutions and the payback. One must examine tradeoffs between various test and rework strategies for MCM designs. Some of these strategies incorporate various DFT options at both the MCM and IC levels. However, test is still treated as an afterthought in the design process, and designers of such complicated systems have little or no MCM testability guidance at this time. This situation can be attributed to the lack of awareness of the importance of test and the lack of tools that can provide the designers with advice and guidance.

The cost and the resultant quality of an MCM are dependent upon (1) the yield of the chips, (2) the number of chips in the module, (3) the yield of the interconnect structure, (4) the yield of the bonding and assembly processes, and (5) the effectiveness of the testing and rework process in detecting, isolating, and repairing those defects. Chip yield plays an important role, as shown in Fig. 2-12. For example, for a 50-chip MCM, if the yield of the incoming chips is 95 percent, the yield of the assembled MCM before testing will be 7.7 percent (without considering any other source of yield loss). Thus more than 90 percent of assembled MCMs have to be diagnosed and repaired, a costly and time-consuming task. Hence, the incoming yield of bare chips must be pushed to nearly 100 percent if the module yield is to be high enough to have a cost-effective MCM process. But while the yield of packaged ICs can be pushed to nearly 100 percent by using sophisticated IC testers, the infrastructure required to produce high-quality known good dice to support MCM manufacturing does not exist today.

An obvious approach to alleviating the need of sophisticated testers at all levels of integration is to incorporate the tester into the circuit under test itself, and hence the notions of design for test and built-in self-test. This eliminates the need for expensive testers and provides a mechanism for accessing and exercising internal design circuitry. BIST, by definition, can be executed at the normal speed of the design under test. However, DFT and BIST options are not free; they require an investment in chip area, and in certain cases they may themselves present additional delays in critical circuit paths.

Ideally, to solve the MCM testing problem, one may require that every chip incorporate DFT and BIST. However, understanding the economics of DFT and BIST is crucial to determine the proper amount of DFT and BIST to include at the IC and module levels. In most cases, only a subset of the chips needs to incorporate testability features, while other cases may not require DFT features. In fact, this situation cannot be avoided since most MCMs contain off-the-shelf or existing components that have varying degrees of testability in addition to custom parts which the MCM builder may have control over. The situation then becomes one of exploiting the partial DFT features in testing the system as much as possible to maintain a high level of test quality and diagnosis.

It is also important to note that, from the IC manufacturer's viewpoint, the economics of incorporating DFT into an IC design may not be favorable for all applications without considering the ramifications at the MCM, board, and system levels.

There are numerous test solutions which have unique advantages and disadvantages. As an example, Fig. 2-15 illustrates various test solutions which might be available through the complete process of producing a fully functional, fault-free, and reliable MCM.

An optimal test strategy must be evaluated concurrently for all stages of the manufacturing process. Numerous factors must be concurrently analyzed in selecting an optimal test strategy for achieving a known good MCM. However, to weigh the benefits of the different approaches, they have to be compared against a common set of metrics. Two primary quantitative metrics should be used as a minimum for comparing the various test methods:

- The impact on the cost of the MCM
- The impact on the quality of the MCM measured by the defect level in parts per million

Selecting cost as one of the comparison metrics reflects the important role that test plays in determining the final cost of an MCM. It also emphasizes the importance of finding cost-effective test solutions, not just test solutions at any cost.

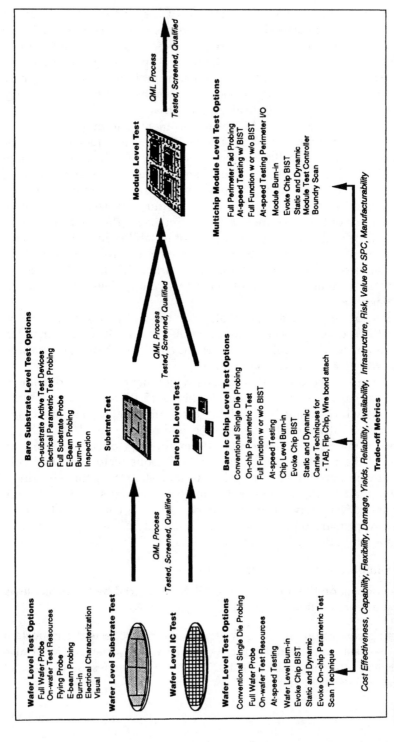

Figure 2-15. Potential MCM test solutions.

The second performance metric serves as a balance against recommending cheap solutions with inferior quality. We will use the defect level (or equivalently, the outgoing yield) of the produced MCM as a measure of quality. Since all alternative test methods should not change the product functionality substantially (some DFT methods result in some performance degradation), cost and quality are the two prime characteristics for evaluating any product.

Besides cost and quality, there are several other evaluation criteria, listed below:

- The impact of a test strategy on time to market. Although variations in design, manufacturing, test, and repair times are always accounted for in the cost, test strategies that result in an excessively long time to market must be avoided.

- Capability, extendibility, and limitations of test strategy with respect to various design complexities, such as circuit size, pin count, technology variations, and maximum speed. Some methods may be eliminated from consideration if they fail to meet state-of-the-art complexities or if they have limitations that would make them obsolete in the near future as complexities continue to increase. For example, some probing methods do not work beyond certain frequencies or for very small contact pads.

- Adverse impact of a test method on the reliability of the product. Certain die test methods can damage the IC pads or may cause thermal overstress. Another example is MCM test strategies that rely on having a high number of rework cycles. The reliability of the product resulting from these test methods might be significantly reduced, and hence they should be avoided if possible.

- Usefulness of a method at higher levels of integration and for field test, diagnosis, and maintenance.

- Electrical performance degradation caused by a test method. Most DFT methods introduce delays into the design. However, proper design of DFT features could avoid affecting critical path, reducing the overall performance impact of DFT.

- Impact of a test method on manufacturability. This covers questions such as these: How easy is it to adapt this method to existing manufacturing processes? How big an infrastructure is required to support the new method? How expensive is the equipment required to support this test method?

2.6.2 Test Economics Modeling Methodology

The test economics modeling methodology described here is based on work done at Microelectronics and Computer Technology Corp.[53] It employs a hierarchical process flow model for representing the complete manufacturing and test process starting at the wafer and going all the way to the MCM level. The tradeoffs among the various test strategies are evaluated by computing the impact of various cost, yield, and test strategy parameters on both the final cost and the quality of an MCM. Some of these key parameters are listed below:

Cost Parameters

- *Wafer cost*
- *Wafer probe cost*
- *Wafer burn-in cost*
- *Die cost*
- *Die burn-in cost*
- *Die test cost*
- *Substrate cost*
- *MCM assembly cost*
- *MCM test cost*
- *MCM diagnosis and repair cost*

Yield Parameters

- *Wafer defect level*
- *Wafer yield*
- *Wafer burn-in mortality rate*   (the fraction of dice that fail during wafer-level burn-in)
- *Die burn-in mortality rate*
- *Substrate and interconnection yield*
- *MCM assembly yield*
- *MCM rework yield*
- *MCM burn-in mortality rate*

Test Parameters

- *Wafer-probe fault coverage.*   Fault coverage is the fraction of the faults detected by the test, relative to all possible faults. Fault coverage could be associated with one specific type of fault (e.g., stuck-at, open, short); however, it is used here as an average coverage for all types of faults.

- *Die-test fault coverage*
- *MCM-test fault coverage*
- *MCM-rework scrap rate.* The fraction of defective MCMs that cannot be repaired and are scrapped.
- *The maximum number of allowed repair cycles per MCM.* An MCM that still appears defective after exceeding the limit of repairs is scrapped.

Various test methods have differing sets of values for some of the parameters listed above. Some of the methods, such as a wafer-level at-speed test method or using a chip carrier for die burn-in, have a local impact on a subset of the parameters, while other test methodologies, such as incorporating BIST into an IC design, have an impact on many parameters. To analyze the relationship among all the above parameters and how they impact the final cost and quality of the MCM, a Hierarchical Test and Economics Advisor tool (Hi-TEA) has been developed.[53] The tool embodies simple generic assembly, test, and rework simulation models. Figure 2-16 illustrates how the three basic process models are used to create a simple MCM manufacturing flow.

Each box in Fig. 2-16 represents a specific process step (assembly, test, and rework). The assembly box takes one or more inputs that correspond to the various types of dice to be assembled and the substrate itself. Each input is characterized by a cost, yield, and a count indicating how many times it is used (for example, 10 SRAM chips). The assembly process itself has a cost and a yield factor, as does the pretested MCM substrate. This module is then tested with a process that has a cost and a certain degree of fault coverage. The modules that pass the test are assumed good and are passed to the next level of assembly. However, since the testing process cannot detect all possible faults, some defective modules may escape and reduce the quality of the output modules.

The modules that fail the test are then diagnosed for possible repair; i.e., some modules may not be repairable and are scrapped. The repairable modules are repaired and are resubjected to the testing process. The diagnosis and rework process has an average cost and also has a yield (i.e., not all repaired modules are good due to rework assembly defects, new-component defects, accidental defects, or misdiagnosis). A module is scrapped if it fails to pass the test after being repaired a maximum allowable number of times.

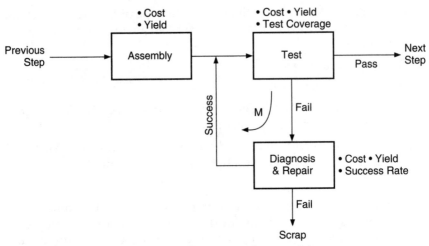

Figure 2-16. A simple MCM manufacturing and test model.

To illustrate the usefulness of the simple model described above, consider the following example. An MCM is to be built containing 50 identical chips. These chips can be procured (purchased or fabricated internally) at different prices and different quality levels, as shown in Table 2-3. According to the table, a chip that costs $2 has an 80 percent yield (that is, 20 percent of those chips are defective), while a chip that costs $9 has a 99.9 percent yield. Note that this is the yield of the die before assembly into the MCM. The table shows the corresponding MCM cost and MCM defect level. The following assumptions were used for the various parameters: Assembly cost = $200, substrate cost = $200, test cost = $150, repair cost = $100, test coverage = 99 percent, 1 repair allowed, repair scrap rate = 5 percent, and repair yield = minimum of 95 percent or chip yield. Both the substrate and the assembly process were assumed to have a perfect yield in this analysis. Hence, the chips were the only source of defects.

Table 2-3. Chip Yield versus MCM Cost and Quality

Die cost, $	Yield, %	MCM defect level (after test), ppm	MCM cost, $
2	80	105,599	1101
3	85	78,073	1120
4	90	51,345	1132
5	94	30,493	1135
6	97	15,143	1119
7	99	5,042	1034
8	99.5	2,533	1025
9	99.9	530	1017

The results shown in Table 2-3 illustrate that although the cost of the 50-chip set more than quadrupled (from $100 to $450), the cost of the MCM decreased by about 10 percent while the quality of the outgoing MCM increased by almost 3 orders of magnitude. The first case (the 80 percent case) results in shipping more than 10 percent defective parts, while the last case results in about 0.05 percent defective shipped parts. Our experience with Hi-TEA has shown that either a bell-shaped curve or in general a two-minima curve relates MCM cost to chip yield. One minimum (one end of the bell curve) corresponds to a high-quality solution using high-quality chips, while the other minimum, corresponds to a low-quality solution that uses inexpensive and low-quality chips.

Having chips with various cost and quality metrics as explained above can be directly attributed to the test strategy used for testing the chip at the wafer level and as a bare die. It may also be attributed to the amount of testing or whether DFT techniques were employed. The above experiment also points to an interesting issue at the business level. From a semiconductor company's point of view, the goal is to produce good-quality chips at the lowest possible price. Adding DFT structures may increase the cost of the chips. However, from a system manufacturer's point of view, the less expensive chips can result in more expensive MCMs, boards, or systems. Thus, it is clear that system houses are the ones that should demand high-quality chips which are easy to test at higher levels of assembly, and they should be willing to pay for it.

The situation is further complicated by the fact that many chips are commodity (off-the-shelf) items and hence are used in a variety of applications that do not demand sophisticated testing requirements.

The simple elements shown in Fig. 2-16 are used to model bigger parts of the manufacturing flow. The user can quickly create a customized manufacturing flow by sequencing the appropriate type of processes and then can use the resulting model to evaluate the cost and quality metrics associated with the design. For example, the model in Fig. 2-17 starts at the wafer-level processing and test steps; followed by die burn-in and test; then MCM assembly, test, and repair; and finally MCM burn-in, final test, and repair. Each of these steps is characterized by various cost, yield, and fault coverage parameters.

Hi-TEA supports a two-level (macro, micro) economics model hierarchy. At the macro level, a manufacturing flow is modeled by using a number of processing steps, as shown in Figs. 2-16 and 2-17. Each processing step, say, MCM assembly, can be broken down further at the micro level into a number of steps. At the micro level the models use parameters such as equipment cost, labor rates, depreciation base, tooling cost, and material cost.

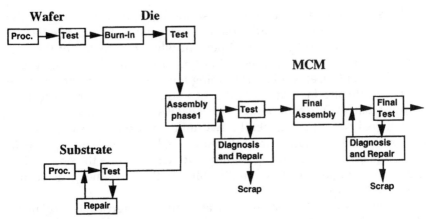

Figure 2-17. An extended test and manufacturing process flow model for MCMs.

Hi-TEA allows the user to model some of the steps at the micro level while retaining the macro-level abstraction of the other steps. Obviously, this provides the user with great flexibility. It also combines the accuracy of detailed micro cost models with the ease of macro modeling and the ease of performing quick "what if" tradeoff analyses on macro models.

2.6.3 A Case Study

An MCM containing a large central processing unit (CPU), a large coprocessor/controller chip, and ten 4-Mbit SRAM chips will be used to illustrate the use of Hi-TEA to explore the economics of various test strategies.[54] The results presented here represent only a summary of the test view of a comprehensive tradeoff analysis study that was performed on this MCM test vehicle which covered electrical, thermal, reliability, and manufacturability aspects. A complete description of the results of the test strategy tradeoffs can be found in Ref. 54.

A testability base case was defined first to serve as the basis for comparing the various test strategies. The base case depicted a favorable scenario for obtaining known good dice. The dice were probed at the wafer level, diced, burned-in, then mounted onto temporary carriers and thoroughly tested to a 99 percent degree of fault coverage. The interconnect substrate was assumed to have been pretested accurately prior to MCM assembly.

After the chips are attached to the substrate, the MCM is tested for assembly defects by using partial boundary scan features on the non-memory chips. Then the MCM is tested functionally by applying test patterns that mimic the normal-use environment and exercise the functionality, electrical characteristics, and speed of the MCM. Very little DFT and BIST were assumed available inside the chips for use during MCM testing (except for boundary scan).

MCMs that fail any of the testing steps are diagnosed for possible repair. The repair may require replacing one or more chips. After rework, the MCM is retested. If it fails again, it is returned to the rework station. Up to five repair cycles per MCM were allowed.

In addition to the base case, the following alternative test strategies were analyzed:

1. Die test variation. This case analyzed the possibility of halving the die test cost by reducing test time and test coverage (95 percent instead of 99 percent).

2. Incorporating BIST into all the chips: 5 percent area overhead at the die level.

3. Incorporating BIST into the two nonmemory chips only.

4. Incorporating a test controller chip into the MCM (cost is $60).

5. Utilizing a two-phase MCM assembly and test approach. In the first phase, the SRAM chips are assembled, and then the MCM is tested and repaired if necessary. In the second phase, the two large chips are added to the MCM, and a second test is applied with repairs as necessary.

Figure 2-18 summarizes the results of the tradeoff analysis obtained by using Hi-TEA. The figure compares the cost and defect level of the MCM (posttested) for the various test strategies. As indicated, using a 95 percent fault coverage test at the die level adversely affects both cost and quality. The two DFT cases at the chip level and the partial-assembly case result in significant cost savings and quality improvements. The full-BIST case has the lowest defect level, but the partial-assembly case has the lowest cost (below $800).

The results clearly indicate that incorporating DFT and BIST techniques in varying degrees at the chip or MCM level is economically justifiable and results in both cost reduction and quality improvement. The results also indicate that the MCM cost could vary by about 10 to 20 percent depending on the test strategy used. However, proper determination of where and how to test, and whether to employ DFT and BIST at the IC or MCM level, requires an evaluation of the economics of the various solutions and the payback. That process is highly dependent on the design under consideration and the parameters associated with the available manufacturing environment(s). Hence, one should be careful about generalizing the lessons learned from specific cases, since it could lead to nonoptimal solutions. Modeling of

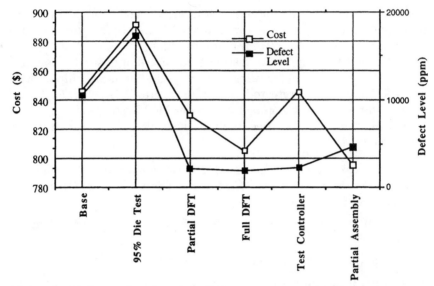

Figure 2-18. Cost and quality tradeoffs for various test and DFT strategies.

design, test, and manufacturing economics and the advisory tools are clearly what the designers of complex multichip modules and systems need in order to produce cost-effective high-quality products.

Acknowledgments

The authors thank their colleagues Hector Moreno and Larry Smith for helpful comments and suggestions during the preparation of this chapter.

References

1. P. A. Sandborn and H. Moreno, *Conceptual Design of Multichip Modules and Systems,* Kluwer Academic Publishers, Boston, 1994.

2. P. A. Sandborn, K. Drake, and R. Ghosh, "Computer Aided Conceptual Design of Multichip Systems," *Proceedings of the Custom Integrated Circuits Conference,* San Diego, CA, May 1993, pp. 29.4.1–29.4.4.

3. J. K. Hagge and R. J. Wagner, "High-Yield Assembly of Multichip Modules through Known-Good IC's and Effective Test Strategies," *Proceedings of the IEEE,* **80** (12): 1965–1994, December 1992.

4. J. Isaac, "MCM Design Tools Filling the Gaps," presented at IEEE/Components, Hybrids, and Manufacturing Technology Symposium on High Density Integration in Communications and Computer Systems, Waltham, MA, October 1991.

5. J. W. Rozenblit, J. L. Prince, and O. A. Palusinski, "Towards a VLSI Packaging Design and Support Environment (PDSE) Concepts and Implementation," *Proceedings of the IEEE International Conference on Computer Design*, Los Alamitos, CA, September 1990, pp. 443–448.

6. G. Choksi, B. K. Bhattacharyya, S. Stys, and B. Natarajan, "Computer Aided Electrical Modeling of VLSI Packages," *Proceedings of the 40th Electronic Components and Technology Conference*, New York, May 1990, pp. 169–172.

7. P. A. Sandborn and H. Hashemi, "A Design Advisor and Model Building Tool for the Analysis of Switching Noise in Multichip Modules," *Proceedings of the International Society of Hybrid Microelectronics (ISHM)*, Chicago, October 1990, pp. 652–657.

8. International Standards Organization (ISO), *EXPRESS Language Reference Manual, External Representation of Product Definition Data*, ISO TC184/SC4/WG5 N14, 1991.

9. S. Drobac, "Design Tools: Necessary, but not Sufficient," *Computer Design*, August 1992, p. 63.

10. D. K. Wilson, "Partitioning—The Payoff Role in Telecommunication System Design," *Proceedings of the International Conference on Advances in Interconnections and Packaging*, Society of Photo-Optical Instrumentation Engineers, vol. 1390, Boston, November 1990, pp. 537–547.

11. A. C. Hartmann, "Software or Silicon? The Designer's Option," *Proceedings of the IEEE*, **74**(6): 861–874, June 1986.

12. D. D. Gajski and R. H. Kuhn, "Guest Editors' Introduction: New VLSI Tools," *Computer*, **16**(12): 11–14, December 1983.

13. W. P. Birmingham, A. P. Gupta, and D. P. Siewiorek, *Automating the Design of Computer Systems, The MICON Project*, Jones and Bartlett Publishers, Boston, 1992.

14. A. W. Wymore, *Model-Based System Engineering*, CRC Press, Boca Raton, FL, 1993.

15. D. Gajski, N. Dutt, A. Wu, and S. Lin, *High-Level Synthesis—Introduction to Chip and System Design*, Kluwer Academic Publishers, Boston, 1992.

16. B. S. Landman and R. L. Russo, "On a Pin versus Block Relationship for Partitions of Logic Graphs," *IEEE Transactions on Computers*, **C-20**: 1469–1479, December 1971.

17. I. E. Sutherland and D. Oestreicher, "How Big Should a Printed Circuit Board Be?" *IEEE Transactions on Computers*, **C-22**(5): 537–542, May 1973.

18. W. R. Heller and W. F. Mikhail, "Package Wiring and Terminals," in R. R.

Tummala and E. J. Rymaszewski, eds., *Microelectronics Packaging Handbook*, Van Nostrand Reinhold, New York, 1989, pp. 65–109.

19. K.-Y. Khoo and J. Cong, "An Efficient Multilayer MCM Router Based on Four-Via Routing," *Proceedings of the 30th Design Automation Conference*, Dallas, June 1993, pp. 590–595.

20. Y. Tsukada, S. Tsuchida, and Y. Mashimoto, "Surface Laminar Circuit Packaging," *Proceedings of the 42d Electronic Components and Technology Conference*, San Diego, CA, May 1992, pp. 22–27.

21. T. S. Steele, "Terminal and Cooling Requirements for LSI Packages," *IEEE Transactions on Components, Hybrids, and Manufacturing Technology*, vol. 4, June 1981, pp. 187–191.

22. L. W. Schaper, "Design of Multichip Modules," *Proceedings of the IEEE*, **80**(12): 1955–1964, December 1992.

23. A. D. Kraus and A. Bar-Cohen, *Thermal Analysis and Control of Electronic Equipment*, McGraw-Hill, New York, 1983.

24. V. W. Antonetti, S. Oktay, and R. E. Simons, "Heat Transfer in Electronic Packages," in R. R. Tummala and E. J. Rymaszewski, eds., *Microelectronics Packaging Handbook*, Van Nostrand Reinhold, New York, 1989, pp. 167–223.

25. H. B. Bakoglu, *Circuits, Interconnections, and Packaging for VLSI*, Addison-Wesley, Reading, MA, 1990.

26. K. C. Gupta, R. Garg, and R. Chadha, *Computer-Aided Design of Microwave Circuits*, Artech House, Norwood, MA, 1981.

27. B. Rubin, "The Propagation Characteristics of Signal Lines in a Mesh Plane Environment," *IEEE Transactions on Microwave Theory and Techniques*, **MTT-32:** 522–531, May 1984.

28. A. Iqbal, M. Swaminathan, M. Nealon, and A. Omer, "Design Tradeoffs among MCM-C, MCM-D and MCM-D/C Technologies," *Proceedings of the IEEE Multi-Chip Module Conference (MCMC)*, Santa Cruz, CA, March 1993, pp. 12–17.

29. I. Catt, "Crosstalk (Noise) in Digital Systems," *IEEE Transactions on Electronic Computers*, **EC-16:** 743–763, 1967.

30. A. J. Rainal, "Computing Inductive Noise of Chip Packages," *AT&T Bell Laboratories Technical Journal*, **63:** 177–195, January 1984.

31. R. Kaw, R. Liu, K. Lee, and R. Crawford, "Effective Inductance for Switching Noise in Single Chip Packages," *Proceedings of the 10th Annual International Electronics Packaging Conference*, Marlborough, MA, September 1990, pp. 756–761.

32. S. T. Baba, W. D. Carlomagno, D. E. Cummings, and F. Guerrero, "Bonded Interconnect Pin (BIP) Technology, A Bare Chip Connection Process for Multichip Modules," *Proceedings of the National Electronic Packaging and Production Conference (NEPCON West)*, Anaheim, CA, February 1991, pp. 1167–1177.

33. E. E. Davidson, "Electrical Design of a High Speed Computer Packaging

System," *IEEE Transactions of Components, Hybrids, and Manufacturing Technology,* **CHMT-6:** 272–282, September 1983.

34. G. A. Katopis, "Delta-*I* Noise Specification for a High-Performance Computing Machine," *Proceedings of the IEEE,* **73:** 1405–1415, September 1985.

35. P. A. Sandborn, H. Hashemi, and B. Weigler, "Switching Noise in a Medium Film Copper/Polyimide Multichip Module," *Proceedings of the SPIE International Conference on Advances in Interconnection and Packaging,* SPIE vol. 1390, Boston, November 1990, pp. 177–186.

36. P. A. Sandborn, H. Hashemi, and K. Ziai, "Switching Noise in ECL Packaging and Interconnect Systems," *Proceedings of the Bipolar Circuits and Technology Meeting,* Minneapolis, MN, September 1990, pp. 148–151.

37. H. Hashemi, P. A. Sandborn, D. Disko, and R. Evans, "The Close Attached Capacitor: A Solution to Switching Noise Problems," *IEEE Transactions on Components, Hybrids, and Manufacturing Technology,* **15:** 1056–1063, December 1992.

38. R. Senthinathan and J. L. Prince, *Simultaneous Switching Noise of CMOS Devices and Systems,* Kluwer Academic Publishers, Boston, 1994.

39. U.S. Department of Defense, Rome Air Development Center, *Reliability Prediction of Electronic Equipment,* Mil-Hdbk-217F, March 1991.

40. R. J. Allen and W. J. Roesch, "Reliability Prediction: The Applicability of High Temperature Testing," *Solid State Technology,* September 1990, pp. 103–108.

41. M. Pecht, M. Dasgupta, D. Barker, and C. Leonard, "A Reliability Physics Approach to Failure Prediction Modeling," *Quality and Reliability Engineering International,* **6:** 267–273, September-October 1990.

42. J. W. Balde, "Crisis in Technology: The Questionable U.S. Ability to Manufacture Thin Film Multichip Modules," *Proceedings of the IEEE,* **80**(12): 1995–2002, December 1992.

43. G. Messner, "Price/Density Tradeoffs of Multichip Modules," *Proceedings of the International Symposium on Hybrid Microelectronics (ISHM),* Seattle, WA, October 1988, pp. 28–36.

44. C. Lassen, "Integrating Multi-Chip Modules into Electronic Equipment: The Technical and Commercial Trade-Offs," *Proceedings of the 10th Annual International Electronics Packaging Conference,* Marlborough, MA, September 1990, pp. 3–15.

45. D. G. Kelemen, "Cost Considerations in High-Density Packaging," *Proceedings of the Ninth Annual International Electronics Packaging Conference,* San Diego, CA, September 1989, pp. 1216–1222.

46. L. H. Ng, "MCM Package Selection: Cost Issues," in D. A. Doane and P. Franzon, eds., *Multichip Module Technologies and Alternatives—The Basics,* Van Nostrand Reinhold, New York, 1993, pp. 133–164.

47. R. O. Carlson, C. W. Eichelberger, R. J. Wojnarowski, L. M. Levinson, and J. E. Kohl, "A High-Density Copper/Polyimide Overlay Interconnection,"

Proceedings of the Eighth International Electronics Packaging Conference, Dallas, November 1988, pp. 793–804.

48. B. T. Preas and P. G. Karger, "Placement, Assignment and Floorplanning," in B. Preas and M. Lorenzetti, eds., *Physical Design Automation of VLSI Systems,* Benjamin/Cummings, Menlo Park, CA, 1988, pp. 87–155.

49. W.W-M. Dai, "Multichip Routing and Placement," *IEEE Spectrum,* November 1992, pp. 61–64.

50. M. J. Lorenzetti and D. S. Baeder, "Routing," in B. Preas and M. Lorenzetti eds., *Physical Design Automation of VLSI Systems,* Benjamin/Cummings, Menlo Park, CA, 1988, pp. 157–210.

51. H. Bollinger, D. Larson, and P. Arana, "Autorouter Technology Update— To Grid or not to Grid Revisited," *Circuit Design,* 6(11): 73–77, November 1989.

52. W. W.-M. Dai, T. Dayan, and D. Staepelaere, "Topological Routing in SURF: Generating a Rubber-Band Sketch," *Proceedings of the 28th Design Automation Conference,* New York, June 1991, pp. 39–44.

53. M. Abadir, A. Parikh, L. Bal, P. Sandborn, R. Ghosh, and C. Murphy, "High Level Testability Economics Advisor (Hi-TEA)," *Proceedings of the Second International Workshop on the Economics of Design, Test and Manufacturing,* Austin, TX, May 1993.

54. M. Abadir, A. Parikh, P. Sandborn, K. Drake, and L. Bal, "Trade-off Analysis of MCM Testing Strategies," *Proceedings of the 13th Annual International Electronics Packaging Conference,* San Diego, CA, September 1993, pp. 1083–1110.

3

Substrate Designs and Processes

This chapter describes the major interconnect substrate designs and processes being used for multichip modules (MCMs). All have in common the multilayering of low-k dielectrics, be they organic polymers or inorganic ceramics, with metal conductor traces and various options for generating z-direction interconnects. A partial list of companies using or producing MCMs and the prevalent process technologies is given in Table 3-1.

3.1 Thin-Film Multilayer Processes

The generic process employed by most MCM-D manufacturers consists of sequentially depositing thin films of metals and dielectrics to form a multilayer structure. Photolithography and selective etching are generally used to pattern fine-line conductors from the metal layers and to generate vias in the dielectric. Many materials and process variations are possible. For example, alumina, silicon, aluminum nitride, or aluminum metal may be used as the supporting substrate; sputtering, vacuum evaporation, metalloorganic decomposition, and electroplating are options for depositing metal films; aluminum, copper, silver, or gold may serve as conductor traces; polyimide, benzocyclobutene, epoxy, or silicon dioxide can be used for the dielectric; and vias may be formed by plasma etching, wet etching, or laser ablation. There are also variations on the design and construction of the interconnect substrate such as the unique high-density interconnect (HDI) process developed by General Electric of forming the thin-film multi-

Table 3-1. Manufacturers and Users of Multichip Modules and Associated Processes*

Firm	MCM type	Process	Dielectric features	Metallization	Interconnects
ACSIST	MCM-L	Multilayer printed-wiring board fine line	Polyimide and BT laminates	Copper: inner layers Nickel/gold: top layer	Wire bond or TAB
Alcoa	MCM-D	Multilayer thin film on silicon or cofired ceramic	Polyimide	Copper, etched vias, electroless Ni plated	Wire bond or TAB
AT&T, Bell Labs	MCM-D	POLYHIC, multilayer thin film on alumina substrates. Integrated thin-film resistors, etched vias	Photodefinable triazine, spray-coated and cured	Sputtered and electro-plated copper	Wire bond
	MCM-D	Silicon wafers, integrated Ta-Si resistors, SiO_2/Si_3N_4 capacitors	Polyimide (low-stress)	Sputtered Al; Ni plating	Flip-chip, solder bumps
Boeing	MCM-D	Multilayer thin film on 5-in-diameter silicon wafers, etched vias	Photosensitive polyimide	Sputtered copper	Wire bond or fiber optics
Computing Devices (formerly Control Data)	MCM-D	Multilayer thin film, plated-up posts	Polyimide	Sputtered and plated copper, barrier metallizations, gold top pads	Wire bond or solder attachment
	MCM-L	Multilayer printed-wiring board	Epoxy laminates	Copper	Wire bond or solder attachment
CTS	MCM-C	Low-temperature cofired alumina ceramic, multilayer	LTCC	Gold	Wire bond
General Electric	MCM-D	Unique overlay process (signal layers formed over chips), chips last approach, laser-etched vias	Polyimide film	Copper	Batch-formed, thin-film metal over die with polyimide separation
Honeywell	MCM-C	High-temperature cofired alumina ceramic	Alumina ceramic	Refractory tungsten inner conductors; gold or solder-plated top conductors	Al wire or flip-chip bonding
Hughes Aircraft	MCM-D	Multilayer thin film on alumina, silicon, or AlN wafers; etched vias	Low-CTE polyimide, spin-coated and cured	Sputtered Al inner layers; top layer Al/Ti-W/Ni/Au	Wire bond
	MCM-C	LTCC, laser-drilled holes	LTCC, alumina ceramic	Gold or silver	Wire bond

Firm	MCM type	Process	Dielectric features	Metallization	Interconnects
IBM	MCM-C/D	Low-temperature cofired ceramic transitioned to thin film for top die attachment layer; thermal conduction module	Cordierite glass ceramic composition matching CTE of Si; top redistribution layers are polyimide	Copper	Bumped die, flip-chip attached
IBM, Japan	MCM-L/D	Epoxy multilayer laminates with thin-film epoxy solder mask for top layers; photoformed vias	Epoxy laminate and photosensitive epoxy	Copper	Wire bond, flip-chip, or solder attachment
Microelectronics and Computer Technology Corp. (MCC)	MCM-D	Thin-film polyimide/copper multilayer, plated-up posts	Polyimide	Copper with barrier and adhesion metallization such as Ti-W, Cr, Ni	Wire bond or TAB
MicroModule Systems (MMS)	MCM-D	Multilayer thin film on Al, Si, or alumina wafers	Polyimide	Copper; top layer: Cu/Ni/Au	TAB or wire bonding
Motorola	MCM-L/D	Multilayer laminate; fine-line top layer, drilled vias	Epoxy-glass (FR-4); BT-glass laminate	Plated copper, plated Ni/Au top layer	Wire bond, solder, or flip-chip
nCHIP	MCM-D	Thin-film multilayer integrated resistors and capacitors on Si wafers; etched vias	Stress-free, silicon dioxide	Sputtered Al or Cu	Wire bond
NTK	MCM-C/D	Low-temperature or high-temperature cofired ceramic transitioned to fine-line thin-film polyimide/Cu	Alumina ceramic for inner layers, polyimide outer layers	Sputtered and electro-plated copper	Solder-attached or flip-chip
Rockwell International	MCM-D	Thin-film multilayer; etched vias	Polyimide	Aluminum	Flip TAB
LTCC	MCM-C	LTCC	LTCC alumina	Gold or silver	Wire bonding
Toshiba	MCM-C/D	Cofired alumina or AlN transitioned to thin film for top layer	Cofired alumina, polyimide	Copper	—

*This is not intended as a complete list of manufacturers and users. The materials and processes given are those felt to be prevalent for the company but do not exclude other materials and processes. The reader is encouraged to contact each firm for its latest technologies.

layer circuitry over the chips ("chips first" or "overlay" process) versus the more conventional "chips last" approach in which the chips are assembled after the interconnect substrate has been produced. Still another variation involves processing the thin-film multilayer structure on a mandrel and then lifting it off and rebonding it to another substrate, desirably one having a high thermal conductivity. This decal or liftoff approach also results in conformal circuits when processed on irregularly shaped parts.

3.1.1 Conventional High-Density Multichip Interconnect Processes

The more conventional thin-film process—employed by companies such as MicroModule Systems/DEC, Rockwell International, Hughes Aircraft, Microelectronics and Computer Technology Corp. (MCC), Boeing, AT&T, and many others—involves the sequential multilayering of thin organic dielectric films with thin metal films that have been photoetched, connecting the metal traces between layers by etching vias in the dielectric and subsequently metallizing them. In these processes, 4-, 5-, or 6-in-diameter wafers approximately 25 mils thick and having a surface smoothness and camber similar to those of silicon wafers used in semiconductor manufacture are processed in a semiconductor line in class 100 clean rooms or better. Wafers generally are silicon or a high grade of highly polished alumina, although the higher thermally conductive aluminum nitride, metals such as aluminum and copper, and metal matrix composites (Lanxide) are also used where high-heat-dissipating integrated circuits (ICs) are employed.

In fabricating a multilayer thin-film interconnect substrate, the first step consists in planarizing the surface by spin-coating, spraying, or flow-coating a dielectric such as polyimide over the wafer and then curing it. The dielectric is next metallized to form the ground plane. However, if the wafer is sufficiently smooth, the wafer may be metallized directly. Aluminum and copper are the two most widely used metals. Silicon as the substrate material has several advantages over other wafer materials. By heavily doping the silicon, the top surface can function as a ground plane, and by thermally oxidizing the silicon, decoupling capacitors can be formed and integrated within the silicon. After the ground plane is completed, a second layer of dielectric is deposited and cured. Vias are then formed in the cured dielectric by plasma or wet etching after a conformal mask has been formed on the dielectric. The conformal mask may simply be a thick photoresist, pat-

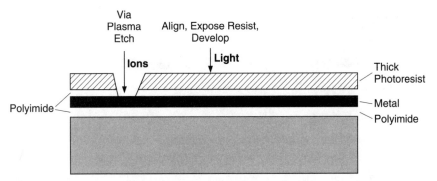

Figure 3-1. Formation of vias by plasma etching.

terned to open the vias (Fig. 3.1), or a metal mask that is formed by a photolithographic process, with the latter requiring more process steps. Vias may be etched in the dielectric by an oxidizing plasma, by reactive ion etching (RIE), or by laser ablation. After etching of the vias, the conformal mask is removed and the next metal layer is deposited. This deposition covers the entire surface including the vias. The photoresist steps are then repeated to pattern the power plane and to generate the z-direction interconnects between layers (Fig. 3-2). The process steps are repeated to form at least two signal layers and finally a top metallization layer that is used to attach and interconnect the bare dice. Most digital circuits require only five conductor layers, although some analog circuits with eight conductor layers have been designed and fabricated. Additional ground layers may also be included between signal layers to ensure controlled impedance.

To obtain consistently high yields of high-density circuits (a line pitch of 100 μm or less), meticulous in-line process controls must be imposed and particulate contaminants must be kept to a minimum, since yield is an inverse function of defect density. Although wafers are processed in class 100 clean rooms, the critical photolithographic steps should be performed in laminar-flow stations or in enclosures meeting the stricter requirements of class 10 clean rooms or better. Even so, a key source of particles is the processing equipment and the robotic movement of cassettes and wafers.[1] Simply operating the equipment in a laminar-flow hood does not suffice. Selection of equipment that generates few particles, routine monitoring of particles generated by the equipment, frequent maintenance and cleaning, and requalifying to a particle count specification prior to the release of equipment back to production are all considered important.[2]

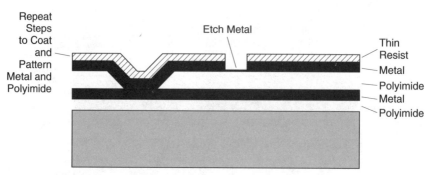

Figure 3-2. Metallization and patterning.

AT&T's POLYHIC Process.   AT&T's polymer hybrid interconnect process, called POLYHIC,* is a good example of the sequential thin-film multilayer process (MCM-D). The process starts with 3.75 × 4.5 in flat panels of high-grade (99.6 percent) alumina from which multiple smaller interconnect substrates are batch-fabricated. If resistors are required, tantalum nitride or nichrome thin-film resistors may be sputter-deposited onto the substrate, annealed, and laser-trimmed, if necessary. A low-temperature curing (200 to 250°C) triazine polymer (PHP-92) compatible with the resistors is used as the dielectric. The more conventional polyimides could not be used because their higher processing temperatures cause resistor values to drift. The triazine dielectric was formulated to be photopatternable, eliminating numerous steps in forming the vias. The triazine prepolymer is spray-coated, dried to remove solvents, and exposed to ultraviolet light through a via pattern mask, after which the unexposed via areas are wet-etched. The remaining polymer is then heat-cured and metallized by sequentially sputtering Ti, Pd, and Cu. The copper is then thickened by electroplating and is photoetched to form the conductor lines. Vias are plated with copper. The sequence is iterated to form several layers (Fig. 3-3). Wire-bonding pads of 70 μm at 105-μm pitch are characteristic of this low-cost, high-production process. Dielectric thicknesses range from 25 to 50 μm, vias are 75 to 100 μm square, and conductor line widths[3] range from 50 to 100 μm. Figure 3-4 is a photograph of a 2 × 2 in, two-conductor-layer MCM fabricated by the POLYHIC process. The POLYHIC process is being used by AT&T for high-volume production of several circuit types used in telecommunication, switching, and transmission functions.

AT&T is developing even lower-cost MCMs for the consumer market. The process produces small substrates from 5-in-diameter silicon wafers, uses polyimide dielectric (5 μm thick) and aluminum metal-

---

* Trademark of AT&T.

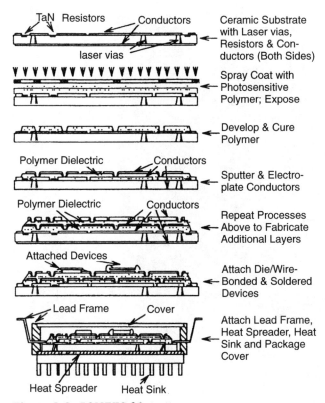

Figure 3-3. POLYHIC fabrication process.

lization (three levels), and integrates thin-film tantalum-silicon resistors and silicon nitride capacitors within the structure.[4]

Hughes HDMI Process. Hughes established a class 100 facility for producing high-density multichip interconnect (HDMI) substrates in 1987 and is among a few companies producing high-performance MCM-Ds for major military weapon systems. At Hughes Aircraft, a 1.8 × 3.8 in interconnect substrate consisting of five conductor layers has become standard for the F-22 radar, F-18 radar, and many other military programs.[5] The approximately 2 × 4 in module size was selected based on an optimum partitioning for the next level assembly on either standard electronic module (SEM-E) or standard avionics module (SAM) cards. For example, the SEM-E can hold two 2 × 4 in MCMs on each side while the SAM card can hold three 2 × 4 in modules per side. Figure 3-5 shows two 2 × 4 in MCMs assembled on an SEM-E card, one unsealed and the other hermetically sealed. Hughes Aircraft established a 6-in-diameter

Figure 3-4. Multichip module produced by POLYHIC process (2 × 2 in, 330 MHz). (*Courtesy of AT&T.*)

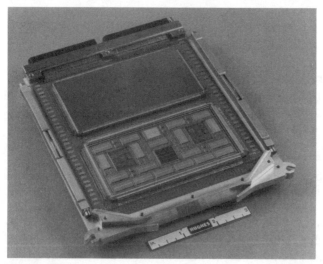

Figure 3-5. Microprocessor MCMs assembled on a SEM-E frame. (*Courtesy of Hughes Aircraft Co.*)

wafer automated line capable of processing two 1.8 × 3.8 in, four 1.8 × 1.8 in, or one 3.8 × 3.8 in HDMI substrate from one 6-in-diameter wafer.

The basic Hughes process starts with 25-mil-thick polished alumina wafers and deposits, by spin-coating, a 5-μm-thick layer of polyimide which serves as a planarization layer. A 5-μm-thick aluminum layer is next sputtered as the ground plane. If highly polished alumina wafers are used, the aluminum may be deposited directly on the wafer, eliminating several polyimide processing steps. After the ground plane has been processed, a second polyimide layer is deposited. Each polyimide layer is cured in three progressively higher-temperature steps, culminating in a 425°C cure. Step curing ensures complete cyclization and elimination of water from the condensation polymerization reaction. Du Pont's PI-2216 polyimide was chosen as the dielectric based on its low dielectric constant (2.9), low coefficient of thermal expansion (CTE) (3 ppm/°C) closely matching that of the silicon die, and low water absorption (0.3 percent). Vias that are 35 μm in diameter are then plasma-etched in the polyimide through a patterned photoresist as a mask. Both staggered staircased vias and stacked vias have been produced.

After the photoresist mask has been removed, a second layer of aluminum is sputtered over the entire surface, simultaneously conformally coating the via walls. This layer is photopatterned to create the power plane. The polyimide deposition, via etching, metallization, and photoetching steps, are then repeated to form the signal layers and the top layer which provides the sites for die attachment and wire bonding. The top aluminum layer is plated with gold by first sputtering Ti/W as an adhesion and barrier layer. The thin sputtered gold is subsequently electroplated with more gold to provide a thickness suitable for wire bonding or TAB attachment (Fig. 3-6). Five conductor layers (one ground, one power, two signal, and one top layer) have been found sufficient to accommodate most designs, although some analog circuits have required eight conductor layers.

Conductor lines of 25 μm and spacings of 75 μm (4-mil pitch) have been found optimum for reduced crosstalk. Though not possessing the electrical conductivity of copper, aluminum was adequate for most circuit functions. Aluminum also has an advantage over copper in being compatible with polyimide, not requiring the additional steps of encapsulating it with barrier metals. Furthermore, sputtered aluminum is a material more familiar to semiconductor processors than copper.

An extension of the basic process was developed by Hughes Aircraft under a Navy manufacturing program.[6] Demonstration modules as large as 4 × 4 in containing up to seven conductor layers were designed, built, and tested. Line widths of 15 μm and spacings of 35 μm were demonstrated on four signal layers comprising 500 lines per

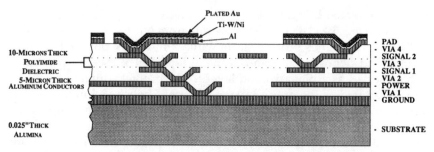

Figure 3-6. Cross section of HDMI substrate (not to scale). (Specifications: 6-in-diameter alumina, 2 × 4-in substrates, five conductor layers, special low-CTE polyimide, plasma-etched vias, aluminum inner layers, Ti-W/Ni-Au top layer.) (Courtesy Hughes Aircraft Co.)

linear inch per layer. Via continuity and line continuity were shown to be reliable after 500 temperature cycles (− 65 to 155°C), 500 temperature shocks, and 1000-h aging at 150°C. Over 15,000 vias on three 4 × 4 in HDMI substrates were tested. The modules displayed superior thermal performance because the thermal vias were designed into the polyimide and because aluminum nitride was used as the package material. Gate array ICs functioning as ripple counters assembled onto the substrates were tested at 350-MHz clock speeds through the package pins.[7]

MCC Plated-up Post Process.   The multichip interconnect substrate process developed by the Microelectronics and Computer Technology Corp. (MCC) and implemented by Control Data Corp., although a sequential process similar to the other thin-film multilayering processes, differs in the manner in which the z-direction interconnects are produced. Solid posts, also called *pillars,* are formed by pattern-electroplating instead of by the customary etching of vias and metallizing.

The base substrate may be silicon, alumina, or other support wafer. An initial layer of polyimide serving as a planarizing layer is spin-coated and cured. Chromium and copper are sequentially deposited by sputtering followed by electroplated copper. The posts are then formed by electroplating (pattern plating) through a thick patterned photoresist mask. After stripping away of the photoresist, the exposed tops and sides of the copper posts are plated with nickel to protect the copper from oxidizing or corroding during subsequent high-temperature curing of the polyimide. Nickel also acts as an adhesion promoter. Other effective barrier layers include titanium, alloys of titanium and tungsten, and chromium. The plating interconnect layer is then etched away, and polyimide is spin-coated over the surface thickly enough to slightly cover the tops of the posts. After curing, the excess polyimide is removed by lapping and polishing the surface flat, exposing the tops of the posts (Fig. 3-7). The entire surface is then metallized with copper and photoetched, and the

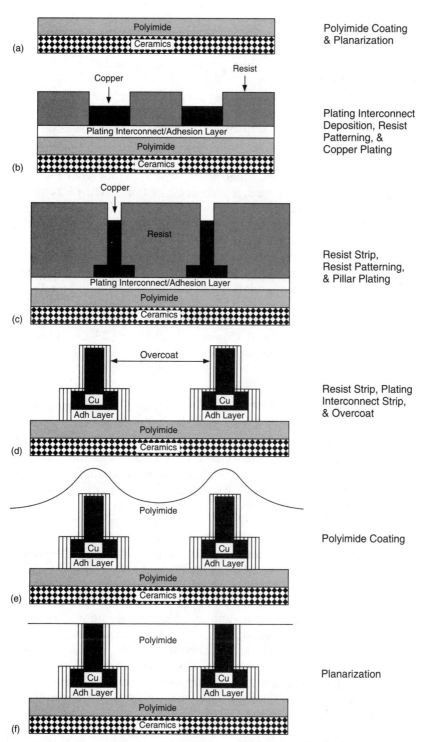

Figure 3-7. Process sequence for plated-post interconnect substrate.

processes are repeated in order to form multiple layers. As with other copper/polyimide processes, the copper is "encapsulated" by an alloy of titanium and tungsten, chromium, or nickel acting as diffusion barriers and adhesion-promoting layers.[8]

A key advantage of the plated-post process is that it ensures the planarity of the surfaces prior to metal deposition; thus many more layers can be fabricated without sacrificing yield. On the other hand, the lapping process generates particles and must be well controlled to avoid particle migration, which causes defects and reduces yields. The lapping process should be performed in an area separate from the photolithographic processing area. A second benefit of plated posts is that they are formed by an additive process, thus resulting in straight-walled posts that can be much smaller than etched vias. The capability to stack posts vertically is a third feature important in bringing electrical connections from the interior layers to a bonding pad on the top layer, where wiring tends to be most dense. Stacked solid posts, besides forming electrical connections, can also serve as thermal conduits transferring heat from the backs of the die through the polyimide to the substrate. It is reported that the process is capable of generating 10-μm lines and spacings and 15-μm vias, although 15-μm lines, 35-μm spacings, and 25-μm vias are more typical.

3.1.2 The nCHIP Silicon Dioxide Process

High-density multichip interconnect substrates produced by nCHIP also belong to the MCM-D family and are formed by sequential thin-film multilayering. However, the nCHIP process is distinct from others in that it uses an inorganic dielectric (thin-film silicon dioxide) instead of organic dielectrics (polyimide or other polymers). Although, over the years, many attempts have been made to deposit silicon oxides over large surfaces, these attempts have not been successful because of the difficulty of forming pinhole-free and stress-free films. Stresses induced in thin oxide films have previously caused cracking and interlayer shorting. However, nCHIP has developed a process that produces low-stress integral silicon dioxide films. Amorphous silicon dioxide is deposited by plasma-enhanced chemical vapor deposition (PECVD). The deposition conditions produce a film that is preloaded under a biaxial compressive stress at room temperature. This stress level compensates for subsequent tensile stresses generated by the aluminum or copper metallizations, resulting in a flat, stress-free structure.[9] The benefits of silicon dioxide compared with polyimide or other organic dielectrics include the following:

- No blistering, outgassing, or dimensional changes.
- Hard surface with greater mechanical stability, ensuring greater reliability of wire, TAB, or other interconnections.
- Improved thermal conductivity. For many applications, using silicon dioxide obviates the need for designing and fabricating thermal vias or wells in the dielectric.
- Compatibility with silicon die and silicon substrates
- Compatibility with copper conductors, eliminating the need for barriers.
- Lower risk of damaging the substrate during rework.
- Improved radiation stability.

Additional features are given in Table 3-2.

Table 3-2. Features of Silicon Dioxide Dielectric

Features	Benefits
Improved thermal conductivity	No thermal vias High wiring density
Low moisture absorption	Reliability; no corrosion with Cu Low-cost, nonhermetic packagng Precise capacitance/impedance control No leakage between conductors No outgassing during processing Low loss tangent
Chemical inertness	No corrosion of Cu (e.g., from polyamic acid) Ni cladding not required, so lower-cost All-inorganic construction (with flip chip)
Fewer process steps compared to polyimide	Lower fabrication cost
Outstanding mechanical properties - Very high Tg - High modulus - Controlled compressive stress	Reliability Nothing moves during high-temperature steps (soldering, etc.) Low deflection under stresses High-yield wire bonding Flat, stress-balanced substrates Maximum strength due to pre-load

SOURCE: Courtesy nCHIP.

The nCHIP nC-1000 series of interconnect substrates is processed on 5-in-diameter silicon wafers onto which are sequentially deposited layers of 7-μm-thick silicon dioxide and 2-μm-thick sputtered aluminum. Vias as small as 15 μm are photoetched in the silicon dioxide, then aluminum is deposited and photodelineated to create the signal, ground, and power lines (Fig. 3-8). Interconnect densities may be as high as 25-μm pitch for signal traces and 75 μm for chip-to-chip spacings. Capacitance is reported to be 1.2 pF/cm, and the inductance is 3.2 nH/cm or 1.2 nH per/wire bond.[10] The nC-1000 substrates are suited to high-density moderate-performance CMOS applications.

The nC-2000 and nC-3000 series substrates combine copper conductor traces with aluminum. Electroplated copper, 3 to 5 μm thick, is used to produce the internal signal lines, while 2-μm-thick aluminum

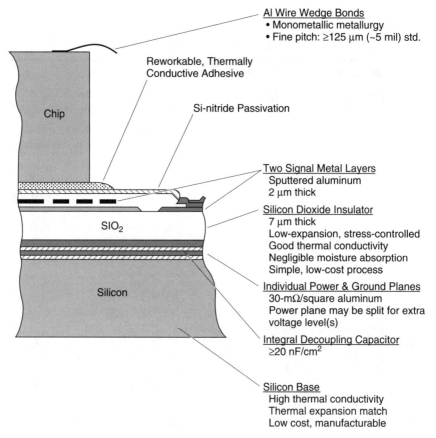

Figure 3-8. Cross section of nC-1000 series silicon substrate (schematic, not to scale). (*Courtesy nCHIP.*)

is used for the power and ground planes and for the top pad metallization to provide an all-aluminum wire-bonding system. The signal layers are separated by 3.7 μm of SiO_2 while 10.5-μm thick SiO_2 is deposited between the power planes and the first signal plane. Applications for the nC-2000 and nC-3000 substrates include high-performance, high-power ECL applications and high-speed GaAs circuits. In both substrate series, thin-film decoupling capacitors and thin-film resistors (tantalum nitride) can be integrated within the multilayer structure at the time of fabrication. For example, by anodizing the aluminum, aluminum oxide capacitors 1500 Å thick can be batch-fabricated and sandwiched between the power planes. Capacitor values of 50 nF/cm^2 are reported to be effective in providing low-inductance decoupling from 0 to >1 GHz and in eliminating package-induced ground bounce and switching noise. Substrates as large as 1.5 inches square have been fabricated to date.[9–11]

Reliability tests performed by nCHIP have shown no mechanical or electrical damage to the substrate when tested in the wafer stage. Tests included 100 thermal shock cycles from − 196 to + 300°C and 100-h exposure to high-pressure salt water and steam. Also the reliability of MCMs containing integral decoupling capacitors was demonstrated[12] by monitoring capacitance and leakage currents during Mil-Std-883C accelerated testing up to 2000 thermal shocks and 2000 temperature cycles from − 65 to + 150°C.

3.1.3 General Electric's HDI Process

The high-density interconnect (HDI) process developed by General Electric is a sequential thin-film multilayer process unique in that the interconnect structure is fabricated over the dice after the dice have been attached to a substrate. The dice are batch-interconnected without requiring wire, TAB, or flip-chip bonding.[13–15] The HDI process is also referred to as the *overlay*, or *chips-first, process* since the bare chips are first attached to the substrate before any connections are made.

Chips are attached in a close-packed array on the substrate either through a frame that is secondarily bonded to the substrate or into milled-out cavities in the substrate. The cavities (chip wells) are formed by laser ablation or by mechanical milling to different depths in order to accommodate both thin ICs and thicker capacitor chips (Fig. 3-9). The substrate-frame combination is produced from ceramic such as alumina or aluminum nitride which is premetallized with aluminum to form ground and power planes. Chips are attached with ULTEM, a polyetherimide thermoplastic adhesive, also developed by

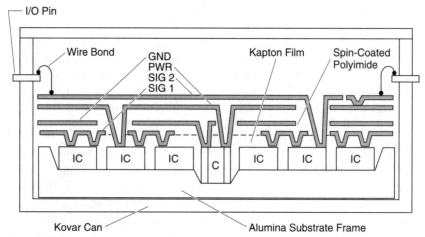

Figure 3-9. Cross section of HDI substrate showing deeper recess for capacitor chips. (*Courtesy GE.*)

General Electric. A Kapton (polyimide) film is then laminated over the dice by using the same thermoplastic adhesive that was used to attach the dice. Vias, coinciding with the die I/O pads, are laser-ablated through the polyimide. An excimer laser, programmed to form the vias in the exact I/O locations, can generate as many as 50 vias per second with 1-μm accuracy. After the vias are formed, the polyimide surface is metallized by sputtering Ti or Cr as a barrier layer followed by copper. The copper is thickened by electroplating, then encapsulated with a barrier of Cr or Ti. Negative-acting photoresist is then applied and exposed to a programmed scanning laser beam. A 2 × 2 in substrate can be laser-scanned in approximately 1 min. The unexposed areas are developed (removed by dissolving in appropriate solvents), and the unprotected metal is etched to form the conductor traces. The generation of the copper traces by using the scanning laser to selectively harden the photoresist is called *adaptive laser lithography,* and it has the advantage of being a direct-write process, i.e., one that does not require a hard mask. Through a computer-aided design (CAD) interface, changes in circuit design can be accommodated, rapidly making this a fast-turnaround process especially for prototypes and small to moderate quantities. To produce subsequent conductor layers, a liquid polyimide is spin-coated or sprayed and cured, and the processes for via formation, metallization, and photoetching are repeated (Fig. 3-10). Typical reported dimensions are 25-μm-diameter vias, 25-μm line widths, and 50-μm spacings.

The key advantages of the HDI process include these:

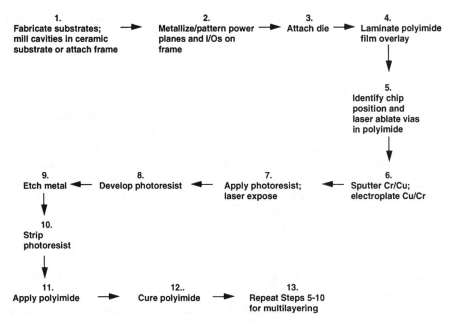

Figure 3-10. General Electric's HDI process steps.

- The batch interconnection of all the dice eliminates the need for conventional wire bonding.

- Interconnect paths are considerably shortened, and inductances are reduced.

- Chips are closely packed, providing very high silicon-to-substrate densities.

- The thermal plane is separate from the electrical plane, providing greater routing density and direct thermal transfer.

In spite of these many advantages, there have been concerns about (1) the ability to easily and economically rework dice once they have been embedded below the polyimide layers and (2) the reliability of the assembled dice after being exposed to HDI processing chemicals and temperatures. General Electric reports rework procedures in which all the interconnect layers can be peeled away after the thermoplastic adhesive is softened. The faulty dice are then removed by using an adhesive-attached thermode approach (see Chap. 9) and are replaced with new dice, and the overlay interconnect structure is refabricated. Reliability after extensive accelerated stress testing has also been reported to be good. Prototypes have been tested for mechanical, thermal, and radiation levels beyond Mil-Std-883 levels. The prototypes are reported to have survived thermal shock from − 200 to + 150°C, centrifuge to 178,000g, and high total-dose radiation levels.[16]

3.1.4 Parallel Processes

In parallel processing of interconnect substrates, the ground and power planes are processed together on an expendable wafer or carrier, and the signal layers are also processed separately on another wafer. The completely processed layers are then removed from the wafers either by a liftoff process or by etching away the wafer. The separate layers are processed by thin-film deposition and photolithographic techniques. The two halves are then bonded with an epoxy adhesive, and vias are drilled and plated to electrically connect the two sections. The structure is then adhesively bonded (laminated) to a rigid permanent support such as copper-molybdenum or aluminum, which metals also provide a good thermal conduction medium. The main benefit of parallel processing is the ability to visually inspect and test each layer or each section, avoiding scrapping of substrates at the finished stage where there has been a high value-added cost. Digital Equipment Corp. has successfully used parallel processing to produce MCMs for their VAX 9000 mainframe computer (Fig. 3-11). Starting with 6-in-diameter silicon or aluminum wafers, 4 × 4 in, high-density substrates having nine conductor layers were processed. Copper was sputtered and electroplated to form the via connections and conductors.[17]

MicroModule Systems (MMS), a spin-off company of Digital Equipment Corp. (DEC), now uses the more conventional approach of

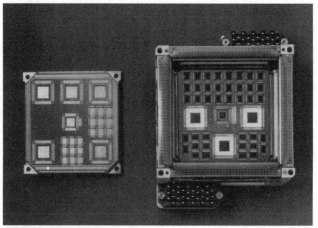

Figure 3-11. Multichip module used in VAX 9000 computer. (*Courtesy Digital Equipment Corp.*)

sequential multilayering to fabricate its interconnect substrates. The MMS process consists of spin-coating the polyimide, plasma-etching vias in the cured polyimide, sputtering copper as a seed layer, pattern-plating copper to the desired thickness, and etching away the seed layer. Chromium is deposited as a barrier between the copper and polyimide to prevent the well-known diffusion problem. A unique feature of the MMS process is the use of aluminum wafers which combine three desirable functions: a rigid, unbreakable, low-cost support; a ground plane; and an excellent thermal conductor.[18] MMS reports line widths as narrow as 17 μm and line pitches of 75 μm. Apertures may be cut in the polyimide down to the aluminum by using an excimer laser. Heat-dissipating dice can then be mounted through these apertures directly onto the aluminum, resulting in excellent thermal conduction.

3.1.5 z-Direction Interconnections in Thin-Film Dielectrics

Besides very fine-pitch conductor lines and spacings, a key to high-density MCMs is the ability to form miniature z-direction electrical connections so that the several circuit layers of a multilayer structure can be connected in the shortest, most direct paths. It is highly desirable that these connections be formed in the center of the conductor line widths instead of through expanded fan-out patterns that consume considerable surface area. Thus for an IC having 4-mil-pitch I/Os (2-mil-wide lines), it is desirable to produce vias of 1-mil diameter or less, centered within the line widths.

Two of the most widely used methods of forming z-direction interconnections in thin-film dielectrics involve (1) etching vias in the dielectric and subsequently metallizing them and (2) plating solid posts. Etching may consist of dry etching, as with plasma or laser, or wet etching, by using photosensitive dielectrics or soluble polyimide precursors. Both staircased and vertical stacked geometries are possible. In all cases, very fine vias (35-μm diameter or smaller) can be produced (Fig. 3-12).

Via Formation. The finest vias are produced by photolithographic processes followed by plasma etching, reactive ion etching, or wet etching of an organic dielectric such as polyimide or of an inorganic dielectric such as silicon dioxide. Many firms opt for dry etching to avoid the risk of chemical solutions becoming entrapped on the surface and later causing corrosion, leakage currents, or metal migration. Dry processing also avoids the expense associated with the handling

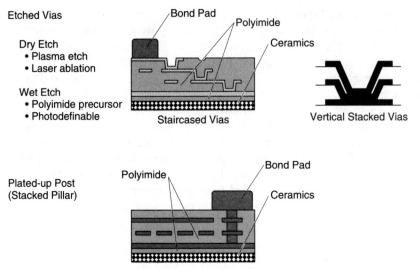

Figure 3-12. The z-direction interconnect configurations.

and disposal of hazardous wet chemicals. The added expense associated with chemical solutions, however, must be balanced against the lower processing and equipment costs for wet etching. In either dry or wet etching, photoresists are used to define the via pattern. The photoresists may be used as direct in-contact masks or to produce a hard in-contact mask from aluminum or spun-on glass, which is removed after the vias have been etched. If used as a direct mask, the photoresist must be applied thickly enough since it, too, will be plasma-etched. Some photoresist must remain to protect the rest of the dielectric until the vias are completely etched.

All fully cured organic dielectrics can be etched with plasma or reactive ions. Vias are formed by chemically reacting or oxidizing the polymer in a highly energetic ion plasma, then decomposing the polymer and volatilizing it. Pure oxygen can be used in the oxidizing plasma, but in practice, mixtures of oxygen with sulfur hexafluoride or with carbon tetrafluoride increase the etch rate and control the sidewall slope of the vias. In reactive ion etching (RIE), the dielectric is removed by a combination of chemical reaction and physical ionic bombardment.[19–21] Differences between these two methods lie in the rates of etching and in the geometries of the vias produced; plasma etching produces sloped vias (Fig. 3-13) while reactive ion etching produces either vertical sidewalled vias or tailored shaped vias.

Wet etching is also widely used and is somewhat more economical than dry etching. Silicon dioxide dielectrics are easily etched by using solutions of hydrofluoric acid. Fully cured polymer dielectrics are dif-

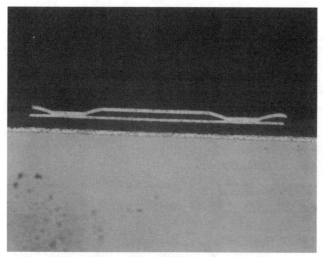

Figure 3-13. Sloped vias produced in plasma-etched poly-imide. (*Courtesy Hughes Microelectronics Division.*)

ficult to chemically wet-etch because very strong chemicals are required to decompose and solubilize the polymer. However, poly-imides may be wet-etched at the precursor stage prior to imidization. The poly(amic-acid) precursor resin is baked to remove solvents, coated with a positive photoresist, exposed to the via pattern, and then developed with an alkaline aqueous solution. The alkaline solution serves a dual role, simultaneously hardening the photoresist (masking off all portions except the vias) and dissolving the exposed precursor, thus forming the vias. After the photoresist is removed, the precursor is heated to 350 to 450°C to effect imidization of the polyimide. This wet-etch mechanism is specific to polyimides and is based on the solu-bility of the carboxylic acid portions of the precursor in alkaline solu-tions such as sodium hydroxide, ammonium hydroxide, or tetram-ethylammonium hydroxide (Fig. 3-14). Wet etching produces sloped vias because of its isotropic nature, etching laterally and vertically at the same rate.

Approximately half the steps in processing vias can be eliminated by using inherently photosensitive dielectrics. With photosensitive dielectrics the vias are formed directly, avoiding the separate photore-sist application, exposure, and developing steps that are normally required (Fig. 3-15; see also Chap. 4). However, in rendering poly-imides photosensitive through modification of their molecular struc-tures, some of their desirable properties have been compromised. Among these are decreased tensile strength and elongation, increased

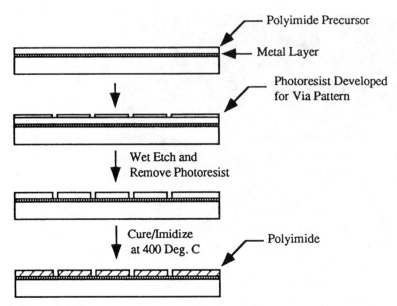

Figure 3-14. Process steps for wet-etching vias in polyimide.

shrinkage during final cure, and an increased coefficient of thermal expansion. In spite of these drawbacks, photosensitive polyimides, triazines, and epoxies are commercially available and are being applied successfully.[22,23] An example of a photosensitive dielectric used in the high-volume production of telecommunication modules is the triazine-based resin PHP-92 developed by AT&T for its well-known POLYHIC process.[3] Boeing also uses a photosensitive polyimide in producing a fiber-optic receiver (Fig. 3-16).[24]

Laser ablation is yet another method of producing miniature vias. The dielectric is volatilized in selected areas by a focused laser beam. The excimer laser has been extensively studied to etch polyimide and other polymer coatings. Thin polyimide films[25] are efficiently etched in air at pulsed excimer laser wavelengths of 248, 308, and 351 nm. There are three variations of laser etching: direct-write, *scanning laser ablation* (SLA), and projection. An excimer laser can be used in either a scanning mode or a direct-writing mode (programmed focused beam). The SLA is a batch process. A metal or inorganic mask having a via pattern is first formed on the dielectric by the normal photolithographic steps. A laser beam is then allowed to flood the entire surface through the mask, followed by removal of the mask.[26] Although SLA is a batch process, the types and number of steps required are almost identical to those used in forming vias by the RIE process. In the direct-write method, however, the laser beam is focused and programmed to form

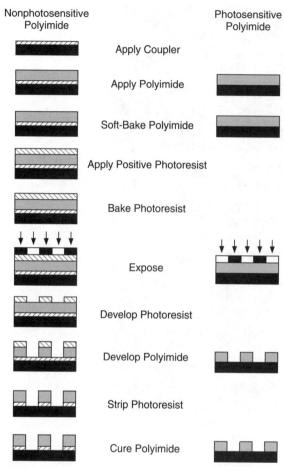

Figure 3-15. Comparison of process steps for non-photosensitive polyimide with a photosensitive polyimide. (*From Ref. 21.*)

each via separately, without the need for either a hard mask or a conformal mask. Projection laser ablation utilizes a separate hard mask to project the via pattern onto the substrate, analogous to conventional semiconductor photolithographic steppers. Projection laser ablation is especially amenable to high production.[27] Using a holographic phase mask and ultraviolet excimer laser equipment, Litel Instruments reports[28] the formation of 125,000 vias of 15-µm diameter in 1-mil-thick polyimide in less than 60 s.

In all three generic processes for etching vias—plasma, wet, and laser etching—vias of 35-µm diameter or smaller are easily produced

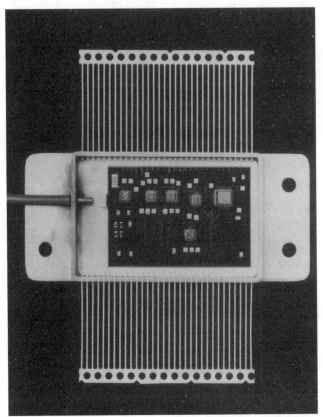

Figure 3-16. Fiber-optic receiver/transmitter, an MCM-D processed with photosensitive polyimide. (*Courtesy Boeing Aerospace.*)

in dielectric thicknesses of 25-μm or less.[29] These methods are extensively used in fabricating MCM-D interconnect substrates.

Metallization of Vias.   To connect circuit layers electrically, the vias must be metallized. This can be done by sputtering or vapor-depositing aluminum, copper, gold, or other metallization—simultaneously coating the walls of the vias and connecting them to pads which then serve as the base for forming other vertical vias. This process results in staircased vias within the multilayer structure. In an alternate method, the vias may be completely filled with metal by electroplating the vias with nickel or copper. Filled vias lend themselves to planar surfaces and to vertically stacked interconnections whose benefit lies in greater routing density and shorter electrical paths.

Plated Posts.   In the plated-post (pillar) process, the interconnections are formed by electroplating solid metal posts and then encapsulating them with the dielectric and lapping the surface flat so that the tops of the posts are flush with the dielectric, thus ready for deposit of the next metal layer (Figs. 3-17 and 3-18), also see Fig. 3-7).[30] The miniature posts, generally consisting of copper, are formed by either an additive or a

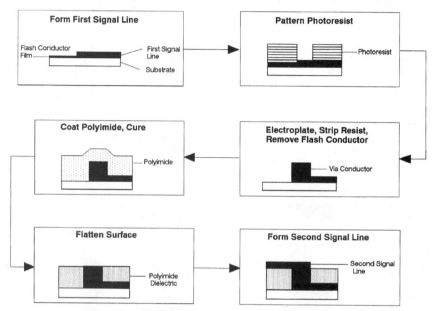

Figure 3-17. Plated-post process steps.

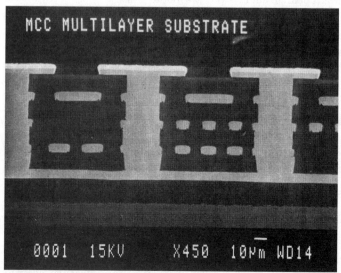

Figure 3-18. Cross section of multilayer substrate showing plated posts. (*Courtesy Microelectronic and Computer Technology Corp.*)

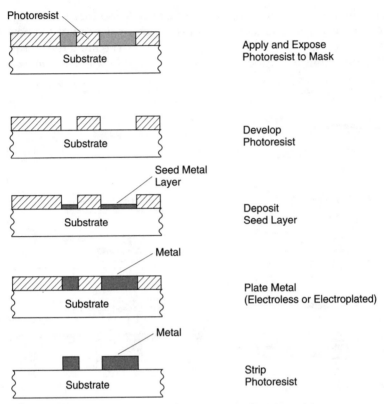

Figure 3-19. Pattern plating: additive process (not to scale).

semiadditive pattern-plating process (Figs. 3-19 and 3-20). A key advantage of the plated-post process is its capability of producing vertically stacked connections.

3.2   Thick-Film Multilayer Processes

As distinguished from thin-film processes, thick-film processes employ thixotropic paste compositions of conductors, dielectrics, and resistors that are screen-printed and fired onto a prefired ceramic substrate or screen-printed and cofired with a "green" unfired ceramic tape. Besides screen printing, thick-film pastes may be applied by direct writing, a technique in which the material is forced through a miniature nozzle and programmed to produce a circuit pattern. There are two variations of the cofired approach: high-temperature cofired ceramic (HTCC) and low-temperature cofired ceramic (LTCC). Thick

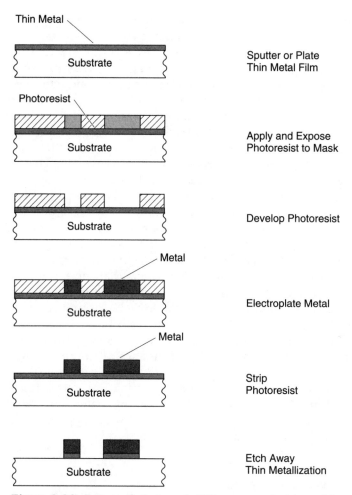

Figure 3-20. Pattern plating: semiadditive process (not to scale).

films may range from 0.5 mil to several mils thick after firing while thin films are less than 0.5 mil and are often measured in angstroms.

3.2.1 Screen-Printing Thick-Film Circuits

Sequential screen printing of thick-film pastes is among the earliest methods employed and still the most widely used method to produce interconnect substrates for hybrid microcircuits. It consists in sequentially screen-printing, drying, and firing dielectric, conductor, and resistor pastes on a ceramic substrate.[31,32] Screen-printing the dielectric through a screen mesh having the via pattern simultaneously generates the vias and produces the interlayer insulator. Vias are typically 6 to 10 mils in diameter and are filled with metal paste at the same time as the next conductor layer is screen-printed. However, to ensure surface planarity and improve yields, separate via-fill screen-printing steps may be used for multilayers having three or more conductor layers. Line widths and spacings are typically 10 mils, although some as low as 2 mils have been reported by using fine mesh screens and specially formulated pastes. Generally, this thick-film process is limited to circuits having less than five conductor layers, although, with some loss in yield, up to nine layers have been produced. Above nine layers, the cofired ceramic technologies should be used. A flow diagram for multilayer substrate fabrication using thick-films is given in Fig. 3-21.

The sequential thick-film process, although a low-cost, mature process with an extensive infrastructure, has not made a significant entry into the very high-density MCM arena. Screen-printing of thick films is limited by the dimensions of conductor lines, spacings, and vias that can be produced; in the number of layers that can fabricated without affecting yields and dimensional tolerances; and in the electrical characteristics of the commercially available dielectric and conductor pastes. Inserting some thin-film processing with thick films, however, shows promise of combining the low-cost attributes of thick-films with the high-density and performance attributes of thin films. For example, the top layers of a thick-film multilayer substrate can be processed by sputtering or vapor-depositing thin metal films and photodelineating the conductors and die bonding pads. The thick-film surface interface may have to be polished or otherwise planarized to achieve the fine-line definition required.

3.2.2 Direct Writing of Thick-Film Circuits

Besides being screen-printed, thick-film pastes may be dispensed by direct writing, thus avoiding the need for hard tooling such as screens and providing fast-turnaround fabrication of prototypes and small to moderate quantities of circuits. Direct-write systems employ a fine nozzle through which the paste is programmed to flow in selected patterns, controlled from a computer database. Thus a direct link exists between computer-aided design (CAD) and fabrication of the interconnect substrate.[33,34] Figure 3-22 shows commercially available equipment. Separate cartridges are used for each paste composition (conductors, dielectrics, and resistors of various sheet resistances). Typical conductor widths are 4 mils for gold conductors and 6 mils for silver. With certain pastes even finer lines of 3 mils are achievable.

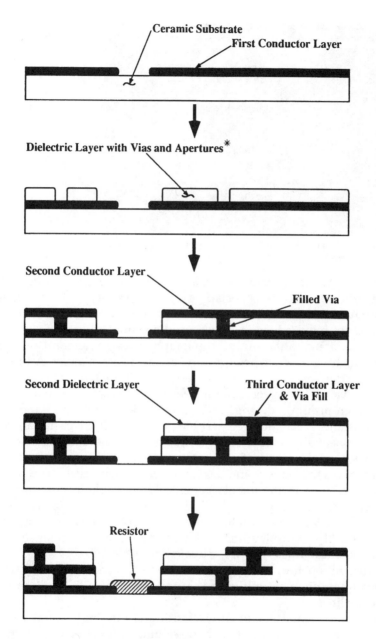

Ceramic Substrate
First Conductor Layer

Dielectric Layer with Vias and Apertures*

Second Conductor Layer

Filled Via

Second Dielectric Layer

Third Conductor Layer & Via Fill

Resistor

* Screen print dielectric twice to avoid pinholes

Figure 3-21. Thick-film multilayer fabrication steps (not to scale).

Figure 3-22. Direct-write equipment for thick films.
(*Courtesy MicroPen.*)

There are several benefits derived from direct writing compared to screen-printing that are especially suited to prototype development and production of small to moderate quantities of substrates. Flat surfaces are not required, as in screen-printing, since direct writing conforms and adapts to the surface topography. Design iterations may be made quickly to produce new prototypes, even within a day. A further advantage derives from dispensing resistor pastes having several sheet resistance values. In screen-printing, separate masks (screens) and separate screen-printing, drying, and firing steps are necessary for each sheet resistance paste. In direct writing, however, all the pastes are dispensed sequentially, then dried and fired at the same time in one operation. Closer tolerances of the as-deposited resistors are achieved which reduces the amount and degree of subsequent resistor trimming.

Direct writing has the further benefit of being able to program and vary the thickness as well as the width of the film. Thus for high-frequency microwave and analog circuits, direct pattern writing produces precisely controlled line widths, spacings, and thicknesses and even curvilinear lines. Finally, direct writing is valuable as a repair procedure to touch up screen-printed conductor lines that are discontinuous open lines by applying the paste in controlled amounts to bridge the lines.

Among the limitations of direct writing are its high initial equipment cost and its slow speed for high-volume production quantities. It is especially adapted to producing, testing, and reiterating prototypes during the initial phases of a program and to producing small to moderate quantities of parts. Direct writing should be considered complementary to screen-printing as a program evolves from engineering prototypes to production.

3.2.3   Fine-Line, Thick-Film Processes

Although the conductor line widths and spacings and via sizes that can be produced by screen-printing thick film pastes are limited, several approaches to producing thick-film circuits having dimensions approaching those of thin-films are available and others are being explored, especially for lowering the cost of high-density multichip modules. By using finer-mesh screens (400 mesh), controlling the screen emulsion thickness, and using paste formulations containing a finer, more controlled size of metal particles, line widths and spacings as low as 2 mils have been reported.[35] Fritless (non-glass-containing) gold pastes especially can provide fine-line definition. Given their high electrical conductivities, fritless gold pastes are suitable for microstrip transmission lines and for low-loss microwave circuits.

Photoetchable thick-film pastes that combine thick-film with thin-film processing have also been developed and result in even finer lines. The paste is first screen-printed over the entire top surface of a multilayer thick-film substrate, then dried and fired. Photoresist is applied next, exposed to ultraviolet light through the desired via mask, and developed. The exposed thick-film, for example, gold, is then etched by using an aqueous solution of potassium iodide and iodine. After etching, the photoresist is stripped away. By this method, line widths and spacings of 15 μm (0.6 mil) have been reported.[36]

Another variation of the photoetching process avoids the need for separate photoresist application and removal steps. It employs thick-film pastes that are inherently photosensitive.[37] Photopatternable pastes[38] with tradename Fodel were first introduced by Du Pont in 1972. Those early compositions were negative-acting; i.e., the exposed portions were hardened while the unexposed portions were washed away. Several drawbacks hindered the wider use of this technology. First, polymerization of the exposed paste was sensitive to oxygen and thus had to be performed in vacuum. Second, 80 to 90 percent of the gold paste was lost or had to be recovered after developing because the pastes were negative-acting. Third, organic solvents were required to develop the patterns. Recently, improved versions of photosensitive pastes that obviate these problems have been introduced by Du Pont. Positive-acting conductor and dielectric pastes that can be developed with aqueous alkaline solutions such as of sodium carbonate are now available. Three-mil-diameter vias, 25-μm line widths, and 50-μm spacings when fired to a thickness of 7 to 9 μm are reported.[39]

Metalloorganic Deposition (MOD).   Still another option to produce fine-line circuitry utilizes metalloorganic inks. Unlike thick-film pastes, metalloorganic inks are true solutions that may be spin-coated, sprayed, or screen-printed onto a substrate. After drying to remove the solvent, the inks are fired at temperatures generally no higher than 850°C, which decomposes and removes the organic portion, leaving the thin metal film. Then photoresist is applied, exposed through a mask, and developed. Next the conductor lines are pattern-electroplated, the photoresist is stripped away, and the original thin-metal film is removed by etching (Fig. 3-23). Metalloorganic films may also be decomposed to the free metal in selected areas by exposure to a computer-programmed laser beam. This variation of MOD is a form of direct writing; but because of the very thin films produced (several hundred angstroms), they must be thickened by subsequent electroless plating. Gold, silver, palladium, and nickel have been deposited from their respective organo salts or complexes. Typical metalloorganic complexes from which gold, silver, or copper may be deposited are shown in Fig. 3-24. Silver, e.g., can be deposited from solutions of sil-

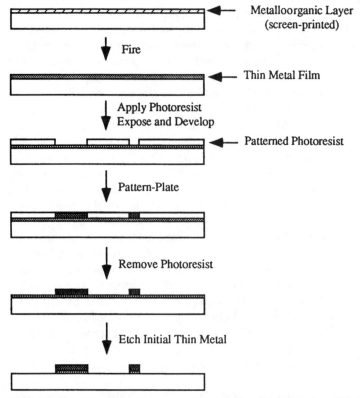

Figure 3-23. Fine-line circuitry using metalloorganic decomposition and pattern plating.

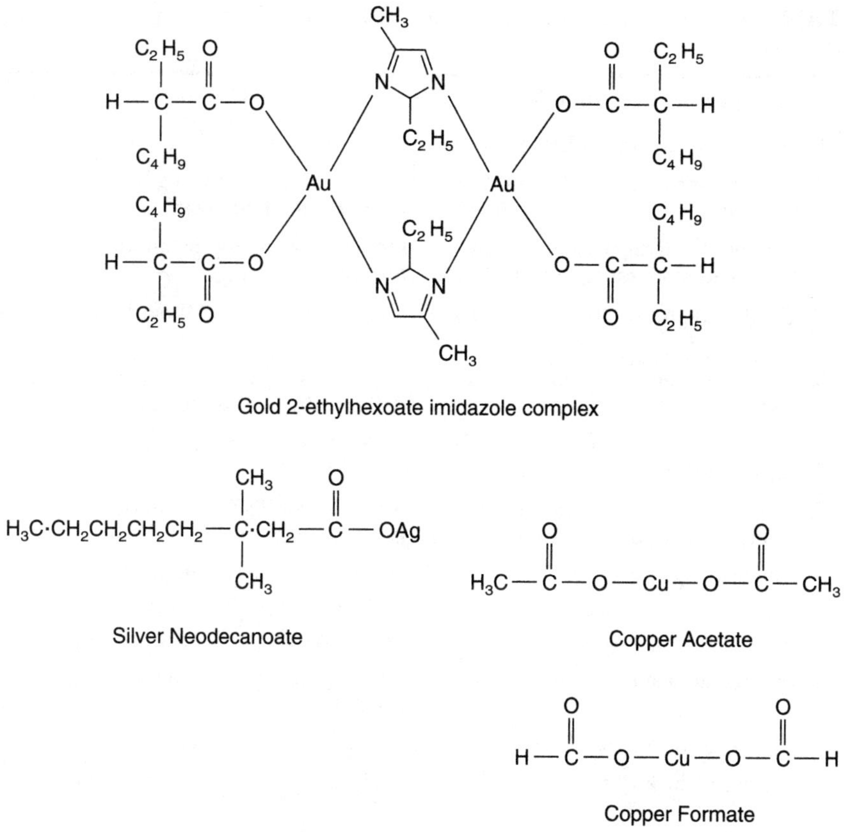

Gold 2-ethylhexoate imidazole complex

Silver Neodecanoate

Copper Acetate

Copper Formate

Figure 3-24. Examples of metalloorganic compounds.

ver neodecanoate.[40] Metalloorganic solutions are commercially available from Electro-Science Laboratories (ESL) and Engelhard Corp. Resistors and dielectrics have also been deposited by MOD.

3.2.4 High-Temperature Cofired Ceramic

High-temperature cofired ceramic (HTCC) is a thick-film process differing significantly from the thick-film process just described, even though both processes involve the screen-printing and firing of thick-film pastes (Table 3-3). Instead of sequentially screen-printing and creating multilayer thick-film pastes on a prefired ceramic substrate, HTCC involves parallel processing of individual layers of unfired (green) ceramic tape. The tape is cast from a slurry of a mixture of

Table 3-3. Comparison of Sequential Thick Film with HTCC Processes

Sequential thick film Process	HTCC process
1. Sequential screening, drying, firing	Batch firing
2. Requires air-firing 850 to 1000°C furnaces	Requires very high firing temperatures (>1500°C in H_2 atmosphere)
3. Low nonrecurring cost, excellent for low to moderate quantities	High nonrecurring cost more suitable for high production
4. Quick turnaround for design iterations	Costly and long lead times for design iterations
5. Permits variety of conductor and dielectric pastes	Limited to refractory metal conductors (W) and high-glass-content dielectrics
6. Permits wide range of batch-fabricated resistors	No screened/fired resistors
7. Yields drop with high number of layers (>7 conductor layers)	Excellent for high number of layers (>20)
8. No shrinkage problem	High shrinkages during firing (20–30%) must be taken into account
9. No plating required	Top W conductor layer must be plated
10. Limited adhesion of conductors	Excellent conductor adhesion (good for brazing lead frames)
11. Vias: 7- to 10-mil diameter	Vias as small as 4 mils
12. Dielectric thickness: 2 to 5 mils	Dielectric thickness as high as 25 mils

ceramic powder, generally alumina, glass, other oxides, organic binders, additives, and solvent, in thicknesses ranging from 4 to 25 mils. First the green tape is punched or drilled to form registration holes, vias, and cavities; then it is screen-printed with conductor pastes to fill the vias and to form the conductor traces. Next the individual layers are aligned and laminated by processes analogous to those used for multilayer epoxy circuit boards. Typical lamination conditions are 100°C and pressures between 1000 and 4000 lb/in². After being cut to shape, the laminates are prefired at 800°C to burn off the organic binders, then sintered at 1500°C or higher (Fig. 3-25). During sintering the ceramic shrinks approximately 12 to 15 percent in the xy direction, 20 percent in the z direction, and 40 to 50 percent in volume. These shrinkages must be taken into account in the design to achieve the desired final via and line tolerances.

Because of the high firing temperatures involved, only refractory metals can be used for the conductors and z-direction interconnections. Gold, silver, and copper, though far better electrical conductors, would

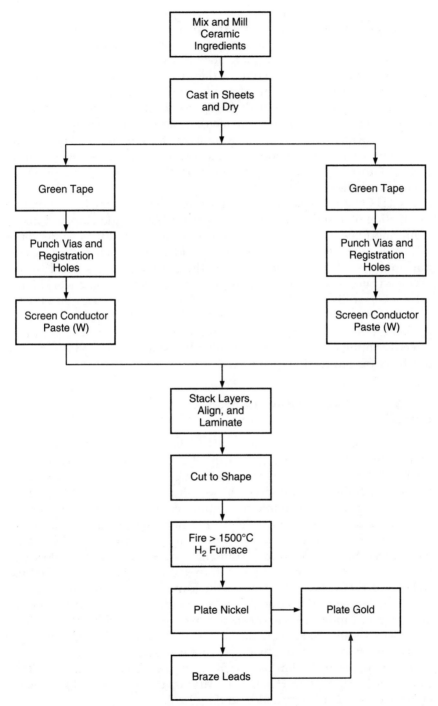

Figure 3-25. High-temperature cofired ceramic (HTCC) process (used for manufacture of packages and substrates).

melt under these high-temperature firing conditions. Furthermore, cofiring of refractory metal pastes with the green ceramic must be carried out in a reducing atmosphere such as hydrogen or forming gas to avoid oxidation and destruction of the metal. Refractory metal pastes generally consist of tungsten, molybdenum, molybdenum manganese, or combinations. Because of their higher sheet resistances (approximately 15 mΩ per square for tungsten), designers must compensate by laying out wider and/or thicker conductors for the internal layers of a multilayer substrate. The top layer, however, can be easily plated with nickel followed by gold to augment electrical conductivity and provide a wire-bondable or -solderable surface.

The HTCC process has been used for many years in the high-volume production of alumina ceramic packages for both single-chip and multichip modules. It has also been used to produce multilayer interconnect substrates used in high-density hybrid circuits such as for cardiac pacemakers. However, because of the severe process conditions involved, users have shied away from producing these parts themselves. Generally, the electronics manufacturer works with an HTCC firm in designing the part, then purchases the parts. Often a circuit or package is first prototyped by using an in-house low-temperature screen-printing and firing process, then transitioned to an HTCC manufacturer for production quantities.

3.2.5 Low-Temperature Cofired Ceramic

Low-temperature cofired ceramic (LTCC) is a relatively new thick-film process technology[41] first reported in 1982 and commercialized by Du Pont[42] in 1985. LTCC has many engineering and manufacturing advantages over both the sequential thick-film process and HTCC (Table 3-4), making it more suitable for high-density, high-speed multichip modules. Like HTCC, the LTCC process starts with a "green" unfired ceramic tape, generally based on alumina and glass. However, the LTCC tape differs significantly in composition, its higher glass content allowing much lower firing temperatures (850 to 950°C compared with 1500 to 1600°C for HTCC tape). Because of the lower firing temperatures, low-sheet-resistance metal conductors such as gold, silver, or copper (instead of tungsten) can be used and can be air-fired. Vias, registration holes, and apertures are produced in the tape by mechanical drilling, punching, or laser ablation. The individual layers are next screen-printed with gold, silver, or copper thick-film pastes, filling in the vias and patterning the conductor lines. As with HTCC, after drying, the layers are aligned and laminated, then cut to size and fired (Fig. 3-26).

Table 3-4. Comparison of Cofired Ceramic Technologies

	HTCC	LTCC
Electrical		
Conductor resistance, mΩ/square	15 (tungsten)	5 (gold) 3 (silver)
Dielectric constant	9.7	4.0–8.0
Dissipation factor, %	<0.03	0.03–0.20
Dielectric strength, V/mil	550	>1000
Volume resistivity, Ω•cm	>10^{14}	>10^{14}
Thick-film resistors	No	Yes
Thermal/Mechanical		
Thermal conductivity, W/(m•K)	15–20	2.0–5.0
CTE, ppm/°C	6.5	5.0–8.0
Flexural strength, Klb/in^2	60	20–30
Density, g/cm^3	3.7	2.9
Physical		
Conductor width, mils	5 or less	5 or less
Via size, mils	4.0–6.0	4.0–6.0

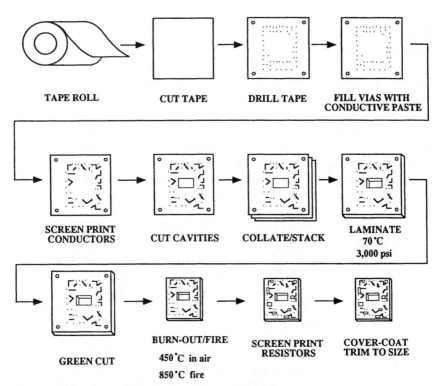

Figure 3-26. Process flow for low-temperature cofired ceramic.

Conventional screen printers and air-firing furnaces, already available at most hybrid circuit manufacturers, are used to process the green tape. Additional equipment required includes lamination and via drilling equipment. Thus the equipment and processing of LTCC come close to what thick-film hybrid manufacturers are familiar with. The processing of LTCC multilayer substrates, integral-lead packages, and substrate-package combinations can be performed in house by the microcircuit manufacturer, resulting in faster turnaround, lower tooling costs, and greater versatility in designs. Ceramic tapes in various thicknesses and dielectric constants are commercially available from several suppliers, including Du Pont, Electro-Science Labs, EMCA-REMEX, and Ferro. The tapes are sold in sheets or rolls supported on a Mylar film.

Besides the advantages of being compatible with high-conductivity metals and firing at low temperatures in air, the LTCC process permits the cofiring and integration of thick-film resistors, capacitors, and inductors. These passive components which are based on various metal oxides cannot be processed in HTCC since the combined high temperatures and reducing atmosphere of that process would reduce the oxides and destroy their electrical functions. The dielectric tapes have dielectric constants lower than those for HTCC or thick-film pastes due to their higher glass content. Some tapes have dielectric constants as low as 4.8. High-dielectric-constant tapes, however, are also available so that one can mix both low-capacitance and high-capacitance layers in the same structure. LTCC structures, also because of their high glass contents, have lower expansion coefficients, rendering them more compatible with large silicon dice and with flip-chip bonding. The expansion coefficients can be tailored to match those of silicon or gallium arsenide devices by varying the composition of the tape. The tapes can be purchased in different thicknesses; thus the separation between conductor layers is easily controlled and varied in order to achieve controlled and matched impedances. Cofired tape also provides greater electrical isolation between conductor layers than sequentially screen-printed pastes, because its dielectric breakdown voltage is over 1000 V/mil. This electrical integrity is essential for high-current high-voltage circuits.

The design versatility of LTCC can be easily shown. First, integral-lead, hermetic-cavity package-substrate combinations can be produced. This significantly reduces volume, weight, and cost since separate metal or ceramic packages are not required. Second, cavities can be formed in the ceramic simultaneously with its fabrication such that IC chips or other components can be inserted and recessed, generating a low-profile package. Metal heat sinks can also be bonded or brazed, if necessary, to the bottom of the package, and the "hot" devices can be mounted directly onto the heat sink through the apertures. Furthermore, thermal plugs or thermal vias can be designed and cofired in the ceramic to draw heat away from the devices. Another perceived advantage of LTCC is the use of low-loss conductors for microwave applications for both microstrip and stripline configurations. The HTCC process with its refractory metal conductors was too lossy, and thick-film paste dielectrics could not be deposited in the required thicknesses to provide controlled impedance transmission lines.

Although, in view of the above, LTCC may seem an ideal packaging material and packaging approach, there are several limitations that a design engineer must consider.

- After firing, the tape shrinks as with HTCC. The amount of shrinkage can be accurately measured for each material and for each set of processing conditions and can be controlled and compensated for in the design of the artwork and masks.

- The high glass content (50 percent or more) results in very low thermal conductivity [2 to 3 W/(m•K)], but this can be obviated by constructing thermal vias or plugs through the tape or, if flip-chip devices are used, by removing the heat from the backs of the devices, as in the case of IBM's thermal conduction module. The thermal efficiency of thermal vias formed in the LTCC has been demonstrated by Schroeder,[43] who showed that the thermal resistance of LTCC with silver- or gold-filled thermal vias was reduced to almost that of beryllia (Table 3-5).

- The high glass content can embrittle the structure; however, there are several reports demonstrating the reliability of LTCC under high vibration and mechanical shock conditions.[43,44] Hermetically sealed LTCC MCMs (1.8 in. square) survived forces of 10,000g acceleration,

Table 3-5. Thermal Resistances of LTCC with and without Thermal Vias

Test Structure	Average thermal resistance, °C/W
LTCC gold, no thermal vias	10.8
LTCC silver, no thermal vias	10.7
LTCC gold, thermal vias ¾ through*	4.99
LTCC silver, thermal vias ¾ through*	4.59
LTCC gold, through thermal vias*	3.02
LTCC silver, through thermal vias*	2.54
96% Alumina, no vias	4.31
99% Beryllia, no vias	2.37

*Thermal vias had 20-mil diameter on 50-mil centers.
SOURCE: Ref. 43.

temperature cycling from − 65 to 150°C, 1500*g* mechanical shock, and 20 to 2000 Hz of vibration performed according to Mil-Std-883.

■ LTCC, like other thick-film processes requiring screen-printing and where shrinkage after firing occurs, is limited in the dimensions of the conductor lines and spacings, the size of the vias, and the overall circuit densities that can be produced. Thus more layers and a thicker substrate are needed to equate to the same density as thin-film structures.

3.2.6   Via Formation in Thick-Film Ceramic Dielectrics

The difficulty in achieving high resolution via patterning has limited the wider use of thick-film processes for multichip modules. The conventional direct process of screen-printing the dielectric paste and simultaneously creating the vias is limited by the size of the screen mesh, the thixotropic properties of the paste, and the thickness of the layers. For optimum yields, design guidelines specify minimum via diameters and spacings of 15 mils. In attempting to screen-print very small vias, the inward flow of the paste after screen printing can actually close up the via prior to filling with metal paste and firing.

In both LTCC and HTCC, vias may be mechanically punched, drilled, or laser-formed in the "green" ceramic tape, then filled with metal paste and fired. The final dimensions of vias in cofired ceramic tape depend on the degree of shrinkage that occurs in the ceramic during firing; however, the amount of shrinkage is reproducible and can be controlled. Although most design guidelines for cofired ceramic specify 6-mil-diameter vias as a minimum, finer vias and tolerances have been obtained by better controlling shrinkage during firing.

Vias that are photolithographically formed combine both thick- and thin-film processes and produce smaller (4-mil-diameter or less) vias, albeit at the expense of extra processing steps and higher cost. Photolithographically formed vias have been investigated for over 20 years but are not yet in wide use, although they have been used in the production of some high-resolution applications.There are several variations of the photoimaging process. In one, the ceramic is formulated with a photosensitive ingredient. After screen-printing the entire surface and drying, the dielectric is exposed to ultraviolet light through a via mask pattern. The unexposed material is removed by a chemical developer, thus generating the vias. In an indirect method the screen-printed and dried ceramic paste is coated with photoresist, the photoresist is exposed to a via mask pattern and developed, whereupon the vias are etched and the photoresist is removed.

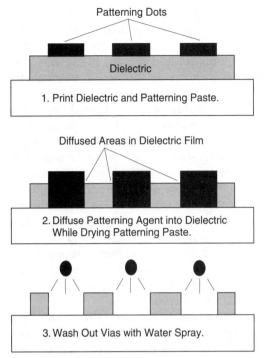

Figure 3-27. Diffusion-patterning process steps. (*From Ref. 45.*)

A recent innovation in forming vias in thick-film ceramic, called *diffusion patterning,* has been reported by Du Pont.[45] In this process the dielectric ceramic layer is printed with a low-viscosity paste and dried, and a via pattern is overprinted with a special paste to form a dot pattern (Fig. 3-27). The overprint paste contains a solubilizer that diffuses into the sublying dielectric during the drying step. The solubilized via pattern is then washed away with a water spray. Vias as small as 4 mils are reported, but typically they are 6 to 12 mils for higher yields.[46] The dielectric constant and dissipation factor for the fired dielectric paste at 10 MHz are 9 to 11 and 0.3 percent, respectively. Thicknesses of 35 μm can be produced by screen-printing and drying two layers. Silver, gold, and silver palladium pastes compatible with the dielectric are commercially available.

3.3 Multilayer Laminates

MCM-L substrates are really multilayer printed-circuit boards (PCBs) scaled to produce finer lines and spacings and smaller vias. Multilayer

PCB technology is an old, mature technology for which an infrastructure of numerous manufacturers and materials and equipment suppliers already exists.[47,48] Copper-clad, glass-reinforced epoxy, primarily FR-4, is the most widely used laminate and among the lowest-cost printed-wiring board materials. It is, however, limited in thermal stability to temperatures below 250°C. For higher-temperature applications, the more thermally stable polyimides and BT (bismaleimide triazine) laminates are available. Polyimides, though somewhat more expensive than epoxies, have better handling strength and much higher thermal resistance—up to 500°C—before decomposition begins. Hence polyimides are better able to withstand solder temperatures and repeated rework cycles. The extra cost of polyimide laminates is reported to be more than offset by lower costs due to the greater speed of wire bonding and assembly. Fluorocarbon laminates have also been used for many years, more for high-frequency, radio-frequency, and microwave applications because of their excellent electrical properties and the stability of these properties over a wide temperature and frequency range. All these laminates have a history of successful and reliable use in both military and commercial applications.

A key advantage of MCM-L substrates is the ability to process the individual layers in parallel, permitting inspection and testing of each layer prior to lamination. The laminates may also be processed as large panels (approximately 18×24 in), then cut into multiple small substrates. The laminates, either single-clad or double-clad, are processed separately (drilled, plated, and photoetched), then bonded together with B-staged prepreg, aligned and stack-laminated under heat and pressure. Typical lamination conditions[49] are 175°C for 30 to 60 min and 250 to 1000 lb/in^2. Interconnections can be made by drilling holes through the stack followed by electroplating with copper.

Although MCM-L is the lowest-cost process for fabricating multichip interconnect substrates, it is limited by the extremely low thermal conductivities of plastic laminates [0.2 $W/(m \cdot K)$ for FR-4 epoxy], their high CTEs (approximately 4 to 5 times higher than that of silicon dice), and the moderate line densities that can be achieved. Typical PCBs have 5- to 10-mil line widths and spacings and 18- to 30-mil plated through-holes. Now that the need for high-density conductor lines and low-cost MCMs has been defined, printed-wiring board manufacturers are exploring modifications to their basic processes that will provide the higher routing densities and improved electrical performance required. Plastic multilayer circuit boards that have been used for decades as second-level assembly boards for prepackaged single chips or for sealed hybrid circuits are now being used as first-level interconnect substrates for bare-chip devices. Finer conductor lines are being achieved by additive pattern plating combined with using thinner laminates and prepregs and thinner copper clad (0.35 mil and less). Processes such as laser ablating or photolithographic forming of the vias instead of drilling them are also capable of forming smaller vias. By controlling the thickness of the copper foil and the etching or pattern-plating process parameters, line widths and spacings of 3 mils and plated-through holes of 6 mils have been achieved. Furthermore, it is reported that 2-mil lines and spacings and 4-mil drilled holes are possible in processing small panels where alignment and registration can be better controlled. The practicality of drilling holes of 4 mils or less, however, is currently limited because of the poor strength and short wear life of miniature drill bits.

To extend the savings resulting from the use of MCM-L materials and processing, they are generally plastic-molded instead of being hermetically sealed in separate-cavity packages. The *quad flat pack* (QFP) module (Fig. 3-28a) in production at Motorola is an example in which the lead frame has been integrated with the substrate and the entire assembly has been plastic-molded.[50] In the plastic-molded version, either an extended lead frame emanating laterally or pads formed on the bottom are used for the external connections. Other MCM-L modules may be enclosed in ceramic pin grid array packages (Fig. 3-28c).

A somewhat different process for producing MCM-L interconnect substrates has been reported by Rogers Corp.[51] It is based on parallel-processing individual layers of thin fluorocarbon films (RO2800). The single layers are processed on metal carriers by pattern-plating the conductors and plated posts, then planarizing with the fluorocarbon film. After removal of the carrier (etching it away), the layers are stack-laminated, simultaneously fusing the thermoplastic film and diffusion-bonding the gold interconnects (Fig. 3-29). The salient features of this process are as follows:

- The fluorocarbon has a very low dielectric constant of 2.6 to 2.8 and excellent thermal stability.

- Individual layers are processed in parallel on expendable carriers.

- Gold-plated posts are diffusion-bonded in the same lamination step used to bond the dielectric layers.

- The fluorocarbon layers, being thermoplastic, are fusion-bonded together under heat and pressure without requiring a separate adhesive or prepreg. Typical lamination conditions are 370°C for 1 h under 1700-in^2 pressure.

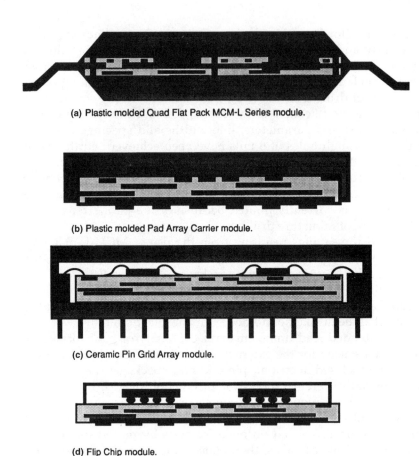

(a) Plastic molded Quad Flat Pack MCM-L Series module.

(b) Plastic molded Pad Array Carrier module.

(c) Ceramic Pin Grid Array module.

(d) Flip Chip module.

Figure 3-28. MCM-L packaging designs. (*From Ref. 50.*)

Interconnect densities approaching those of MCM-D thin-film multi-layers are reported to have been produced from 50-μm-thick RO2800 dielectric. Dimensions reported are 25-μm conductor lines widths at 75-μm pitch and 50-μm vias.[51]

3.4 Combinations

In order to strike a compromise between density and performance, on one hand, and cost and yield, on the other hand, combinations of MCM-D/L and MCM-D/C have emerged. Both approaches utilize the fine-line higher-cost thin-film processing, but only for the top several layers where it is required for either high density or high signal speed.

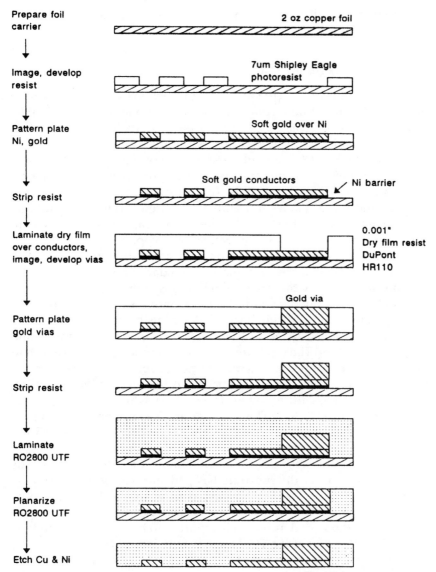

Figure 3-29. Rogers Corp. fluorocarbon process for MCM-L. (*From Ref. 51.*)

The bulk of the multilayer structure consists of the more conventional, less costly laminated plastic or cofired ceramic. Both the laminated plastic and cofired ceramic result in coarser geometries, but they can be compensated for by designing several more layers for routing. The sequence of steps for processing the thin-film multilayer structure on either cofired multilayer ceramic or multilayer plastic laminate is given in Fig. 3-30.

3.4.1 MCM-D on MCM-C

The combination of MCM-D and MCM-C technologies has been used primarily by mainframe computer companies, notably IBM, Fujitsu, and NEC. IBM 's *thermal conduction module* (TCM) is an example that has been abundantly described in the literature and at conferences.[52] MCMs for the ES/9000 computer required very high packaging densities of ECL devices, faster speeds, and improved methods for heat removal. To achieve these goals, IBM designed most of the power, ground, and signal planes in multilayer cofired ceramic. A cordierite glass-ceramic composition was developed that can be fired at low temperatures (950 to 1000°C), is compatible with copper conductors, has a CTE closely matching that of silicon, and has a dielectric constant of 5.2, much lower than the 9.5 typical of standard cofired alumina. The only desired property that the ceramic lacks is high thermal conductivity, which is not a problem with the TCM since heat is conducted from the back sides of the flip-chip bonded dice by using copper pistons.

The TCM is constructed of 63 layers of cofired glass-ceramic and copper conductors. The top layer, however, is processed by using thin-film polyimide/copper as a redistribution layer compatible with the bumped flip-chip device bonding pads.[53] A top view of the redistribution layer for one chip is shown in Fig. 3-31 while a cross-sectional view is given in Fig. 3-32. The modules are 5 × 5 in and contain sites for flip-chip bonding 100 to 121 chips. Logic chip sizes are 255 mil square, and memory chips are 275 × 335 mils. The chips are processed with area array bumps (C4), some containing up to 648 bumps. The top thin-film layer of the substrate has pad sites corresponding to the bumps on the chip. Heat is conducted from the backs of the chips through copper pistons and springs to a water-cooled heat exchanger. The maximum power dissipation is reported to be 2000 W per module. Other characteristics of the TCM are given in Table 3-6.

Another example of an MCM-D on MCM-C combination is NTK's multichip module. These modules are fabricated from 4 × 4 in alumina green tape cofired with refractory metal pastes to produce up to 20 layers. The conductor line widths are 4 to 8 mils, while the vias are 8 mils in diameter or greater. The electrical resistance of the metal is 10 to 15 per mΩ square, which is typical of refractory metal compositions. Four to six layers of thin-film copper/polyimide are then processed on top of the cofired ceramic to provide catch-pad electrical connections, fine-line signal layers, and device bonding sites. Tantalum nitride thin-

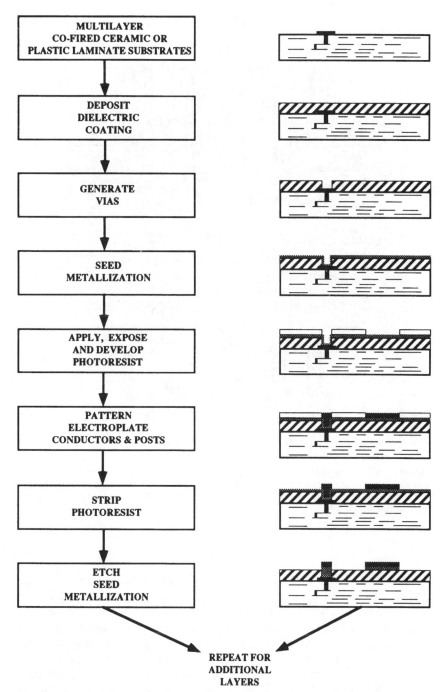

Figure 3-30. Flowchart for combining MCM-D with MCM-C or MCM-L.

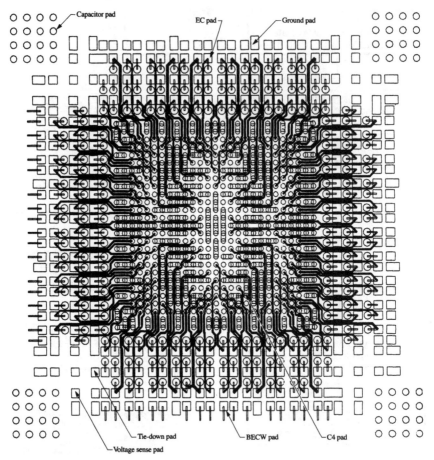

Figure 3-31. Top-surface view showing the thin-film redistribution wire pattern and the engineering change pad pattern. (BECW = buried engineering change wires, EC = engineering change.) (*Copyright 1992 by IBM Corporation. Reprinted with permission.*)

film resistors, 10 to 50 Ω per square, can be integrated within the thin-film structure. Thicknesses of the copper and polyimide are 6 to 10 and 25 μm, respectively, while the line widths are 25 μm and spacings are 50 μm. The CTE of the polyimide is 20 to 70 ppm/°C, and its dielectric constant is 3.2 at 1 MHz. The top metallization layer (for bonding pads) consists of nickel gold-plated to a sheet resistance of 7 mΩ per square. Vias have 30-μm diameter on 100-μm pitch. In the final step, up to 5000 external I/O pins are brazed to pads on the bottom of the ceramic substrate.

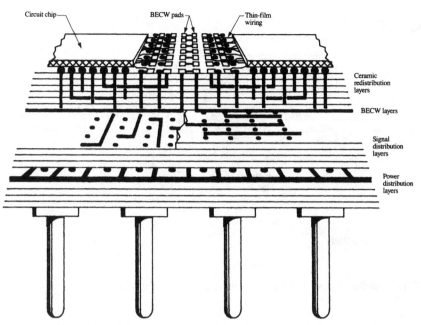

Figure 3-32. View of the ES/9000 module showing the top-surface features and the key regions of the cross section. (*Copyright 1992 by IBM Corporation. Reprinted with permission.*)

Table 3-6. Physical and Electrical Parameters for ES/9000 Module

Physical	
Number of layers	63
Number of signal layers	28
TFW dimension $W \times H$, µm	28×5.5
TFW pitch, µm	55
C_4 pad pitch, µm	225
Thick-film line dimensions. $W \times H$, µm	89×25
Thick-film line pitch, µm	450
Thick-film via pitch, µm	450
Pin pitch, mm	2.5 staggered

Electrical	
Relative dielectric constant	5.2
Thick-film characteristic impedance, Ω	58
Nominal thick-film resistance, mΩ/mm	25
Nominal thin-film resistance, mΩ/mm	150
Nominal thin-film capacitance, fF/mm	60
Simultaneous switching per chip	50–100
Saturated near-end crosstalk, %*	8

*For nine coupled lines.

SOURCE: Ref. 52, courtesy of IBM

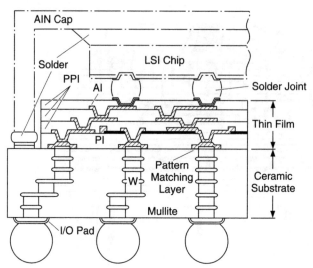

Figure 3-33. Cross section of microcarriers for Hitachi
M-880.

3.4.2 MCM-D on MCM-L

IBM's surface laminar circuitry (SLC) is an example of an MCM-D on
MCM-L interconnect substrate.[54] This structure, developed by IBM in
Japan, uses standard low-cost multilayer PCB processes to fabricate
most of the ground, power, and signal layers. The top fine-line redistribution layer is then formed by using thin-film MCM-D processes. A
photosensitive epoxy solder mask is deposited in which vias are
formed in a single step. The epoxy is then cured, metallized, and photoetched to generate the top fine-line conductors and bonding pads.

In subsequent work, Motorola is investigating a similar construction
except that benzocyclobutene (BCB) is used instead of epoxy for the
top layers. BCB offers better planarization and improved thermal stability. Lower costs were achieved by processing the printed-circuit
boards from large 18 × 24 in panels, then cutting them into single substrates, and processing the thin film on top of each substrate.[55] The
MCM-D/L approach restricts the high-density, high-cost steps to only
where needed, such as for the top layers, and therefore provides a
good compromise between performance and cost.

3.4.3 MCM-D/C Carrier
Packages on MCM-C

A unique variation of MCMs is used by Hitachi for its HITAC M-880
mainframe computers. Individual large-scale integration (LSI) chips

are flip-chip-attached to small carriers consisting of cofired ceramic (mullite) for the inner layers and are thin-film deposited polyimide/aluminum for the outer signal layers, thus producing a miniature MCM-D/C structure. The thin-film portion consists of five sputtered aluminum conductor layers separated by polyimide, while the cofired ceramic portion is a seven-layer mullite/tungsten cofired structure. The I/Os for these carriers consist of bumps on the bottom sides. The carriers are then prepackaged by soldering an aluminum nitride cover as in Fig. 3-33. These single-package ICs, called MCCs (microcarriers for LSI chips), are then flip-chip-attached to an MCM-C motherboard, resulting in a double flip-chip situation. The motherboard is formed by cofiring 44 layers of mullite/tungsten. Among the advantages of this approach are the removal of heat through the high thermally conductive aluminum nitride lid, the ability to test each die at the carrier stage before assembly, and the higher yields and lower costs associated with processing small carriers.[56] However, there is a penalty due to lower die packing densities.

3.5 Low-Cost, Fast-Turnaround Substrates

Though high performance and high densities have been achieved with the MCM-D technology, costs are still too high to broaden its application to commercial and consumer electronics. Reducing the cost an order of magnitude, from approximately $200 per square inch to $20 per square inch or less for a five-conductor-layer interconnect substrate, is the focus of many investigations. There are tradeoffs, however, between cost, interconnect density, and electrical performance. The greater the line density, the greater the probability of defects and the lower the yield. The finest line and via pitches can only be produced by using thin-film processes under the super-clean conditions of a semiconductor facility. The elements of substrate fabrication that have the greatest impact on cost include the yield, number of process steps, material costs, equipment cost and depreciation, equipment maintenance, equipment downtime, and levels of labor skills required.

The most dramatic effect on cost is the yield, especially for the larger substrates. One defect in a 2×4 in substrate, where only two substrates can be processed from a 6-in-diameter wafer, immediately reduces the yield to 50 percent. Yields can be increased by designing smaller substrates. In the same 6-in-diameter wafer designed for processing four 2×2 in substrates, one defect increases the yield to 75 percent. Therefore a primary approach to improving yields and reducing costs is to process small substrates from large panels. Large panels

(12 × 12 in or larger) such as are used to produce printed-wiring boards and flat panel displays are being investigated.

Production volume is a second factor affecting the cost. Some contend that in high volumes there is little difference in cost among the three MCM types when similar designs are compared.

A third cost element is the number and complexity of process steps. It is not unusual for a five-conductor-layer thin-film interconnect substrate to require over 100 steps and 30 to 60 days for completion. The greater the number of steps, the longer the turnaround time. The number of steps, however, may be reduced in several ways. Foremost is by using photosensitive polyimides or other photosensitive dielectrics so that the vias can be patterned directly, obviating the normal photoresist application, baking, exposure, and removal steps required to process each via layer. Photosensitive polyimides and epoxies having low dielectric constants are commercially available and are used by several companies. Several more steps may be eliminated by starting with an electrically conductive wafer which can double as a substrate support and as a ground plane. Heavily doped silicon, aluminum, or other metals may be used.

Using dielectrics that do not require step curing to high temperatures and that planarize to a thick layer in one pass also reduces the number of steps. Some widely used polyimides require three to four steps to effect a complete cure and may have to be applied two or three times to produce a sufficiently thick layer. Dielectric coatings having a high solids content and low viscosity are therefore desirable. Finally, finding a compatible metal-dielectric combination avoids the added steps required to deposit barrier metal layers to prevent copper from diffusing into the polyimide. Polyimide metallized with aluminum and BCB metallized with copper are examples where barriers are not needed.

Material costs become a significant factor in scaling to high production. In spin-coating polyimide, over 90 percent of the material is lost and not recoverable. Alternate spray-coating or flow-coating processes are more cost-effective. A variation of flow coating is liquid extrusion coating in which pressurized liquid resin is programmed to flow over the entire surface or onto selected portions of the surface. In contrast to spin coating, very little material is wasted (5 to 10 percent) in flow coating, and large panels can be continuously coated.[57]

Thick-film multilayer substrates such as low-temperature cofired ceramic and high-temperature cofired ceramic are potentially lower in cost than their thin-film multilayer counterparts. Analogous to epoxy printed-circuit boards, each layer of LTCC or HTCC is processed and inspected separately, then stacked and laminated together. There is less chance for pinholes or other defects since the dielectric is purchased and used in tape form. Thicknesses are also better controlled and can be easily varied within the same structure by purchasing the green tape in various thicknesses. The investment in equipment and facilities to process cofired ceramic substrates is lower than that for thin-film substrates since the latter require semiconductor-type equipment and clean rooms. Equipment maintenance and downtime are also lower for ceramic processing. However, to get the full cost benefit of LTCC, silver or copper conductors instead of gold need to used. The price of the commercially available ceramic green tape also needs to drop, or the user would need to formulate and cast the tape in-house. Any cost comparison between ceramic and thin-film multilayering should take into consideration the extra layers required for routing in cofired ceramic due to its coarser line and via dimensions.

The lowest-cost substrates, if high performance is not the main issue, are epoxy multilayer laminates. MCM-L employs printed-wiring board technology that is mature and has a longstanding infrastructure. As a result of the multichip module revolution, process improvements have been made in achieving finer line pitches and higher densities in FR-4 epoxy multilayer circuit boards. In fact, the MCM-D and MCM-L technologies are drawing closer to each other as some of the printed-wiring board processes are being applied to MCM-D.

3.6 Rapid Prototyping of Interconnect Substrates

Rapid prototyping of interconnect substrates and MCMs is essential in assessing the suitability of a particular design and technology, in debugging device and electrical problems, and in generally establishing proof of design before proceeding to larger quantities. Rapid prototyping is best done by computer-aided interactive design and fabrication. Direct-write approaches are therefore ideal since they avoid the need to generate hard masks and tooling. The conductor lines and vias are routed directly from the computer database that was derived from the design. Design changes can thus be made rapidly by reprogramming and transferring the data to fabricate the prototype. There are several variations to direct writing of both thick- and thin-film conductors.

1. A programmed laser beam may be used to decompose organometallic coatings or adsorbed vapors on a dielectric substrate. For example, copper formate or copper acetate films are decomposed to metallic

copper, carbon dioxide, and water, of which the last two volatilize as gases.[58] Organometallic complexes of gold and nickel may also be used to form gold or nickel conductors, respectively. Tungsten hexafluoride (WF_6) as a gas, adsorbed on a substrate, can also be decomposed to metallic tungsten by using a focused laser beam.

2. A programmed ultraviolet laser beam can be used to directly expose a photoresist, again avoiding the need for fabricating a mask.

3. A laser beam may be used to decompose or reduce a palladium salt to free palladium as a seed layer for the additive electroless plating of copper or other metals.

4. Thick films can be directly written by forcing thick-film pastes (conductor, resistor, or dielectric) through a fine nozzle. The paste flow is controlled from a computer database and is programmed to flow in selected patterns. Precision line widths of 4 mils for gold and 6 mils for silver conductors have been reported, and with certain pastes, even finer lines of 3 mils are achievable.

Another approach to rapid prototyping is the *quick-turnaround interconnect* (QTAI) design approach developed by Microelectronics and Computer Technology Corp. Substrates are designed and fabricated so that all the inner layers are generic to many applications while the upper layer can subsequently be customized for a specific application. A special software tool has been developed to automatically route the interconnections. This tool is an integral part of the QTAI design and is necessary to take full advantage of the approach. The generic substrates can be fabricated and stored, then personalized for specific functions. Microelectronics and Computer Technology Corp. has demonstrated this approach by fabricating and testing a three-chip MCM consisting of a Motorola 88100 RISC processor and two 88200 cache controllers. The generic interconnect substrate consisted of a 2.5 $\times$ 5 cm alumina base on which three layers of thin-film copper/polyimide were processed. The underlying conductors consisted of broken segments 2 mm long with vias at both ends extending through the multilayer structure to the top surface. This permits both links and cuts to be formed. The links or cuts may be made between vias connecting the underlying segments, between surface conductors, or between bond pads. The substrate was customized by making 1624 laser cuts and creating 3970 links. The links were formed by laser-irradiating a palladium seed layer followed by electroless copper and gold plating. The integrity of the interconnections was reported to be established by capacitance testing.[59]

References

1. S. Titus and P. Kelly, "Defect Density Reduction in a Class 100-lab Utilizing the Standard Mechanical Interface," *Solid State Technology,* November 1987.

2. S. Harris, S. Radley, and P. A. Trask, "Advancements in HDMI Substrate Fabrication," *Proceedings of the National Electronic Packaging and Production Conference (NEPCON West),* Anaheim, CA, February 1991.

3. E. Sweetman, "Characteristics and Performance of PHP-92," *Proceedings of the International Conference on Multichip Modules,* Denver, CO, 1992.

4. *Manufacturing Market Insider,* January 1993.

5. D. A. Doane and P. D. Franzon, eds., *Multichip Module Technology and Alternatives,* Van Nostrand Reinhold, New York, 1993, chap. 15, pp. 695ff.

6. Final Report, Naval Command, Control, and Ocean Surveillance Center, "Manufacturing Technology for VLSIC Packaging," contract no. 66001-88-C-0181, April 1993.

7. P. Sullivan and J. Chernicky, "High Density MCM Testing," *Proceedings of the Second International Conference on Multichip Modules,* Denver, CO, April 1993.

8. J. T. Pan, S. Poon, and B. Nelson, "A Planar Approach to High Density Copper-Polyimide Interconnect Fabrication," *Proceedings of the Eighth International Electronics Packaging Conference,* 1988.

9. B. Randall, et al., "GaAs Digital MCM in $Si/Cu/SiO_2$ Technology for Clock Rates up to 1 GHz," *Proceedings of the International Electronics Packaging Society,* San Diego, CA, 1993.

10. T. Horton and B. McWilliams, "MCM Driving Forces, Applications, and Future Directions," *Proceedings of the National Electronic Packaging and Production Conference (NEPCON West),* Anaheim, CA, February 1991.

11. B. McWilliams, "Comparison of Multichip Interconnect Technologies," *Proceedings of the International Electronics Packaging Society,* San Diego, CA, September 1991.

12. P. F. Marella and D. B. Tuckerman, "Prequalification of nCHIP SiCB Technology," *Proceedings International Conference on Multichip Modules,* Denver, CO, 1992.

13. L. M. Levinson, et al., "High Density Interconnects Using Laser Lithography," *Proceedings of the National Electronic Packaging and Production Conference (NEPCON West),* 1989.

14. *Compendium on Multichip Modules,* IEEE, New York, 1989, p. 177.

15. R. O. Carlson, et al., "A High Density Copper/Polyimide Overlay Interconnection," *Proceedings of the International Electronics Packaging Society,* Dallas, 1988.

16. R. A. Fillion and C. A. Neugebauer, "High Density Multichip Packaging," *Solid State Technology, Hybrid Supplement,* September 1989.

17. U. Deshpande, S. Shamouilian, and G. Howell, "High Density Interconnect for VAX 9000 System," *Proceedings of the International Electronics Packaging Society*, 1990, p. 46.

18. C. W. Ho and U. A. Deshpande, "A Low Cost Multichip Module-D Technology," *Proceedings of the International Conference on Multichip Modules*, 1993.

19. C. Chao, et al., "Multilayer Thin Film Substrates for Multichip Packaging," *IEEE Transactions of Components, Hybrids, and Manufacturing Technology*, **CHMT-12:** 180, 1989.

20. G. Turban and M. Rapeaux, "Dry Etching of Polyimide in O_2/CF_4 and O_2/SF_6 Plasmas," *Journal of the Electrochemical Society*, **130:** 2231, 1983.

21. T. G. Tessier, W. F. Hoffman, and J. W. Stafford, *Proceedings of the EIA/IEEE, Electronic Components Conference*, Atlanta, 1991, p. 827.

22. K. K. Chakravorty, et al., "High Density Interconnection Using Photosensitive Polyimide and Electroplated Copper Conductor Lines," *IEEE Transactions of Components, Hybrids, and Manufacturing Technology,,* **13,** 1990.

23. M. T. Pottiger, "Second Generation Photosensitive Polyimide Systems," *Solid State Technology*, December 1989.

24. D. K. Smith, et al., "A 1 Gbps Fiber Optic Transmitter/Receiver Pair for Military Avionics Network Applications," *DoD Fiber Optics Conference*, 1990.

25. J. H. Brannon, et al., "Excimer Laser Etching of Polyimide," *Journal of Applied Physics*, **58** (5), September 1985.

26. T. G. Tessier, et al., "Via Processing Options for MCM-D Fabrication: Excimer Laser Ablation vs. Reactive Ion Etching," *Proceedings EIA/IEEE Electronic Components Technology Conference*, Atlanta, 1991.

27. E. Myszka, et al., "The Development of a Multilayer Thin Film Copper/Polyimide Process for MCM-D Substrates," *Proceedings of the International Conference on Multichip Modules*, Denver, CO, 1992.

28. A. H. Smith, "Phase Mask Machining for High Throughput Via Formation," *Proceedings of the Second International Conference Multichip Modules*, Denver, CO, April 1993.

29. G. Adema, M. Berry, and I. Turlik, "Via Formation in Thin Film Multichip Modules," *Electronic Packaging and Production*, November 1992.

30. L. Smith, "High Density Copper Polyimide Interconnect," *Proceedings of the 3d SAMPE Electronics Conference*, Los Angeles, June 1989.

31. J. J. Licari, "Thin and Thick Films," chap. 8, in R. Sampson and C. Harper, eds., *Electronic Materials and Processes Handbook*, 2d ed., McGraw-Hill, New York, 1993.

32. J. J. Licari, "Hybrid Microelectronic Packaging," chap. 7, in C. Harper, ed., *Electronic Packaging and Interconnection Handbook*, McGraw-Hill, New York, 1991.

33. W. J. Havey, "The Formation of Conductive Patterns on Hybrid Circuits and PWBs Using Alternative Techniques," *Hybrid Circuit Technology*, August 1985.

34. M. Shankin, "Write It, Don't Screen It," *Proceedings of the International Symposium on Hybrid Microelectronics*, Minneapolis, MN, 1978.

35. Johnson Matthey, *Thick-Film Products Bulletin*, JM1301-Fine Line Gold Conductor.

36. Johnson Matthey, *Thick-Film Products Bulletin*, JM1202-Ultra Fine-Line Etchable Gold Conductor.

37. J. J. Licari and W. S. DeForest, British Patent 1256344, Photographic manufacture of printed circuits, December 8, 1971.

38. D. H. Scheiber and R. M. Rosenberg, "Circuitry from Photoprintable Paste—A New Technology," *Proceedings of the International Symposium on Hybrid Microelectronics*, Washington, DC, 1972.

39. Du Pont Experimental Data Sheets, PhotoPatterning System-6050D Dielectric Paste and 1056-2 Gold Conductor.

40. R. W. Vest, *Metallo-Organic Materials for Ink Jet Printing*, Final Report, Navy contract N00163-81-C-0350, 1982.

41. W. Vitriol and J. I. Steinberg, "Development of a Low Temperature Cofired Multilayer Ceramic Technology," *Proceedings of the International Symposium on Hybrid Microelectronics*, San Francisco, CA, 1992.

42. J. I. Steinberg, S. J. Horowitz, and R. J. Bacher, "Low Temperature Cofired Tape Dielectric Material Systems for Multilayer Interconnections," *Solid State Technology*, January 1986.

43. D. Schroeder, "The Use of Low Temperature Cofired Ceramic for MCM Fabrication," *Proceedings of the International Conference on Multichip Modules*, Denver, CO, 1992.

44. R. Brown and P. Polinski, "Manufacturing of Low-Temperature Cofired Ceramic Modules for Advanced Radar Applications," *Digest of Government Microcircuit Applications Conference (GOMAC)*, Las Vegas, NV, 1992.

45. J. J. Felten, C. B. Want, and J. Collins, "High Resolution Via Patterning by Aqueous Diffusion Patterning," *Inside ISHM*, July/August 1993.

46. J. J. Felten and J. Collins, "Small Via Generation by Diffusion Patterning," *Proceedings of the International Symposium on Microelectronics*, Chicago, IL, 1990.

47. C. F. Coombs, *Handbook of Printed Circuits*, 3d ed., McGraw-Hill, New York, 1988.

48. L. M. Higgins, "Laminate-Based Technologies for Multichip Modules," chap. 5, in D. A. Doane and P. D. Franzon, eds., *Multichip Module Technologies and Alternatives*, Van Nostrand Reinhold, New York, 1993.

49. D. P. Seraphim, et al., "Printed Circuit Board Packaging," chap. 12 in R. Rummala, Ed., *Microelectronics Packaging Handbook*, Van Nostrand Reinhold, New York, 1989.

50. W. Blood and A. Dixon, "MCM-L: Cost Effective Multichip Module Substrate," *Inside ISHM*, **20**(2), March/April 1993.

51. G. S. Swei, W. David Smith, and D. J. Arthur, "Ultrathin Fluoropolymer Dielectrics for MCM-L," *International Journal of Microcircuits and Electronic Packaging*, **16**(2), 2d quarter, 1993.

52. E. E. Davidson, et al., "Physical and Electrical Design Features of the IBM Enterprise System/9000 Circuit Module," *IBM Journal of Research and Development*, **36**(5), September 1992.

53. T. F. Redmond, C. Prasad, and G. A. Walker, "Polyimide Copper Thin Film Redistribution on Glass-Ceramic/Copper Multilevel Substrates," *Proceedings of the 41st Electronic Components and Technology Conference*, Atlanta, May 1991.

54. Y. Tsukada, S. Tsuchida, and Y. Mashimoto, "Surface Laminar Circuit and Flip Chip Attach Packaging," *Proceedings of the International Microelectronics Conference*, 1992.

55. T. G. Tessier and E. Myszka, "High Performance MCM-L/D Substrate Approaches for Cost Sensitive Packaging Applications," *Proceedngs of the Second International Conference on Multichip Modules*, Denver, CO, April 1993.

56. K. Takeda, Y. Naritsuka, and M. Harada, "High Density Module with Small Packaged LSIs for Mainframe Computers," *Proceedings of the Second International Conference on Multichip Modules*, Denver, CO, 1993.

57. T. Snodgrass and G. Blackwell, "Advanced Dispensing and Coating Technologies for Polyimide Films," *Proceedings of the International Conference on Multichip Modules*, Denver, CO, 1992.

58. R. Miracky, et al., "Laser Customization of Multichip Modules," *Digest of Government Microcircuit Applications Conference (GOMAC)*, November 1991.

59. T. Hirsch, R. Miracky, and S. Sommerfeldt, "Laser Customization of a Multichip Module Substrate," *Transactions of the IEEE*, **1**(3), September 1993.

4

Materials for Interconnect Substrates

The materials employed to fabricate high-density interconnect substrates fall into three categories: ceramics, metals, and plastics. Often, because all three types are used in the same structure, care is required in selecting materials that are mutually compatible mechanically, thermally, electrically, and chemically. The properties of materials most widely used for electronic packaging are given in Table 4-1.

4.1 Substrates

All major classifications of solid materials—metals, ceramics, metal-ceramic composites, glasses, plastics, and reinforced plastic laminates—are being used as substrate materials upon which interconnect circuits are formed and active and passive chips are assembled. Substrates for thin- and thick-film circuits basically serve four functions:

- Mechanical support to interconnect and assemble bare or packaged dice
- Electrical isolation medium on which conductor traces and passive components are deposited and patterned
- Thermal medium to conduct and transfer heat from the dice
- Dielectric layer to control impedance, crosstalk, and signal propagation in high-speed circuits

Table 4-1. Properties of Electronic Packaging Materials

	Melting point °C	Electrical resistivity, 10^{-6} Ω•cm	Dielectric constant	Coefficient of thermal expansion, ppm/°c, 0 to 300°C	Thermal conductivity, W/(m•K), at 25°C	Density, g/cm³
Metals						
Copper	1083	1.7	N/A*	17.0	393	8.93
Silver	962	1.6	N/A	19.7	418	10.49
Gold	1063	2.2	N/A	14.2	297	19.3
Tungsten	3415	5.5	N/A	4.5	160	19.3
Molybdenum	2610	5.2	N/A	5.0	146	10.2
Platinum	1772	10.6	N/A	9.0	71	21.5
Palladium	1552	10.8	N/A	12.0	70	12.0
Nickel	1455	6.8	N/A	13.3	92	8.9
Chromium	1875	13.0	N/A	6.3	66	7.2
Invar	1425	80.0	N/A	3.1	11	8.0
Kovar	1450	50.0	N/A	5.3	17	8.3
Silver/Palladium	1145	20.0	N/A	14.0	150	—
Gold/Platinum	1350	30.0	N/A	10.0	130	—
Aluminum	660	2.7	N/A	25.0	247	2.7
Au/20% Sn	280	16.0	N/A	15.9	57	16.9
Pb/5% Sn	310	19.0	N/A	29.0	63	11.1
Cu/W (20% Cu)	1083	2.5	N/A	7.0	248	17
Cu/Mo (20% Cu)	1083	2.4	N/A	7.2	197	—
Silicon	1415	2.3×10^5	11.8	3.0–3.5	84–135	2.3
Ceramics and Glasses						
92% Alumina	1500†	>10^{14} Ω•cm	9.2	6.0	18	3.5
96% Alumina	1550†	>10^{14} Ω•cm	9.4	6.4	20–35	3.8
99% Alumina	1600†	>10^{14} Ω•cm	9.9	6.6	37	3.9
50% Alumina (LTCC)	850–950†	>10^{14} Ω•cm	7.3–7.9	—	2.4–3.0	2.9–3.1
Silicon nitride	1600†	>10^{14} Ω•cm	7.0	2.3	30	3.44
Silicon carbide	2000†	>10^{14} Ω•cm	42.0	3.7	270	3.2
Aluminum nitride	1900†	>10^{14} Ω•cm	8.8	4.4	140–230	3.3
Beryllia	2000†	>10^{14} Ω•cm	6.8	6.8	240	2.9
Boron nitride	>2000†	>10^{14} Ω•cm	6.5	3.7	600	1.9
Borosilicate glass	1075	>10^{14} Ω•cm	4.1	3.0	5	—
Silicon dioxide (quartz)	—	>10^{14} Ω•cm	3.8	0.5	7	—
CVD diamond	—	>10^{14} Ω•cm	5.7	0.8–2.0	1500–2000	3. 52
Plastics						
Epoxy Kevlar (xy) (60%)	200†	>10^{14} Ω•cm	3.6	6.0	0.2	—
Polyimide/quartz (x axis)	200†	>10^{14} Ω•cm	4.0	11.8	0.35	—
Epoxy FR-4 (xy plane)	175†	>10^{14} Ω•cm	4.7	15.8	0.2	1.85–1.95
Polyimide	300–400†	>10^{14} Ω•cm	3.0–3.5	3–50.0	0.2	1.42–1 .61
Poly-benzocyclobutane	250†	>10^{14} Ω•cm	2.6	35.0–60.0	0.2	1.05
Teflon	400†	>10^{14} Ω•cm	2.2	20.0	0.1	2.15
Silicone (high-purity)	150†	>10^{14} Ω•cm	2.7–3.0	330	0.157	1.05–2.5
Parylene N	vapor-deposited room temp	>10^{14} Ω••cm	2.65	69	0.125	—

*Not applicable.
†Approximate processing temperatures.

Although, for many years, the sole functions of hybrid microcircuit substrates have been mechanical support and electrical insulation, the electrical and thermal functions now assume critical roles for multichip modules, high-density interconnect substrates, and three-dimensional packaging. In these cases, closely packed very large-scale integrated circuit (VLSIC) chips generate more heat per unit area, thus degrading reliability if the heat is not drawn away from the ICs so that they can operate at low junction temperatures. Some integrated circuits (ICs) dissipate over 30 W/cm². Removing heat and maintaining low junction temperatures are essential for the long-term reliability of the die and circuit. Furthermore, for high-speed performance, the substrate now plays an active role in controlling and matching impedance to the ICs.

4.1.1 Engineering Considerations

The substrate can either improve or degrade the performance and reliability of a circuit. The major engineering considerations for substrate selection are listed in Table 4-2. They fall into four categories: electrical, thermal, physical (mechanical), and chemical.

Electrical Substrates must have high electrical insulation resistance initially and after exposure to various environmental extremes such as high humidity (85 to 100 percent relative humidity) and elevated temperature (up to 200°C). Fortunately, the most widely used substrate materials—alumina and most other ceramics—easily meet this requirement. Generally, ceramic substrates used for thin- or thick-film circuits have ini-

Table 4-2. Substrate Engineering Requirements

Electrical	■ High electrical insulation resistance ■ Low and uniform dielectric constant
Thermal	■ Thermal stability ■ High thermal conductivity ■ Matched expansion coefficients
Physical	■ Low porosity ■ Smoothness ■ Camber ■ Strength
Chemical	■ Chemical inertness ■ Resistor/conductor compatibility

tially high insulation resistances of 10^{14} $\Omega \bullet$ cm or greater. Unless the ceramic has absorbed or adsorbed contaminants, its insulation resistance does not change drastically on exposure to humidity and/or temperature. Plastic substrates, however, should be chosen with care. Although the initial insulation resistance of plastics is also very high, it may drop several orders of magnitude under humid or high-temperature conditions, depending on the polymer's molecular structure, formulation ingredients, and degree of cure. As a general rule, both ceramics and plastics have negative *temperature coefficients of (electrical) resistivity* (TCRs).

Dielectric constants and dissipation factors are two other electrical parameters that have not been critical until now, but with the emergence of high-speed, high-performance multichip modules (MCMs), these parameters have assumed vital importance. Reducing capacitance and controlling impedance are critical to the performance of circuits operating at clock speeds of 100 MHz and higher. The interconnect substrate now assumes an integral role in the electrical functioning of the module. The signal propagation time delay is directly proportional to the square root of the dielectric constant of the material surrounding the conductor trace (Fig. 4-1). The capacitance is also directly proportional to the dielectric constant. Thus substrates and dielectric isolation layers having low dielectric constants (<4) and low capacitance are required. The dielectric constant of high-purity alumina ceramic (9 to 10) is relatively high. However, as the alumina content decreases and

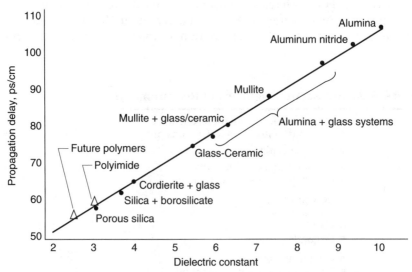

Figure 4-1. Propagation delay as a function of dielectric constant for various substrate materials.

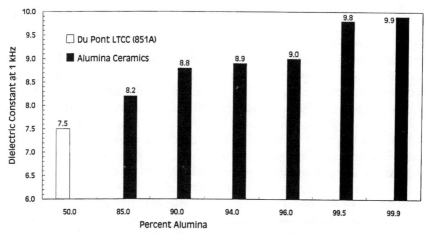

Figure 4-2. Percentage of aluminum oxide in alumina ceramics versus dielectric constant.

the glass content increases, the dielectric constant of the ceramic decreases (see Fig. 4-2). Alumina ceramics having a very low k (4 to 5) have been developed but mostly at the expense of increased porosity, lower thermal conductivity, and reduced strength. Du Pont has developed a low-k (4 to 5), low-temperature cofired ceramic tape whose coefficient of thermal expansion (CTE) is also low (4.5 ppm/°C), making it compatible with circuits having large-area bare silicon dice.

The best low-k materials are high-purity polymers such as polyimides, benzocyclobutene, fluorocarbons, silicones, triazines, and some epoxies. Many of these polymers have dielectric constants ranging from 2 to 3, but, as with low-k ceramics, polymers are extremely poor thermal conductors. Unfortunately, a material that combines high thermal conductivity, high electrical resistivity, low dielectric constant, and ease of processing does not exist. Ultimately, the electrical design engineer must settle on a compromise.

Thermal Although the thermal conductivities of alumina ceramics are relatively low, alumina ceramics are satisfactory for most hybrid microcircuits and multichip modules being produced today. Where high currents or high-heat-dissipating chips are used, the thermal conductivity of alumina substrates can be augmented by using metal, beryllia, aluminum nitride, or diamond heat spreaders beneath the heat-generating devices. Thermal vias or thermal wells may also be designed and incorporated in the ceramic, thus transferring the heat directly to a heat sink on the bottom of the substrate or to the package. Relatively thick copper can be applied to alumina or to aluminum nitride by active brazing, by plating, or through gas-metal eutectic bonding, also known as *direct-bonded copper*.[1–3] Beryllia substrates have been used for many decades for both thin- and thick-film high-power circuits but have always been limited by their high cost due primarily to the safety precautions required in processing. Industry is now looking forward to aluminum nitride and diamond as emerging substrate materials that have a thermal conductivity as high as or even higher than that of beryllia without the toxicity issue.

Besides thermal conductivity, thermal expansion is drawing increased attention, especially as silicon IC chips approach 1 in square in size and contain many hundreds of I/O pads. A close CTE match of the IC material with the substrate material is desirable to avoid fracturing the dice or the interconnections (particularly bumped flip-chip interconnects) during temperature cycling. This is an added reason for the high interest in AlN since its CTE of approximately 4.4 ppm/°C closely matches that of silicon (3 ppm/°C). Of course, silicon substrates are ideal in providing an exact CTE match with silicon chips and in having a fairly high thermal conductivity. Gallium arsenide ICs whose CTE is approximately 6 present a unique situation, especially if they are to be assembled alongside silicon dice on the same substrate.

Glass-ceramic and cordierite-glass materials in which both the CTE and dielectric constants have been tailored to meet specific circuit requirements for mainframe computers have been formulated and patented by IBM. The thermal conductivities of these materials, however, are quite low because of the high amount of glass in the formulations.

Thermal stability is yet another substrate requirement, though one that is easily met. Substrates must not decompose, outgas, change in dimensions, or fracture during any of the thermal processing steps or subsequent accelerated screen tests that a multichip module (MCM) must undergo. Ceramics can withstand the highest firing temperatures that are encountered in thick-film processing (850 to 1000°C). Thin-film processing temperatures are far lower—the highest might occur during extended sputtering of metal films (about 200°C) or in the curing of polyimides (about 400°C). Thus ceramic substrates are ideal and most widely used for both thick- and thin-film circuits. Lower-temperature stable substrates such as plastics can also be used for thin-film circuits and for *polymer thick films* (PTFs) since high-temperature processing is not required in those cases. Low-cost PTFs are processed (cured) at temperatures of 125 to 175°C and thus are compatible with plastic laminate substrates.[4,5] Plastic laminates such as those of epoxies, polyimides, and fluorocarbons, all having low dielectric constants, are ideal in fabricating low-cost multichip modules (MCM-L) for the consumer market.

Chemical. The chemical inertness of the substrate is important in thin-film processing because of the number and nature of chemical solutions used in etching metal conductors, resistors, and dielectrics; in developing and stripping photoresists; and in plating. For example, the hydrofluoric acid used to etch tantalum nitride resistors will also etch the glassy constituent of alumina ceramic substrates, especially if the glass content is high. A porous substrate entraps and retains chemical solutions, subsequently causing electrical leakage currents, metal migration, or corrosion. Even a high-grade, as-fired alumina, used as a substrate for thin-film processing, can absorb and retain ionic contaminants from etching or plating solutions, if it is not thoroughly cleaned. Washing with deionized water and monitoring the electrical resistivity of the water rinses are important in ensuring the complete removal of ionic residues. Vacuum baking or firing the substrates at 850°C prior to thin- or thick-film processing is also effective in removing adsorbed organic residues.

Physical. The key physical attribute of substrates is surface smoothness. Surface smoothness is essential in fabricating thin-film precision circuits. *Smoothness* is defined as the average deviation from an arbitrary centerline that traverses the hills and valleys of a surface profile (measured by using a profilometer). One metric for reporting smoothness (or, conversely, roughness) is microinches of centerline average (CLA). In fabricating precision fine-line conductors 25 μm wide or less, a surface smoothness of 6 μin or less is required. For precision, reproducible resistors, smooth, highly polished surfaces are also essential. If one attempts to use a rough ceramic surface, e.g., one having a CLA of 20 μin (5000 Å), to deposit thin-film nichrome or tantalum nitride resistors that are only 200 Å thick, it is apparent that the thin film will conform to the irregularity of the surface. The length of the resistor will then vary from area to area, and since resistance varies directly with length, both the length and resistance values will be variable and unpredictable according to

$$R = \frac{rl}{A}$$

where R = resistance, l = length, r = resistivity, and A = area. Furthermore, weak spots and discontinuities often occur in thin films at the peaks, resulting in open circuits and in poor-quality resistors. With thick-film resistors, the opposite is true. A rough surface is actually desirable, to achieve some mechanical interlocking and thus enhance adhesion. Assuming a scenario for a surface having a CLA of 20 μin, a thick-film resistor approximately 0.5 mil thick (125,000 Å) would completely fill in and encapsulate the surface. The length of the resistor and its value would therefore be reproducible.

Camber is yet another substrate characteristic that can affect the processing of thin or thick films. *Camber* is defined as the warpage or bowing of a substrate or the total deviation from flatness and is reported in inches per inch deviation from flat. Camber is particularly important in processing large-area substrates where those having a high camber are difficult to attach inside packages. These substrates require thicker adhesive or solder preform which, in turn, reduces thermal transfer. Substrate camber and waviness also affect thick-film screen printing, resulting in poor print resolution. Typically, ceramic substrates have a camber of 0.001 to 0.003 in/in when measured across the diagonal. Both camber and waviness may be reduced by lapping and polishing. Test procedures and further definitions of camber, waviness, and surface roughness may be found in ASTM F865-84 and ASTM F109-73.

Finally, high mechanical strength is required of all substrates. Mechanical strength becomes much more critical as substrates increase in size, complexity, and shape. High-stress areas at corners or nicks at edges can be the foci for fracturing. Generally, ceramics are quite strong, but as more glass is incorporated into their compositions, the flexural strength decreases and brittleness increases (Table 4-3).

4.1.2 Alumina

Ceramics, in particular alumina ceramics, are the most widely used substrates for both thin- and thick-film hybrid circuits and MCMs.

Table 4-3. Strength Properties of Substrate Materials

Material	Flexural strength, MPa	Young's modulus, GPa	Poisson's ratio	Tensile strength, MPa
99% Alumina	570–620	276–372	0.2	214
96% Alumina	365–400	303–310	0.21	193
HTCC	317–413	268	0.22	
LTCC	124–193	220–303	0.23–0.26	1.38–1.72
AlN	276	269–331	0.24	
99.5% BeO	207–241	344	0.2–0.34	137
Silicon		113		
Aluminum (6061)	276	65	0.33	310
Lanxide (MCX-622)	300	265	0.22	
Diamond	1000	1000–1048	0.148	

NOTE:

$$Pa \times 1.45 \times 10^{-4} = lb/in^2$$

The ranges indicated are composites of values reported by several companies.

However, depending on whether the circuit is thin or thick film, different grades of alumina must be used. Chemical composition and surface smoothness are two key factors that affect the performance and the cost. Thin films require a substrate of high alumina content (99 + percent) and a very smooth surface finish (< 6 μin CLA); otherwise the key benefits of fine-line definition and precision derived from thin films are lost. In contrast, thick films actually benefit from a rougher surface (20- to 50-μin CLA) and from a lower alumina content (96 percent), both of which ensure better adhesion of fired thick-film pastes and reduce overall costs. Tables 4-4 and 4-5 give properties of alumina substrates used in hybrid microcircuit and MCM fabrication.

The balance of the composition of alumina ceramics consists of glassy silicon dioxide and other oxide materials. The glass content can vary from very high (approximately 50 percent) in the case of LTCC to very low (less than 1 percent) in the case of alumina used for thin-film circuits. Because of the extremely poor thermal conductivity of glass, the low thermal conductivities of ceramics having a high glass content must be accounted for in designing high-density, high-power circuits. Figure 4-3 shows thermal conductivity as a function of alumina content. The dielectric constant of alumina ceramics is also affected by the amount of glass, varying inversely as the glass content increases (Fig. 4-2). Alumina may be purchased as highly polished wafers up to 6 in in diameter or square panels as large as 6 × 6 in for processing multiple high-density interconnect substrates.

4.1.3 Silicon

Silicon is the basic substrate material for the fabrication of almost all semiconductor devices and integrated circuits. However, attempts at extending single-chip silicon technology to wafer-scale integration of many circuit functions have met with limited success because of low yields and inability to rework at the monolithic level. As a compromise, high-density multichip interconnect substrates have been developed which start with silicon wafers as the substrate and use thin-film multilayering processes to form the interconnect structure. As in hybrid circuits, uncased bare dice are then interconnected to the substrate.

Silicon is commercially available at a relatively low cost. Semi-grade silicon wafers (Ml STD 8-65) of various sizes up to 6 in in diameter and 15 to 40 mils thick are commercially available. These monocrystalline wafers have a smooth polished surface and are flat to 60 μm over a 6-in span. A mature infrastructure already exists in the semiconductor industry for materials, processes, and equipment. Besides its availabil-

Table 4-4. Characteristics of Alumina Substrates for Thin-Film Circuits

Characteristic	Unit	Test method	ADS-995	ADS-996	Substrate 996
Alumina content	Weight %	ASTM D2442	99.5	99.6	99.6
Color	—	—	White	White	White
Nominal density	g/cm³	ASTM C373	3.88	3.88	3.87
Density range	g/cm³	ASTM C373	3.86–3.90	3.86–3.90	3.85–3.89
Hardness, Rockwell	R45N	ASTM E18	87	87	87
Surface finish (working surface— "A" side)	Microinches centerline avg	Profilometer— 0.030-in cutoff, 0.0004-in diameter stylus	6	3	2
Average grain size	Micrometers	Intercept method	<2.2	<1.2	<1.0
Water absorption	%	ASTM C373	Nil	Nil	Nil
Gas permeability	—	*	Nil	Nil	Nil
Flexural strength	klb/in²	ASTM F394	83	86	90
Elastic modulus	Mlb/in²	ASTM C623	54	54	54
Poisson's ratio	—	ASTM C623	0.20	0.20	0.20
Coefficient of linear thermal expansion	$10^{-6}/°C$	ASTM C372			
25–300°C			7.0	7.0	7.0
25–600°C			7.5	7.5	7.5
25–800°C			8.0	8.0	8.0
25–1000°C			8.3	8.3	8.3
Thermal conductivity	W/(m•K)	Various			
20°C			33.5	34.7	35.0
100°C			25.5	26.6	26.9
400°C			—	—	—
Dielectric strength (60 cycles ac avg. rms)	V/mil	ASTM D149			
0.025 in thick			600	600	600
0.050 in thick			450	450	450
Dielectric constant (relative permittivity)	—	ASTM D150			
1 kHz			9.9 (± 1%)	9.9 (± 1%)	9.9 (± 1%)
1 MHz			9.9 (± 1%)	9.9 (± 1%)	9.9 (± 1%)
Dissipation factor (loss tangent)	—	ASTM D150			
1 kHz			0.0003	0.0003	0.0001
1 MHz			0.0001	0.0001	0.0001
Loss index (loss factor)	—	ASTM D150			
1 kHz			0.003	0.003	0.001
1 MHz			0.001	0.001	0.001
Volume resistivity	Ω•cm	ASTM D257			
25°C			$>10^{14}$	$>10^{14}$	$>10^{14}$
100°C			$>10^{14}$	$>10^{14}$	$>10^{14}$
300°C			$>10^{12}$	$>10^{12}$	$>10^{13}$
500°C			$>10^{9}$	$>10^{9}$	$>10^{10}$
700°C			$>10^{8}$	$>10^{8}$	$>10^{14}$

*Helium leak through a plate with 25.4-mm diameter by 0.25 mm thick measured at 3×10^{-7} torr vacuum versus approximately 1 atm of helium pressure for 15 s at room temperature.

SOURCE: Courtesy Coors Ceramics Co.

Table 4-5. Characteristics of Alumina Substrates for Thick-Film
Circuits

Property	Unit		Test method	ADOS-90R	ADS-96R
Alumina content		Weight %	ASTM D2442	91	96
Color		—	—	Dark brown	White
Bulk density		g/cm³	ASTM C373	3.72 min.	3.75 min.
Hardness, Rockwell		R45N	ASTM E18	78	82
Surface finish— CLA (as fired)		μin	Profilometer ▪ cutoff: ▪ 0.030 in ▪ stylus diameter: 0.0004 in	≤45	≤35
Grain size (avg.)		μm	Intercept method	5–7	5–7
Water absorption		%	ASTM C373	None	None
Flexural strength		klb/in²	ASTM F394	53	58
Modulus of elasticity		Mlb/in²	ASTM C623	45	44
Poisson's ratio		—	ASTM C623	0.24	0.21
Coefficient of linear thermal expansion	25–200°C 25–500°C 25–800°C 25–1000°C	10⁻⁶/°C	ASTM C372	6.4 7.3 8.0 8.4	6.3 7.1 7.6 8.0
Thermal conductivity	20°C 100°C 400°C	W/(m•K)	Various	13 12 8	26 20 12
Dielectric strength	0.025 in thick	V/mil	ASTM D116	540	600
(60-cycle ac avg rms)	0.050 in thick			—	—
Dielectric constant (relative permittivity) @ 25°C	1 kHz 1 MHz		ASTM D150	11.8 10.3	9.5 9.5
Dissipation factor (loss tangent) @ 25°C	1 kHz 1 MHz		ASTM D150	0.1 0.005	0.0010 0.0004
Loss index (loss factor) @ 25°C	1 kHz 1 MHz		ASTM D150 ASTM D2520	1.2 0.05	0.009 0.004
Volume resistivity	25°C 300°C 500°C 700°C	Ω•cm	ASTM D1829	>10^{14} 4×10^{8} — 7×10^{6}	>10^{14} 5.0×10^{10} — 4.0×10^{7}

SOURCE: Courtesy Coors Ceramics.

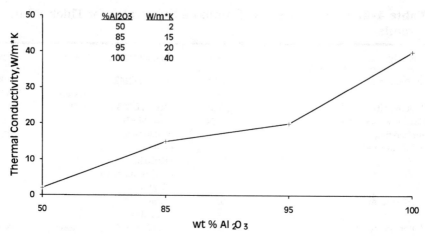

Figure 4-3. Effect of weight percent of aluminum oxide on thermal conductivity of alumina ceramic substrates.

ity and low cost, silicon has several performance advantages over other substrate materials:

- Decoupling capacitors can be batch-fabricated and integrated within the silicon by thermally oxidizing the silicon surface, forming silicon dioxide. In some MCM circuits, the integration of capacitors can free up to 20 percent of the top die bonding area so that more ICs can be interconnected. The batch fabrication of capacitors, besides resulting in increased IC density, reduces costs associated with the purchase, inspection, handling, and assembly of chip capacitors. Reducing the number of interconnections also improves reliability.

- Resistors and some active devices may also be monolithically formed in the silicon.

- By starting with a highly doped (high-electrical-conductivity silicon substrate) the silicon can function as a ground layer, avoiding a metallization step.

- Silicon substrates are compatible (matched CTE) with custom silicon ICs whose sizes are approaching 1 in square.

- The thermal conductivity of silicon is much higher than that of alumina ceramics. It is variously reported between 85 and 135 W/(m•K) depending on its purity and crystalline structure.

- Silicon is easily metallized with aluminum or other metals, although in some cases an adhesion promoter must be used.

Only a few problems have been encountered in using silicon wafers as MCM substrates. Among these are its lower flexural strength compared with alumina and its excessive bowing or warpage after depositing thick layers of dielectric and metallization. The lower mechanical strength requires greater care in handling to avoid chip-outs and breakage. The increased bowing (Young's modulus for silicon is over 50 percent lower than that for alumina—see Table 4-3) results in a threefold increase in warpage.[6] This can stop automatic pick-and-place equipment from further processing the wafers, can induce stresses within the interconnect structure, and, for large substrates, can cause both mechanical and thermal problems in adhesive attachment to the inside of flat packages. Several ways to reduce warpage are using thicker silicon wafers, using dielectrics having a CTE closely matching that of silicon, and depositing dielectric and metal on both sides of the wafer.

4.1.4 Aluminum Nitride

Aluminum nitride does not occur in nature as does alumina; hence aluminum nitride will always be somewhat more expensive than alumina since extra process steps are required for its synthesis. There are two widely used processes for its synthesis. One process starts with alumina and converts it to aluminum nitride by nitriding in the presence of carbon, according to

$$Al_2O_3 + 3C + N_2 \rightarrow 2AlN + 3CO$$

A second process involves the direct nitriding of aluminum:

$$2Al + N_2 \rightarrow 2AlN$$

Although the latter process is referred to as *direct*, it is really indirect since Al is first obtained by the reduction of bauxite, the natural mineral from which alumina is extracted. Aluminum nitride powders obtained by either process have been characterized as having similar properties.[7]

Although first synthesized almost 100 years ago, AlN has not found any major applications until recently. It is now being developed and used in electronic packaging to take advantage of its high thermal conductivity and close CTE match with silicon dice. Its key benefit is a thermal conductivity that is almost as high as that of beryllia, coupled with excellent electrical insulating and dielectric properties. AlN is among the few materials that combine the high thermal conductivities of metals with the electrical insulation properties of other ceramics. Other materials in this category include diamond, beryllia, and cubic boron nitride. A further benefit, also important for the new generation of high-density and three-dimensional electronics, is that the CTE of aluminum nitride closely matches that of silicon dice (4.4 versus 3 ppm/°C, respectively). As silicon integrated circuits become more complex and larger and contain higher numbers of I/Os (some over 500), CTE compatibility becomes critical in minimizing or avoiding stresses that can fracture the die or the interconnect bonds.

Considerable progress has been made in the past 5 years in producing and processing AlN powder into ceramic that has reproducible thermal, mechanical, and electrical properties. Today AlN substrates are readily available from many Japanese and U.S. suppliers. Among the major suppliers of AlN powder are Dow Chemical Co. and Tokuyama Soda. Ceramic suppliers include Carborundum, Coors Ceramics, General Ceramics, Kyocera, NTK, Sumitomo, Tokuyama Soda, and Toshiba.

Both pressureless and hot-press processes are used to sinter aluminum nitride "green" tape. Sintering is performed at 1800 to 1900°C in a reducing atmosphere such as hydrogen or forming gas. The reducing ambient is essential in avoiding the formation of oxides, a small amount of which significantly degrades the thermal conductivity. The pressureless process has the advantage of lower cost; but, as with alumina cofired tape, xy shrinkage must be taken into account to control the final dimensions. In the hot-press process, the tape is constrained in the xy plane during pressurization and sintering; hence virtually no xy shrinkage occurs. Shrinkage occurs totally in the z direction. Hot pressing has successfully been used to produce large 4 × 4 in cofired substrates and package bases having 612 I/Os on 25-mil pitch.[8,9]

Plots of thermal conductivity, thermal expansion, and dielectric constant for pressureless sintered AlN as a function of temperature are given in Figs. 4-4 through 4-6. The properties of hot-pressed AlN ceramic are given in Table 4-6, and design guidelines can be found in Table 4-7.

4.1.5 Plastic Laminates

The low-end consumer market for multichip modules (MCM-L) employs glass-reinforced plastic laminates as the multilayer interconnect structure. Plastic multilayer laminates are not new; they have been used since the early 1960s, and even today the majority of circuit boards for television, radio, computers, toys, etc., are fabricated by using print-

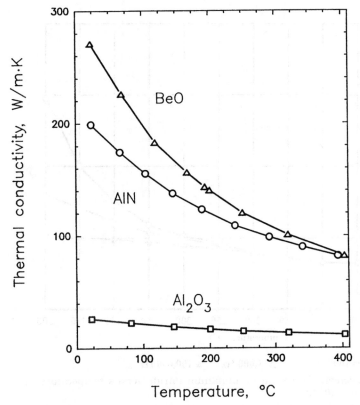

Figure 4-4. Thermal conductivity of aluminum nitride compared with beryllia and alumina versus temperature. (*Courtesy Carborundum.*)

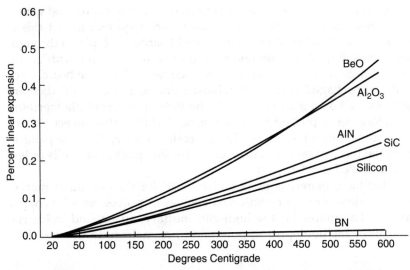

Figure 4-5. Thermal expansion of aluminum nitride compared with other ceramics. (*Courtesy Carborundum.*)

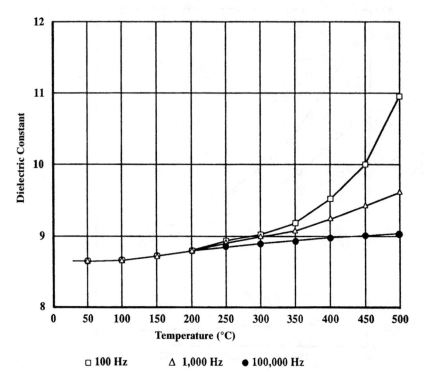

□ 100 Hz   △ 1,000 Hz   ● 100,000 Hz

Figure 4-6. Dielectric constant of aluminum nitride versus temperature. (*Courtesy Carborundum.*)

ed-circuit board (PCB) processes. The main difference between these mature processes and those being used for MCMs is a matter of degree in fine-line definition and density. In the classical printed-wiring boards, prepackaged components are soldered to the board, and generally no more than two interconnect layers are required; in fact this is often achieved by fabricating double-sided boards, with plated through-holes connecting devices on one side to the other.[10,11] Line widths and spacings of 20 mils or greater are not uncommon for these boards. For MCMs, unpackaged bare dice including one or more VLSIC dice are interconnected on a multilayer PCB. The state of the art of photoetching is therefore being pushed to the limits so that the interconnect pattern can accommodate VLSIC dice. This generally involves fine-line patterning for the top two layers so that conductor line pitches of 6 mils or less can be achieved (see Chap. 3).

The laminate materials most widely used for the consumer market consist of glass- or paper-reinforced epoxies or polyesters with copper cladding. Laminates for the high-end military, space, and industrial applications include epoxies, polyimides, bismaleimide triazine (BT),

Table 4-6. Properties of Aluminum Nitride
Package Material*

Mechanical properties (room temperature)	
Young's modulus, GPa (Mlb/in²)	339 (48.4)
Shear modulus, GPa (Mlb/in²)	137 (19.6)
Poisson's ratio	0.24
Modulus of rupture, MPa (lb/im²	280 (40,000)
Vicker's hardness, kg/mm²) in²	200 (100 g loaded)
Thermal properties	
Thermal expansion, ppm/°C	3.2 (RT to 100°C)
	3.7 (RT to 200°C)
	4.3 (RT to 400°C)
	4.7 (RT to 600°C)
Specific heat, J(g•K)	0.74
Thermal conductivity, W/(m•K)	170–190
Physical properties	
Density, g/cm³	3.25–3.26
Microstructure (grain size), μm	5–10
Surface roughness† R_a, μin	30
Surface flatness,‡ in/in	<0.001
Electrical properties	
Dielectric constant @ 25°C	8.7 @ 1 kHz
	8.5 @ 1 MHz
	8.5 @ 10 MHz
	8.3 @ 9.3 GHz
Dielectric loss	0.0002 @ 1 kHz
	0.0001 @ 1 MHz
	0.0001 @ 10 MHz
	0.0012 @ 9.3 GHz
AC breakdown, V/mil	330 (94 mil thick)

*Prepared from AlN powder manufactured by Dow Chemical
Company and processed by W. R. Grace & Co.
†After grinding with resin-bonded diamond wheel.
‡After grinding.
SOURCE: Ref. 7. *Courtesy W. R. Grace & Co.*

cyanate esters, and fluorocarbons reinforced with glass, quartz, or
aramid fibers and cladded with copper. The electrical and physical
properties of some laminate and reinforcement materials are given in
Tables 4-8 and 4-9.

Epoxies, Polyimides, and Fluorocarbons Glass-reinforced epoxy
laminate, especially FR-4, is the dominant material used for single-layer
and multilayer circuit boards. Photoetched copper-clad epoxy laminates

Table 4-7. Design Guidelines for Hot-Pressed AlN

Dimension	Description	Typical	Minimum	Maximum
A	Tape thickness	0.010	0.005	0.030
B	Via diameter	0.010	0.005	0.015
C	Filled-via cover pad diameter*	B+0.002	B+0.002	N/A
D	Via-to-via centerline	0.025	0.0125	N/A
E	Via-to-via centerline with a trace between	0.040	0.027	N/A
F	Line width	0.008	0.005	N/A
G	Line spacing	0.008	0.005	N/A
H	Surface metal pattern to product edge	0.010	0.000	N/A
I	Buried metal pattern to product edge	‖0.030 ⊥0.000	‖0.020 ⊥0.000	N/A N/A
J	Width of isolation ring around cover pad or via	0.015	0.008	N/A
K	Via centerline to ceramic edge	0.035	0.030	N/A
L	Depth of hermetic through-vias	0.030 (0.012 dia.)	0.025 (0.010 dia.)	—
M	Package size	5 × 5	N/A	6.2 × 6.4
N	Number of layers	5	N/A	10
O	Metallization	Refractory	Resistivity: 15 mΩ per square	N/A

All dimensions in inches.
*Not required for via diameters of 0.010 in or greater
XY tolerances ± 0.1% typical.

SOURCE: Courtesy W. R. Grace and Coors Ceramics.

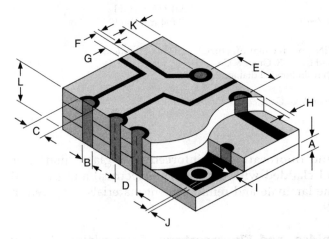

CIRCUIT CROSS SECTION

Table 4-8. Electrical and Physical Properties of Laminate Materials

Material type	$T(g)$, °C	Dielectric constant Resin	Dielectric constant With/E-glass	Propagation delay, ps/cm	Percent of H_2O absorption	Copper peel strength lb/in^2 @ 25°C	Copper peel strength lb/in^2 @ 200 °C
Standard FR-4 epoxy	135	3.6	4.4	70	0.11	11.0	5.7
High-performance FR-4 epoxy	180	2.9–3.6	3.9–4.4	66	0.04–0.20	9.0–11.0	7.5–7.9
Polyimide	220	3.4	4.3	69	0.35	8.5	8.0
BisMaleimide Triazine	195	3.1	4.0	67	0.40	8.7	5.2
Cyanate ester	240	2.8	3.7	64	0.39	8.0	6.3
Silicon-carbon	190	2.6	3.4	61	0.02	5.0	5.0
PTFE (Teflon)	16	2.1	2.5	53	0.01	10.0	8.0

SOURCE: Ref. 12.

Table 4-9. Electrical and Physical Properties of Laminate Reinforcements

Material type	CTE, ppm/°C	Dielectric Constant	Density, g/cm^3	Dielectric constant with cyanate ester	Propagation delay, ps/cm
E-glass	5.04	6.2	2.58	3.7	64
S-glass	2.34	5.2	2.46	3.5	62
D-glass	2–3	3.8	2.14	3.2	59
Quartz	0.54	3.8	2.5	3.2	59
Aramid*	–4	4.1	1.44	3.4	62
PTFE†	0.1	2.1	1.46	2.4	52

*Aramid is a class of aromatic polyamides (nylons).
†PTFE (polytetrafluoroethylene) is a generic name for Teflon, a trade name of Du Pont.
SOURCE: Ref. 12.

have served the electronics industry well for over 30 years as printed-wiring boards for digital, analog, and microwave circuits. FR-4 epoxy laminates have a glass transition temperature of 135°C and a relatively low dielectric constant of approximately 4.4. However, higher-performance FR-4 epoxy resins are available and are better suited to MCM fabrication. The latter have greater thermal stability [$T(g) = 180$°C] and lower z-direction expansion; therefore they are better able to withstand the solder, thermocompression, or thermosonic bonding temperatures employed in MCM assembly.

For the higher-end MCM-L market, polyimide or fluorocarbon laminates offer greater thermal stability and better electrical performance. The $T(g)$ for polyimides is among the highest of all plastics—greater than 220°C. Polyimides are thus better able to withstand extended times at high temperatures, which is important in high-reliability applications. However, polyimide laminates are both costlier than epoxy laminates and more difficult to process.

Fluorocarbon polymers, a notable example of which is polytetrafluoroethylene (PTFE), or Teflon, are especially useful in very high-frequency (gigahertz) and radio-frequency (RF) applications because of their very low dielectric constants (2.1 for virgin Teflon and 2.6 to 2.8 for silica-filled Teflon) and low electrical losses over a wide temperature and frequency range. Rogers Corp. has been a leader in the processing of fluorinated polymer dielectrics for printed-circuit board applications.[13] Marketed under the trade name RO2800 UTF, Rogers Corp.' fluorocarbons are available as films (0.2 to 2 mils thick) or as copper-clad laminates, making them suitable for both sequential and parallel processing. Typical properties are given in Table 4-10.

Table 4-10. Properties of Fluorocarbon RO2800 UTF

Property	Test method	Condition	Units	Direction	Typical value
Dielectric constant	ASTM D150	1 MHz, 23°C	—	Z	2.6–2.8
Dissipation factor	ASTM D3380	10 GHz, 23°C	—	Z	0.003 max.
Volume resistivity	ASTM D257	A	$\Omega \cdot cm$	Z	10^{12}
Surface resistivity	ASTM D257	A	Ω	X,Y	10^{13}
Tensile modulus	ASTM D882	A, 23°C	MPa (klb/in²)	X,Y	690 (100)
Ultimate stress	ASTM D882	A, 23°C	MPa (klb/in²)	X,Y	6.9 (1.0)
Ultimate strain	ASTM D882	A, 23°C	%	X,Y	300
Copper peel strength	Mil-P-13949	After 260°C solder float	kN/m (lb/in)	—	1.4 (8)
Water absorption	ASTM D570	D48/50	%	—	0.07 max.
Thermal conductivity	Rogers' TR2721	100°C	W/(m•k) [Btu•in•ft²/ (h•°F)]	Z	0.44 (3.0)
Specific heat	Calculated	100°C	J/(g•K) [Btu/(lb•°F)]	—	0.93 (0.22)
Coefficient of thermal expansion	ASTM D3386	10 K/min	parts/ (part•°C)	X, Y, Z	17×10^{-6} to 50×10^{-6}
Specific gravity	ASTM D792	23°C	—	—	2.1
Dielectric breakdown	ASTM D149	A, 23°C	kV/mm (kV/mil)	Z	39.4 (1.0)

SOURCE: Ref. 13. (*Courtesy of Rogers Corp.*)

Cyanate Ester Laminates (Polytriazines). Cyanate ester laminates are emerging as substrates for MCMs. They are reported to combine the ease of processing and low cost of epoxies with the excellent thermal stability and electrical characteristics of polyimides.[14] Cyanate ester resins, like epoxies and polyimides, can be reinforced with glass fibers or other fillers and processed as prepregs and laminates. Processibility into multilayer interconnect substrates (drilling, plating, photoetching) is reported to be easy and similar to conventional FR-4 glass-epoxy circuit boards. Cyanate esters, however, have an advantage over epoxies in being better suited to high-speed digital circuits because of their lower dielectric constants. The dielectric constant of the unfilled (neat) resin is 2.8, lower than that of other commercial resins used in printed-wiring boards except for PTFE (Teflon) (Fig. 4-7). The dielectric constants of the corresponding glass-reinforced laminates are higher due to the glass content, but are still in the same relative order as the neat resins (Fig. 4-8). A comparison of the laminate properties of Norplex/Oak's CE-245 cyanate ester with those of FR-4 epoxy is given in Table 4-11.

Conventional cyanate ester resins are synthesized from bisphenol A by forming the dicyanate. On heating the dicyanate resins in the presence of a catalyst, they are converted to thermally stable, highly crosslinked triazines. Dow Chemical Co. has developed cyanate esters in which a cycloaliphatic group is substituted for the bisphenol group. This modification is reported to decrease both the moisture absorption and the dielectric constant of the cured triazines and still retain their high thermal stability.[15]

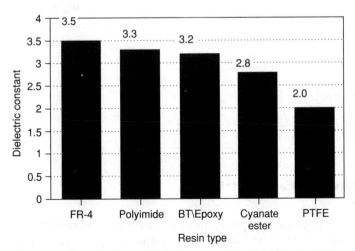

Figure 4-7. Comparison of dielectric constants of neat resins. (*Source: Ref. 14.*)

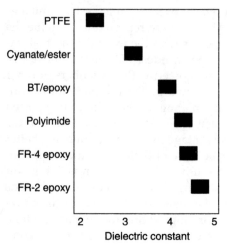

Figure 4-8. Dielectric constants of printed-circuit board laminates. (*Source: Ref. 14.*)

Table 4-11. Properties of Cyanate Ester Laminate Compared with Epoxy Laminate

	Epoxy FR-4	Cyanate ester CE-245
Dielectric constant, 1 MHz		
E-glass	4.3	3.5
S-glass	4.2	3.4
Dissipation factor, 1 MHz		
E-glass	0.02	0.005
S-glass	0.02	0.005
$T(g)$, °C	130	245
CTE, ppm/°C		
x-axis	16	12
y-axis	16	12
z-axis	80	45
Flammability rating, UL	V-O	V-O
Moisture absorption, %	0.9	0.7
Peel strength, lb/in width	10	7
Dimensional stability, mils/in	0.005	0.003

SOURCE: Ref. 14.

4.1.6 Metals and Metal Matrix Composites

Metals (such as aluminum, copper, and molybdenum), alloys of copper, metal matrix composites, and metal laminates can serve as substrates on which thin-film multilayer interconnect structures are processed. The key advantages of using metal substrates include their low cost, availability, high thermal conductivity, ability to serve as ground planes, and ability to process large panels (not limited in size, as ceramic panels are). Copper-clad molybdenum and copper-tungsten have low CTEs, ranging from 5 to 7 ppm/°C, closely matching those of other electronic packaging materials such as ceramic chip carriers, GaAs devices, and alumina packages and substrates. Furthermore the CTEs of metal matrix composites and of some metal laminates can be tailored to match those of other portions of an MCM, the dice, or the package. Besides CTE and thermal conductivity, the weight, stiffness, and cost are important elements in material selection.

Aluminum. Aluminum has a high thermal conductivity [237 W/(m•K) for the pure metal], has high strength, and, relative to other metals, is lightweight (2.7 g/cm^3). It is also inexpensive and can be purchased in or cut to various sizes and shapes. Large aluminum panels may also be useful for the batch processing of large numbers of interconnect substrates, increasing yields and reducing costs.

Now 6-in-diameter aluminum wafers, similar in shape and dimensions to silicon semiconductor wafers, are being used by Multichip Module Systems (MMS) as the basic support substrate for its MCM-D process. Besides its low cost and unbreakable nature, aluminum serves as the initial ground plane, thus reducing the number of processing steps. Otherwise, the MMS process is similar to other sequential thin-film processes in which polyimide is spin-coated as the interlayer dielectric, vias are plasma- or wet-etched, and copper metal is sputtered, electroplated, and photoetched to pattern the conductors.

Copper-Clad Laminates. Molybdenum or Invar with copper cladding on both sides may be used as substrate materials. They provide high thermal conductivity and CTEs closely matching those of other circuit materials and of the dice. The cladded metals may also be incorporated as a core within a plastic laminate so that the entire laminate is mechanically and thermally compatible with the assembled module. The thickness ratios of the core metal to copper cladding determine the CTE of the composite. The CTEs can thus be tailored for specific applications. The thermal and mechanical properties of some copper-clad substrates are given[16] in Table 4-12.

Silicon Carbide–Reinforced Aluminum. Silicon carbide–reinforced aluminum belongs to a general class of materials known as *metal matrix composites*. This unique material, commercially available from Lanxide Electronic Components under the tradename Lanxide, combines four properties desirable for substrates and package materials:

Table 4-12. Properties of Copper-Clad Metals

Material	Vol. ratio, %	Density, g/cm³	Tensile strength, MPa (klb/in²)	Young's modulus, GPa (Mlb/in²)	CTE, ppm/°C	Thermal conductivity, W(m•K)
Cu/Invar/Cu*	12.5/75/12.5	8.33	379–483 (55–70)	140 (20.3)	4.5–4.7	
	20/60/20	8.45	310–413 (45–60)	135 (19.6)	5.1–5.3	164
Cu/Mo/Cu†	5/90/5	10.1	690 (100)	303 (44)	5.1	166
	13/74/13	9.89	600 (87)	269 (39)	5.7	208
	20/60/20	9.7	503 (75)	241 (35)	6.5	242
	25/50/25	9.6	427 (62)	220 (32)	7.9	268

*From IPC CF-152.
†From Climax Specialty Metals.
SOURCE: Ref. 16.

- Very lightweight—only slightly heavier than aluminum but much lighter than copper-clad laminates

- High thermal conductivity—similar to that of aluminum nitride or beryllia but much higher than that for Kovar or alumina

- Closely matched CTE with that of GaAs devices or alumina packages, carriers, or substrates

- High stiffness

This combination of properties distinguishes Lanxide from other metal composites and makes it particularly useful as a substrate for high-power circuits, a heat spreader for devices, an SEM-E frame material, or a constraining core material for epoxy or polyimide printed-circuit boards.[17]

Lanxide is being produced in net shapes, thus obviating one of its early drawbacks—its extreme hardness and consequent difficulty in machining and cutting to a final shape. Lanxide is produced by a pressureless infiltration process in which silicon carbide powder is packed into a graphite mold, topped with an aluminum-alloy ingot, placed in a controlled atmosphere furnace, and heated to the melting temperature of aluminum (660°C). The molten aluminum wets and infiltrates the SiC powder, forming the composite on cooling.[18,19] Ceramic feedthroughs can be incorporated by positioning them in the filler prior to heating. They, too will be wetted by the aluminum, thus forming reliable hermetic joints. When the feedthroughs are removed from the graphite mold, the package is a finished part requiring no further machining. Tolerances are ± 0.003 in for a 2-in span, and the surface finish is 64 μin.

Coefficients of thermal expansion ranging from 6 to 14 can be tailored based on the volume percent of SiC used.[20] For example, a CTE of 6.2 ppm/°C, closely matching that of GaAs devices, is reported for a composite containing 70 percent by volume of SiC (Table 4-13). The thermal conductivity is also quite high, approximating that of aluminum, aluminum nitride, or beryllia. Variations of thermal conductivities and CTEs with temperature are given in Figs. 4-9 and 4-10, respectively.

4.1.7 Cofired Ceramics

Both high-temperature cofired ceramic (HTCC) and low-temperature cofired ceramic (LTCC) based on alumina can be used as interconnect substrates for multichip modules. As already discussed (Chap. 3), LTCC has several engineering and design advantages over HTCC, ren-

Table 4-13. Properties of Silicon Carbide–Reinforced Aluminum

	MCX-622	MCX-693
Coefficient of thermal expansion	$6.2 \times 10^{-6}/°C$	$6.9 \times 10^{-6}/°C$
Density	3.0 g/cm3	3.0 g/cm^3
Thermal conductivity	170 W/(m•K)	180 W/(m•K)
Young's modulus	265 GPa	235 GPa
Flexural strength	300 MPa	300 MPa
Fracture toughness	10 MPa•m$^{1/2}$	9 MPa•m$^{1/2}$
Poisson's ratio	0.22	0.24
% SiC loading	70	65

MCX-622 and MCX-693 are trademarks of Lanxide Electronic Components (LEC).
SOURCE: *Courtesy of Lanxide Electronic Components.*

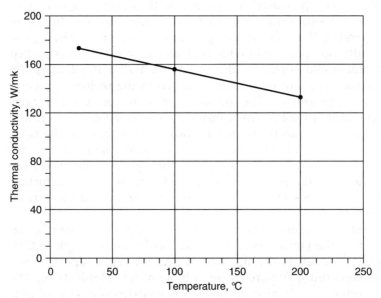

Figure 4-9. Thermal conductivity versus temperature of MCX-622 silicon carbide–reinforced aluminum. (*Courtesy Lanxide Electronic Components.*)

dering LTCC more suitable for high-performance MCMs. The main difference between the two compositions lies in the glass content. For HTCC, the glass content is relatively low, approximately 8 to 15 percent, whereas LTCC has a high glass content—50 percent or more. This difference in the glass contents of the green tapes results in significant differences in the firing temperatures, which in turn affects the types

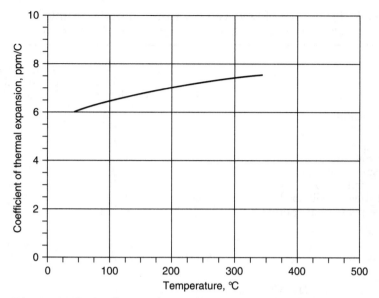

Figure 4-10. Coefficient of thermal expansion of MCX-622 silicon carbide–reinforced aluminum. (*Courtesy Lanxide Electronic Components.*)

of metal conductors that can be used. For HTCC, only refractory metals such as tungsten or molybdenum fired in a reducing atmosphere are compatible with the 1200 to 1500°C sintering temperatures that are required. For LTCC, however, gold, silver, and copper conductor pastes are compatible with the much lower 850 to 950°C processing temperatures. The low firing temperatures of LTCCs coupled with their ability to be fired in air give the engineer added freedom in designing and integrating passive devices such as resistors, capacitors, and inductors within the cofired structure. The large amount of glass in LTCC reduces its dielectric constant, which is favorable for higher-speed circuits. Unfortunately glass also reduces the mechanical strength and thermal conductivity of the cofired ceramic. For these reasons large LTCC substrates often require attachment to a high-strength backing material and incorporation of thermal vias or wells to transfer heat from the devices to a heat sink.

LTCC green tape was developed by Du Pont[21,22] and is commercially available from several manufacturers. Screen-printable thick-film gold, gold-alloy, silver, and silver-alloy pastes compatible with the green tape are also available.[23] General comparisons of the physical and electrical properties of LTCC with those of conventional thick-film circuits and with HTCC are given in Tables 4-14 and 4-15.

Table 4-14. Properties of Low-Temperature Cofired Ceramic Compared to Thick-Film and High-Temperature Cofired Ceramic

	LTCC	Thick film on 96% alumina	High temperature cofired package (92% alumina)
Physical properties			
CTE at 300°C, ppm/°C	7.9	6.4	6.0
Density, g/cm^3	2.9	3.7	3.6
Camber, mils/in	1–4*	1–2	1–4*
Surface smoothness, μin	8.7	14.5	20.0
Thermal conductivity, W/(m·K)	3–5†	20‡	14–18
Flexural strength, klb/in^2	22	40‡	46
Thickness per layer after firing, mils	3.5–10.0	0.5–1.0§	5–20
Dimensional tolerances			
Length and width	± 0.2%	N/A	± 1.0%
Thickness	± 0.5%	N/A	± 5.0%
Electrical properties			
Insulation resistance, Ω at 100 V dc	>10^{12}	>10^{12}	>10^{12}
Breakdown field, V/mil	>1000	>1000	>700
Dielectric constant, 1 MHz	7.1	9.3	8.9
Dissipation factor, %	0.3	0.3	0.03

*Function of firing setter and part design.
†16 to 20 with thermal vias.
‡Property of substrate.
§Property of printed dielectric.

Glass Ceramics   Specifically formulated glass-ceramic compositions have been developed and patented by IBM for its mainframe computer electronic modules.[24] IBM has developed a series of glass ceramics that are based on two compositions: one has beta-spodumene $Li_2O \cdot Al_2O_3 \cdot 4SiO_2$ as its principal crystalline phase, while the other is based on cordierite, a ternary compound of magnesium oxide, aluminum oxide, and silicon dioxide: $2MgO \cdot 2Al_2O_3 \cdot 5SiO_2$. Both materials are sinterable and crystallize below 1000°C, typically and desirably 950°C. The glass ceramics are formulated with organic binders such as polyvinylbutyral and small amounts of oxides such as P_2O_5, TiO_2, and B_2O_3 which act as nucleation promoters, grain growth regulators, and sintering aids. Other glass-ceramic tapes based on cordierite are possible by varying the weight percent ratios of the main constituents and including small amounts of other oxides. In so doing, both the dielectric constant and the CTE can be varied[25] and made to match closely those of sili-

Table 4-15. Properties of Low-Temperature Cofired Tape Dielectric, Du Pont 845

Electrical properties*	
Dielectric constant	≤4.8 (5 kHz to 5 GHz)
Dissipation factor	≤0.3% (5 kHz to 5 GHz)
Insulation resistance (at 100 VDC)	>10^{12} Ω
Breakdown voltage	>500 volts/25-μm (1-mil)-thick sample

Physical properties*	
Thermal expansion (25°C to 300°C)	4.5 ppm/°C
Density	2.4 g/cm^3 (>95% theoretical density)
Camber†	Conforms to setter
Surface smoothness	20 to 25 μcm (8 to 10 μin)
Thermal conductivity	2.0 W/m·K
Flexural strength	240 MPa (35 klb/in^2)

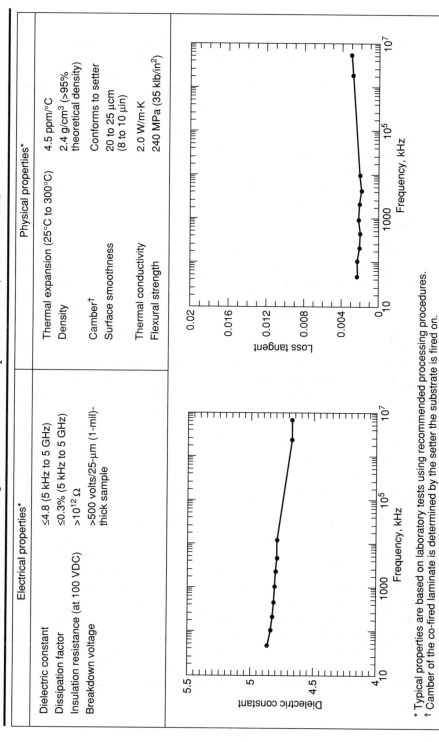

* Typical properties are based on laboratory tests using recommended processing procedures.
† Camber of the co-fired laminate is determined by the setter the substrate is fired on.

SOURCE: *Courtesy of Du Pont.*

con. The compositions are then tape-cast and dried, forming the "green" tape, then processed to form the multilayer interconnect structure. A key advantage of these glass-ceramic tapes, besides the other advantages of LTCCs, is their compatibility with copper thick-film pastes, thus resulting in lower material cost. The copper paste was also specially formulated so that it could be fired below 1000°C in an inert atmosphere without oxidizing.

Multilayer substrates fabricated from glass ceramics having up to 63 layers of copper conductors are being produced by IBM for its thermal conduction multichip modules.[26]

4.1.8 Diamond

Diamond may be the ideal material for the most advanced high-performance microcircuits. It is certainly unique among materials in combining the highest thermal conductivity [over 2000 W/(m•K) for single-crystal natural type IIa diamond] with high electrical resistivity (> 10^{13} Ω•cm), low dielectric constant (5.6), high thermal and radiation resistance, and excellent passivation properties. It is also the hardest and most chemically inert of all known materials. There has been renewed interest in synthetic diamond since the invention of pressureless, low-temperature processes in which hydrocarbon and hydrogen gases are energized by a plasma or an arc to form thin films or thick substrates. Plasma-assisted chemical vapor deposition (CVD), microwave plasma CVD, dc arc jet, hot filament, and other low-pressure methods have been developed for converting methane, acetylene, or other hydrocarbon gases to single-crystal and polycrystalline diamond.[27–29] Large amounts of hydrogen, which in its monoatomic state, suppresses the formation of graphite, and slow rates of deposition have been found necessary in producing high-quality diamond films.

Theoretically, diamond created by CVD should be inexpensive since the starting materials are relatively inexpensive. However, because of the very slow deposition rates, long times required to deposit even several micrometers of film, and high consumption of energy, the cost of the final product remains high. The cost, however, is a function of the quality of the diamond, which in turn is reflected in its higher thermal conductivity. The thermal conductivity of even low-quality diamond is still higher than that of most metals, so a cost versus quality-thermal tradeoff study for each application should be made. To date, the highest-quality optical or electronic-grade diamond has been produced at deposition rates of less than 1 μm/h and at substrate temperatures of approximately 900°C. Increasing the deposition rate or reducing the temperature results in lesser-quality darker-color diamond having lower thermal conductivity. The quality of diamond is generally measured by Raman spectroscopy in which the diamond crystalline structure can be distinguished from the graphitic structure. High deposition rates of good-quality diamond have been reported by Norton Diamond Film while using its proprietary arc jet process. By this process Norton is producing diamond wafers 4 in in diameter and 16 mils thick. These wafers are translucent and have a thermal conductivity of approximately 1400 W/(m•K). Properties of diamond created by CVD compared to those of other materials are given in Table 4-16.

As multichip modules are further developed, diamond wafers and coatings will be essential in removing heat generated, e.g., from very high-density two-dimensional modules and from three-dimensional stacked circuits. Removing heat from gallium arsenide ICs has always been difficult; but when mounted on a diamond substrate or heat spreader, GaAs dice have been shown to run much cooler than when they are mounted on alumina substrates.[30] There is considerable activity in optimizing adhesion, fabricating larger substrates (up to 6-in diameter), increasing the deposition rates, and reducing deposition temperatures in order to extend the usefulness of diamond in electronic applications.[31]

Diamond Ceramics. An interesting development in attempting to reduce the cost of diamond is the formation of composites from diamond created by CVD and high-pressure synthesized diamond. In this process developed by Crystallume, a mold is first packed with low-cost synthetic diamond powder commercially available in volume from General Electric or other sources using high-pressure synthesis. This diamond grit composes the bulk of the substrate. The particles are then consolidated by a process called *chemical vapor infiltration* (CVI) in which diamond created by CVD grows within the interstices of the packed powder, bonding the particles together. Silicon nitride or other ceramic powders may be substituted for the diamond grit. The process is reported to produce near net-shape articles of complex geometries and potentially large sizes exceeding 60 in^2 with little or no warpage. Specifically, a thin layer of the powder is placed on a silicon substrate or in a mold and is consolidated by growing diamond in a microwave chemical vapor deposition reactor. Subsequent layers are applied and consolidated to form thicker sections of diamond ceramic. The silicon substrate is subsequently removed by etching with hydrofluoric acid (Figure 4-11).

Table 4-16. Properties of Diamond Produced by CVD Compared with Other Substrate Materials

	Diamond produced by CVD	BeO	AlN	Alumina (96%)	Copper
Approximate thermal conductivity, W/(m·K) at 100°C	1300–1500	200–250	170–200	18–20	400
Thermal diffusivity, cm^2/s	7.4–10.9	0.67	0.65	0.05	1.2
Coefficient of thermal expansion (CTE), ppm/°C	0.8–2.0	6.4	4.5	8	18.8
Thermal shock (relative to BeO)	926	1	2.1	0.07	—
Dielectric constant	5.2	6.7	8.8	8.9	—
Electrical resistivity, Ω·cm	10^{12}–10^{14}	10^{14}	10^{13}	10^{13}	1.6×10^{-6}
Dielectric strength, V/mil	8750	850	1275	850	—
Density, g/cm^3	3.5	3.0	3.3	3.7	8.92
Young's modulus, Mlb/in^2	145	45	39	52	—

SOURCE: *Courtesy of Crystallume Corp.*

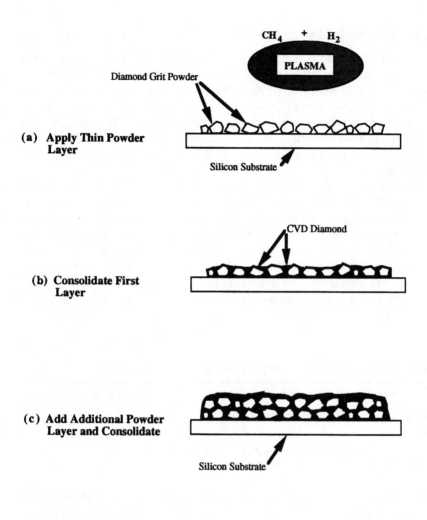

(a) Apply Thin Powder Layer

(b) Consolidate First Layer

(c) Add Additional Powder Layer and Consolidate

(d) Etch Away Silicon Using HF

Figure 4-11. Diamond ceramic produced by chemical vapor infiltration method. (*Courtesy Crystallume Corp.*)

4.2 Dielectrics

Interlayer dielectric materials used in fabricating multilayer interconnect substrates for MCMs must possess a combination of desirable
engineering and manufacturing features, as listed in Table 4-17. Since
no single material is ideal in meeting all these requirements, compromises must be made with parameters ranked based on the intended
application. Two generic dielectric materials are in use: organic (polymeric) and inorganic or ceramic. Both may be deposited as thin or
thick films.

Organic polymers as thin films or in the form of reinforced laminates
are used in MCM-D and MCM-L, respectively. To a lesser extent, some
thin-film inorganic oxide dielectrics, such as silicon dioxide, are also
used in MCM-Ds. Of the ceramic dielectrics, compositions based on
alumina are most widely used for MCM-C interconnect substrates.
Alumina ceramic tapes for both high-temperature and low-temperature
cofiring are available. The tradeoffs in selecting ceramic versus polymer organic dielectrics will be apparent from the ensuing discussions.

Table 4-17. Desirable Properties for MCM Dielectrics

Engineering properties	Manufacturing properties
Low dielectric constant and dissipation factor	Low material cost
High electrical insulation resistance (even under extremes of temperature and humidity)	Low processing cost
Low outgassing	Low equipment and maintenance cost
Low moisture absorption	Low curing/processing temperature
High thermal stability	Short cure/processing time
High thermal conductivity	Minimum steps to form vias
Low percentage of shrinkage during cure/firing	Ease of planarization
CTE matched to die material	Rapid etch rate for vias
Low stresses (no blistering)	Environmentally safe
High adhesive strength to metallization	Reproducible thickness Availability of materials Ability to process large panels Automated processing

4.2.1 Thin-Film Dielectrics

Thin-film organic polymers have played a major role in the development of modern-day electronics. Organic polymers are used as protective coatings for printed-circuit assemblies, hybrid microcircuits, and chip devices; as electrical and wire insulation, flexible cables, capacitors, and particle immobilizers; and recently as multilayer dielectrics for ICs and multichip modules. Polymer films having low dielectric constants (4 or less) are being used to fabricate high-speed digital and analog circuits, generally referred to as *high-density multichip interconnect* (HDMI) or *high-density interconnect* (HDI) substrates. These interconnect substrates are the basis for the MCM-D modules. In one widely used process, the thin-film interconnects are formed on a substrate base such as alumina or silicon by depositing the organic dielectric, etching vias, metallizing, photopatterning the metal, and then repeating the sequence. Dielectric thicknesses may range from 2 to 25 µm, depending on the polymer's dielectric constant and the capacitance and impedance properties required of the circuit.

The most widely used high-performance dielectrics are the polyimides because they combine excellent electrical properties with high thermal and mechanical stability. Polyimides have a history of over 30 years of successful applications in military, space, and commercial electronics and the benefit of an established infrastructure, including extensive technical support from manufacturers and suppliers. In spite of this, considerable activity exists in evaluating other polymers that may provide the same performance attributes as polyimides but require fewer process steps, thus decreasing the turnaround time and cost. Polyimides along with other polymeric dielectrics and one inorganic dielectric are discussed in the following sections.

Polyimides. Polyimides have been used for decades as high-performance electrical insulation, especially for high-temperature applications. Single-layer and multilayer laminates for printed-circuit boards, wire and cable insulation, conformal protective coatings, stress relief coatings for ICs, and alpha particle barriers for memory ICs are among some of the many applications. Although there are hundreds of polyimide formulations on the market, only a few can meet the specific requirements of an interlayer dielectric for MCMs. Of the many properties desired of a dielectric (see Table 4-17), three properties are critical and can be measured for initial screening and narrowing down of the selection: dielectric constant, moisture absorption, and CTE, all of which should be low. An evaluation of 10 polyimide formulations showed that three had the lowest values for these properties (Table 4-18). However, the final selec-

Table 4-18. Polyimide Selection Results

Polyimide type	Dielectric constant, 1 kHz, 25°C	CTE, ppm/°C	Water absorption, %
Ciba Geigy 293	3.3	28	4.48
Du Pont 2525	3.6	40	1.36
Du Pont 2555	3.6	40	2.52
Du Pont 2574D	3.6	40	1.68
Du Pont 2611D	2.9	3	0.34
Hitachi PIQ-L100	3.1	3	1.56
Hitachi PIX-L110	3.4	5	0.55
National Starch Thermid EL	2.8	5–10	0.7
Toray SP-840	3.4	30	1.55

tion is often based on other factors including processability and cost, which may lead to compromises. Three basic types of polyimides, differing in molecular structures, properties, and processing conditions, are commercially available. They are classified according to their curing mechanism as

- Condensation-cured
- Addition-cured
- Photocured

Condensation-Cured Polyimides. The polyimides most widely employed for multichip module substrates cure by a condensation mechanism; i.e., the applied resin, an amide-acid polymer, eliminates water on heating, forming the cyclized imide structures along the polymer chain, a process called *imidization.* One widely used condensation-curing polyimide is referred to by chemists as PMDA-ODA because of its synthesis from pyromellitic dianhydride (PMDA) and oxydianiline (ODA).[32] Polyamic acid or polyamic ester resins, used as the polyimide precursors, must be heated in steps, first to evaporate the solvent (generally *N*-methyl pyrrolidone, NMP), then up to 425°C to effect complete imidization. These resins are converted to polyimides by expelling two molecules of water (or alcohol in the case of the polyamic ester) for every polymer repeating unit (Fig. 4-12). The polyamic ester precursors are reported to offer improved solubility and better control of viscosity and solids content. The alcohols that are released during curing are less corrosive to metals than the water produced from polyamic acids.[33] A comparison of the properties of polyimides produced from these two precursors is given in Table 4-19.

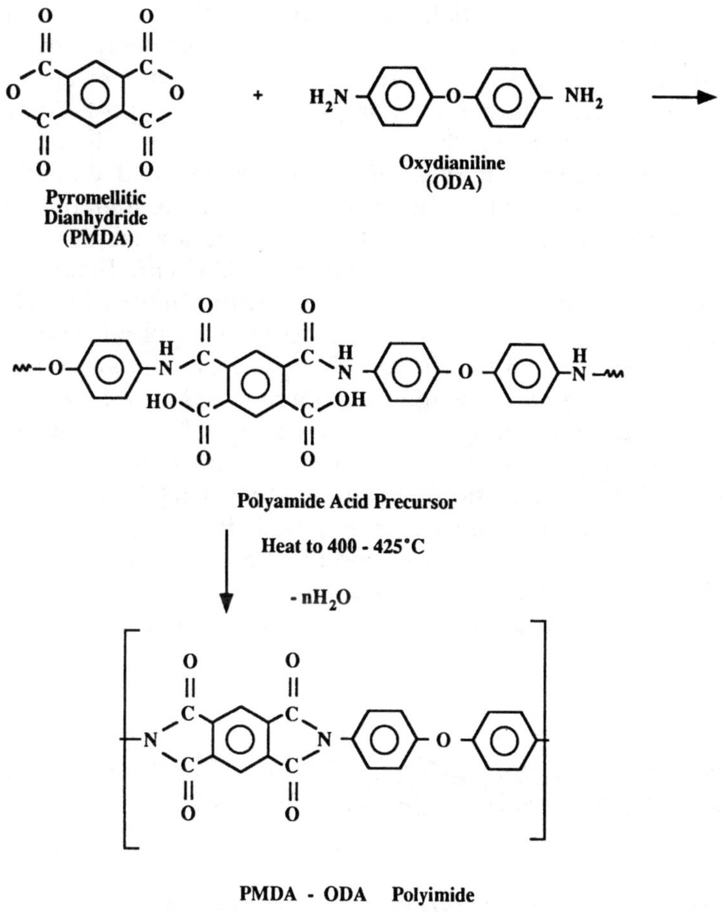

Figure 4-12. Reactions and molecular structures for condensation-cured polyimides.

Table 4-19. Properties of Polyimide Films Prepared from Polyamic Ester and Polyamic Acid Precursors

Material property	Polyamic ester precursor	Polyamic acid precursor
Young's modulus, GPa	3.1	3.3
Ultimate strength, MPa	230	265
Elongation, %	115	80
Coefficient of thermal expansion, ppm/°C	29	32
Coating stress, MPa	21	20
Glass transition temperature, °C	380	380
Dielectric constant, 50% relative humidity	3.3	3.4

SOURCE: Ref. 33.

If the polyimide is only partially imidized (cured), then water (or alcohol) and NMP solvent continue to evolve and affect the reliability of the circuit through loss of adhesion to the metal, blistering, or degradation of electrical and physical properties. This was proved in a study by Bachman and others[34] in which the properties of polyimides cured at 300°C were inferior to those of the same resin cured at 350°C. Outgassing of both NMP and water was shown for the undercured polyimide, and the dielectric constant and dissipation factor were higher—4.37 and 0.017, respectively—while the 350°C cured material had a dielectric constant of 3.88 and dissipation factor of 0.010. Incompletely cured polyimides may also react with copper metallization, causing corrosion and diffusion. Copper has been shown to migrate even in fully cured polyimides, hence the need for interjecting barrier layers of titanium, tungsten, nickel, or chromium to separate the copper from the polyimide.[35] The barrier layer also doubles as an adhesion layer. A three-stage step cure culminating at 400 to 425°C has been found effective in achieving complete imidization.

Du Pont's PI 2525, a condensation-cured polyimide, is a widely used dielectric for MCM-D. Its molecular structure is a variation of the PMDA-ODA molecule (Fig. 4-13). Du Pont has also synthesized and

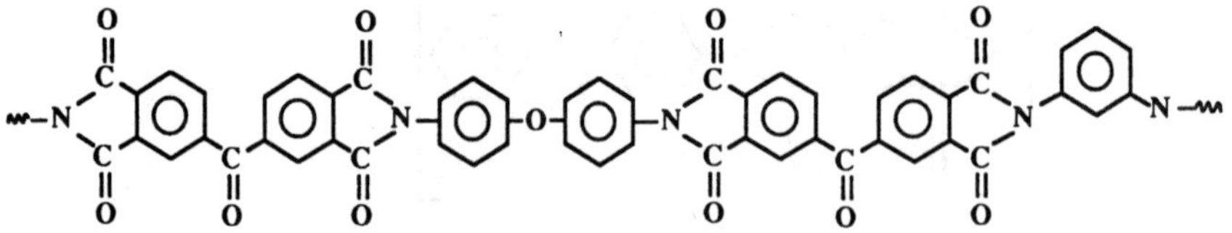

Benzophenone Tetracarboxylic Dianhydride / Oxydianiline / m-phenylene Diamine

BTDA / ODA / MPD   DuPont 2525

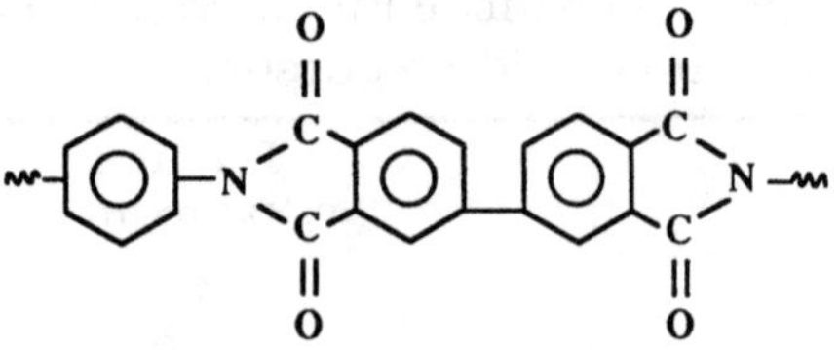

Biphenyl dianhydride   p - phenylene diamine

BPDA / PPD   DuPont 2611D

Figure 4-13. Condensation-cured polyimides: Du Pont 2525 and 2611D.

formulated a polyimide having a more compact molecular structure (PI-2611D) which is gaining popularity because of its low coefficient of thermal expansion of approximately 3 to 5 ppm/°C, closely matching that of silicon, thus obviating thermal mechanical stresses in large silicon wafer substrates and ICs (Fig. 4-13).

An important feature of condensation-curing polyimides is the ease of forming vias (used for z-direction interconnections) either by dry-plasma-etching the cured polyimide or by wet-etching the polyamic acid precursor, then completing the cure. The latter process is possible because the amide-acid structure is readily soluble in alkaline solutions such as ammonium hydroxide, potassium hydroxide, or tetramethyl ammonium hydroxide (Fig. 4-14). The process consists of depositing photoresist over the dried prepolymer, exposing and patterning the photoresist to form the via pattern, then etching the exposed polymer. In fact, if a positive photoresist is used, the same aqueous alkaline solution used to develop the resist simultaneously etches the vias. After stripping away of the photoresist, the prepolymer is imidized at 400°C.[36,37] Once fully cured, all polyimides are thermally stable to at least 400°C for long periods, and they do not start decomposing until 500 to 600°C. This excellent thermal behavior is shown in the *thermal gravimetric analysis* (TGA) curves for Du Pont's PI2611 given in Figs. 4-15 and 4-16. Properties for several condensation-cured polyimides are given in Table 4-20.

Addition-Cured Polyimides. A second generic class of polyimides cures by an addition mechanism. This class is based on polyimide oligomers (low-molecular-weight prepolymers) in which the imidization reaction has already been completed. The oligomers contain acetylenic functional groups (– C ≡ CH), which on heating join "head to tail" to form long-chain high-molecular-weight polymers without eliminating water or other undesirable compounds (Fig. 4-17). Acetylene-terminated

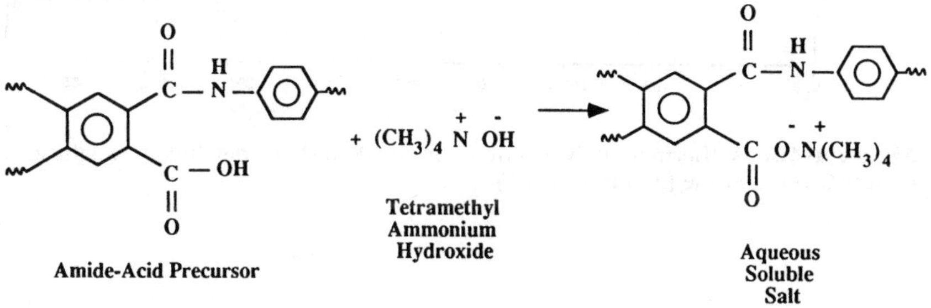

Figure 4-14. Chemistry for wet-etching vias in polyimide.

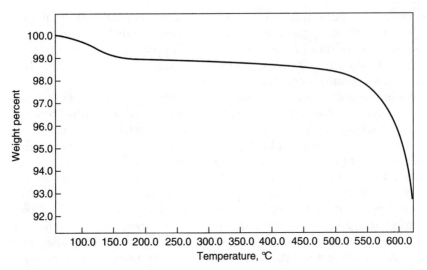

Figure 4-15. Thermogravimetric analysis curve for Du Pont 2611 polyimide (Perkin Elmer 7; temperature rise = 10°C/min).

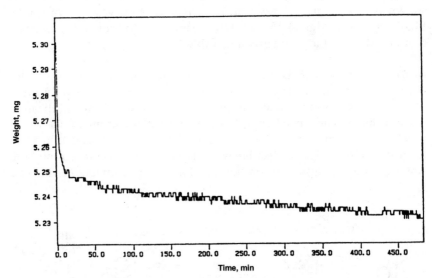

Figure 4-16. Isothermal analysis curve for Du Pont 2611 polyimide, weight loss at 400°C versus time (Perkin Elmer 7).

Table 4-20. Comparison of Properties of Condensation-Cured Polyimides

Property	WE-1111	PI-2611	PI-2525	PI-2555
Solids, %	19–21	12.4–14.5	24–26	18–20
Viscosity, P	90–110	110–135	50–70	12.0–16.0
Modulus, GPa	4.4	7	2.9	3.2
Tensile strength, MPa	300	580	150	180
Elongation, %	55	60	25	25
CTE, ppm/°C	19	5	50	41
TGA, °C for 5% loss	580	610	550	550
$T(g)$, °C	385	350	>310	>320
Dielectric constant, 1 MHz	2.8	2.9	3.5	3.3
Water uptake, %	0.9	<0.8	3	3
Self-priming	Yes	No	No	No
Wet-etch latitude	Broad	None	Narrow	Narrow

SOURCE: Courtesy of Du Pont Electronics.

polyimides were first synthesized by investigators at Hughes Aircraft[38,39] and later licensed and marketed by National Starch and Chemical Corp. under the trade name Thermid. It is reported that these polyimides, unlike the condensation-cured types, are compatible with copper without requiring a barrier metal. Properties of addition-cured polyimides can be found in Table 4-21.

Photocured Polyimides. Photocured polyimides behave like photoresists in that they can be polymerized (hardened) on exposure to ultraviolet light through a mask. Photocured polyimides offer both cost and quick-turnaround advantages over conventional polyimides by eliminating approximately 75 percent of the process steps required to form the vias (Table 4-22 and Fig. 4-18). The first photosensitive polyimides were developed at Siemens[40] in the late 1970s, but only recently have they been modified and reformulated to meet the requirements for inner-layer dielectrics.[41] Most photodefinable polyimides are negative-acting; i.e., they cross-link and harden on exposure to ultraviolet light. In these compositions the basic polyimide molecular structure has been altered to incorporate photoreactive unsaturated ester groups (such as acrylates) and sensitizers. Once the coating has been imaged, the photoreactive ester groups are burned out during the final high-temperature curing/imidization step. The decomposition and removal of these ester groups cause shrinkage of the coating (as high as 60 percent has been reported), generating stresses, marginal via geometries, and some degradation of mechanical properties.[42] A second photocuring mechanism

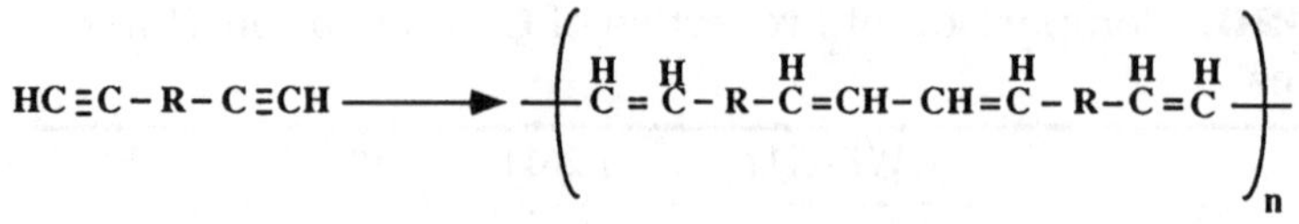

$$HC \equiv C - R - C \equiv CH \longrightarrow \left(\underset{C}{\overset{H}{\underset{\|}{C}}} = \underset{C}{\overset{H}{\underset{\|}{C}}} - R - \underset{C}{\overset{H}{\underset{\|}{C}}} = CH - CH = \underset{C}{\overset{H}{\underset{\|}{C}}} - R - \underset{C}{\overset{H}{\underset{\|}{C}}} = \underset{C}{\overset{H}{\underset{\|}{C}}} \right)_n$$

GENERIC ADDITION CURE MECHANISM

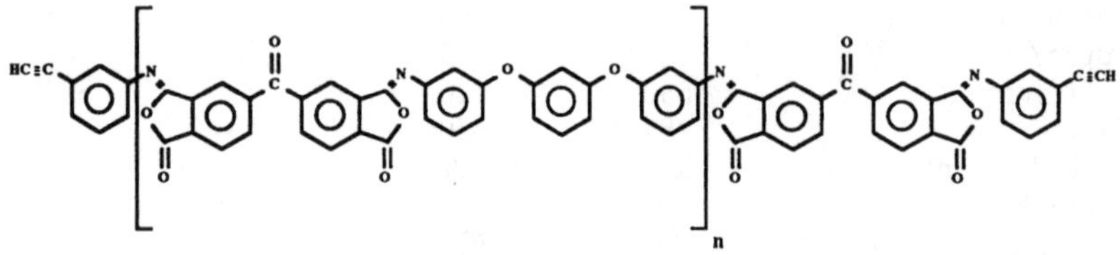

STRUCTURE OF CONVENTIONAL THERMID POLYIMIDE

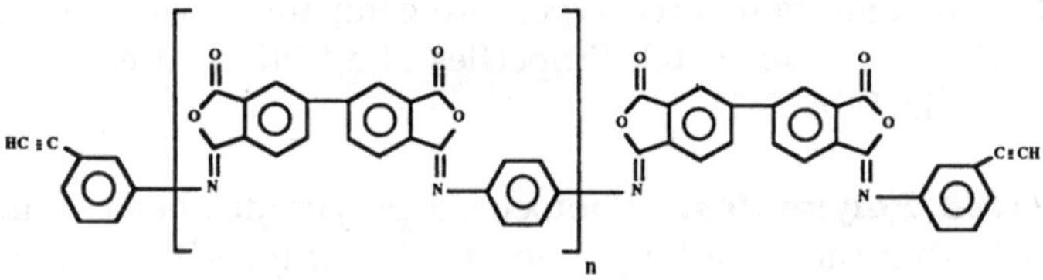

DEVELOPMENTAL LOW-STRESS THERMID 144B

Figure 4-17. Reactions and molecular structures for addition-cured polyimides. (*Courtesy National Starch and Chemical Co.*)

involves ionic salt formations which are also photoreactive. These coatings have reduced shrinkage and much improved mechanical properties,
although at the expense of poorer resolution for the thicker layers
(20 μm). A further drawback to all negative-acting polyimides is the
need to use organic solvents to develop the vias. In spite of these limitations, many firms (Boeing, NTT, NEC, Toshiba, and Mitsubishi) are successfully using photosensitive polyimides to fabricate their high-density
multichip interconnect substrates.[42,43] Positive-acting polyimides are
being developed by major suppliers. When available, these polyimides

Table 4-21. Properties of Addition-Cured Polyimides

Thermid EL-5000 Series	Typical properties		Suggested uses and properties
	Viscosity per % solids	Filtration, μm	
EL-5010	1500 cP per 35% @ 25°C	0.5	General-purpose formulation. Suggested for passivation coating and as an interlayer dielectric; k of 3.2 @ 10 kHz. Excellent planarization. Will adhere direct to microcircuitry oxides and nitrides; also excellent intercoat adhesion.
			Recommended for spin coating; at as-received viscosity, will deposit 15 μm and can be varied to 1 μm with dilution.
EL-5501	1000 cP per 53% @ 25°C	0.5	Fluorinated polyimide formulation. Suggested for passivation coating and as an interlayer dielectric. Forms a rigid, low-interfacial-stress film having a high $T(g)$ and modulus with superior solvent resistance. A k of 2.8 @ 10 kHz and low moisture absorption. Excellent adhesion and planarization.
EL-5512	1100 cP per 47% @ 25°C	0.5	Fluorinated polyimide formulation. Suggested for same applications as EL-5501. Forms solvent-resistant film with good tensile properties. A k of 2.8 @ 10 kHz and low moisture absorption. Excellent planarization and adhesion.
			Both EL-5501 and EL-5512 are recommended for spin coating; at as-received viscosity, will deposit 15 μm and can be varied to 1 μm with dilution.
EL-5330	980 cP per 24% @ 25°C	0.2	Suggested as alpha particle barrier and for heavy deposition protective applications. A k of 3.2 @ 10 kHz. Excellent adhesion.
			Recommended for syringe application on dynamic random access memory IC chips.
Suggested cure schedule* for all Thermid EL formulations (in nitrogen or air)			Soft-bake @ 180°C for 30 min.
			Bake @ 300°C for 60 min.
			Postcure at 400°C for 15 min.

*Cure schedule—i.e., total time at temperature—is for the Thermid material. Heat sink effects due to the substrate may require adjustment of total schedule.

SOURCE: Courtesy of National Starch & Chemical Co.

Table 4-22. Reduction of Via Process Steps by Using Photosensitive Dielectric

In-contact hard mask process	In-contact sacrificial mask	No mask (photosensitive dielectric)
1. Apply dielectric.	1. Apply dielectric.	1. Apply dielectric.
2. Cure dielectric.	2. Cure dielectric.	2. Expose.
3. Metallize cured dielectric.	3. Apply thick photoresist.	3. Develop (form vias).
4. Apply photoresist.	4. Expose.	4. Cure dielectric.
5. Expose.	5. Develop.	
6. Develop.	6. Plasma-etch vias.	
7. Etch via pattern in metal.	7. Remove photoresist.	
8. Remove photoresist.		
9. Laser-ablate or plasma-etch vias.		
10. Etch away metal.		

For a five-conductor-layer HDMI:		
40 steps	28 steps	16 steps

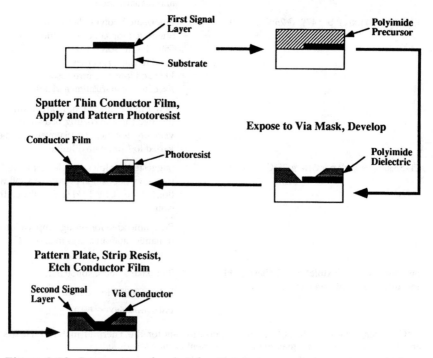

Photoetch First Conductor Layer   **Apply Photosensitive Polyimide, Prebake**

First Signal Layer

Substrate

Polyimide Precursor

Sputter Thin Conductor Film, Apply and Pattern Photoresist

Conductor Film

Photoresist

Expose to Via Mask, Develop

Polyimide Dielectric

Pattern Plate, Strip Resist, Etch Conductor Film

Second Signal Layer   Via Conductor

Figure 4-18. Process steps for photoforming vias.

will be able to be patterned by using environmentally friendly aqueous solutions instead of volatile organic solvents. A comparison[44] of the properties of a nonphotosensitive polyimide (Du Pont PI2611) with those of a photosensitive one (Du Pont PI2741) is given in Table 4-23.

Triazines. The chemistry of triazine dielectrics is based on cyclizing cyanate ester monomers to form triazine polymers on heating. This chemistry has previously been described in the section on cyanate ester laminates. Cyanate ester resins based on bisphenol A cyclize on heating to triazines that have very high thermal stability. Their $T(g)$ is reported to be 250°C, and their decomposition temperature is approximately 450°C. However, this excellent thermal stability is offset by the brittleness of the cured films. In practice, triazines are formulated with materials to add flexibility and other additives in order to yield good films, though this entails some compromise of their thermal properties.

Photodefinable triazines have been used on a production scale since 1987 by AT&T in the POLYHIC process. A cyanate ester resin formulation (PHP-92) was developed by AT&T that is rendered photosensitive by blending with four photosensitive compounds: an acrylonitrile butadiene rubber, an epoxy acrylate resin, N-vinyl pyrrolidone, and trimethylolpropane triacrylate (U.S. patent 4554229). The cyanate ester

Table 4-23. Cured* Film Properties of Du Pont's Photodefinable and Nonphotodefinable Polyimides

	Nonphotodefined (PI-2611)	Photodefined (PI-2741)
Thermal properties		
Glass transition temperature, °C	350	365
Decomposition temperature, °C	620	620
% Weight loss (500°C in air, 2 h)	1.0	1.5
Thermal expansion coefficient, ppm/°C	5	10
Residual stress, MPa	10	20
Mechanical properties		
Young's modulus, GPa	6.6	6.1
Tensile strength, MPa	600	330
Elongation to break, %	60	50
Electrical properties		
Dielectric constant (@ 1 kHz, 0% RH)	2.9	3.0
Water absorption (wt % at 50% RH)	0.8	1.2

*Cured at 400°C in nitrogen.
SOURCE: Ref. 44.

is synthesized from bisphenol A, a common starting material for epoxy resin synthesis. The formulation contains other ingredients to improve adhesion and control flow.[45] PHP-92 is a negative-acting dielectric capable of producing films that are 25 to 50 μm thick when applied by spray coating. Its key attributes include a low-temperature cure of 200°C, compatibility with tantalum nitride thin-film resistors, a low dielectric constant of 2.8, and high dielectric strength of 900 V/mil. The dielectric constant was measured by three test methods over a frequency range of 0.1 Hz to above 1 GHz. The dielectric constant decreased at the higher frequencies, reaching a value of 2.8 in the gigahertz region.[45] The water absorption of approximately 2 percent, though relatively high, is reported to result in impedance changes of less than 5 percent. The $T(g)$ is not sharp because of the many ingredients used in the formulation, and its thermal stability, for the same reason, is not as high as that for the neat resin. Detailed electrical and physical properties are given in Tables 4-24 and 4-25, respectively.

Benzocyclobutene (BCB).   Benzocyclobutene resins, developed under the trademark Cyclotene by the Dow Chemical Co., are a promising series of prepolymers that can be used as interlayer dielectrics for high-speed multichip modules, coatings for flat panel liquid-crystal displays, or as inner-layer dielectrics for gallium arsenide integrated circuits. Compared with polyimides, polymers of Cyclotene 3022, a divinyl silox-

Table 4-24. Electrical Properties of PHP-92* Triazine

Property	Value	Comments
Dielectric constant	2.8	Gigahertz range: Z_0 from reflection coefficient and propagation velocity
tan δ	<0.025	Gigahertz range
Dielectric strength	35 V/μm (900 V/mil)	ASTM D149
Volume resistivity	>3.7 × 10^{12} Ω·cm	ASTM D257 (instrument limited)
Temperature, humidity, bias I (surface) ρ (volume)	~1 pA per square >10^{15} Ω·cm	85°C, 85% RH, 180 V, 1000 h

*Trade name of AT&T.
SOURCE: Ref. 45.

Table 4-25. Physical Properties of PHP-92* Triazine

Property	Value	Comments
Moisture absorption	$\leq 2\%$ $\Delta Z_0/Z_0 < 5\%$	Typical of epoxy systems 25% RH–85% RH
Thermal conductivity	0.2 W/(m·K)	Thermomechanical analysis
TCE	65 ppm/°C	(25–100°C)
$T(g)$	180°C 155–210°C	Differential scanning calorimetry and thermo-mechanical analysis
Thermal stability	300°C $\leq 120°C$	Spikes (i.e., soldering) max. operating temperature
Young's modulus	2–7×10^{10} dyne/cm^2	Mechanical tensile tester
Room-temperature film stress	2.2–2.9×10^8 dyne/cm^2	Warpage test with fused quartz disk
Thermal cycling	>200 cycles	-40 to $+130$°C

*Trade name of AT&T.
SOURCE: Ref. 45.

ane bis-benzocyclobutene (DVS bis-BCB) resin (Fig. 4-19), are reported to have better planarization (>90 percent in a single pass), lower moisture absorption (<2 percent), rapid curing at lower temperatures (<300°C), and better electrical properties. The electrical properties of DVS bis-BCB are among the best in polymer coatings. Its dielectric constant is 2.7 at 10 kHz and remains essentially unchanged up to 40 GHz and over a temperature range of -150 to 250°C. Its dissipation factor is also very low, 8 $\times 10^{-4}$ at 1 kHz, increasing to approximately 0.002 at 10 GHz. Unlike most polyimides, the excellent electrical properties of DVS bis-BCB are preserved under humid conditions because of its low water absorption and the hydrophobic nature of its siloxane structure.[46] Other electrical and physical properties are given in Table 4-26.

BCB resins cure by a thermal rearrangement and addition mechanism which does not require catalysts, hardeners, or other ingredients. Thus water, alcohol, and other volatiles are not emitted as with condensation-cured polyimides. The mechanism and kinetics of curing are somewhat complex, involving the thermal opening of both the vinyl double bonds and the cyclobutene rings to form free radicals (Fig. 4-20). These free radicals then couple in several ways to form long-chain, high-molecular-weight polymers.[47] Curing temperatures are relatively low (200 to

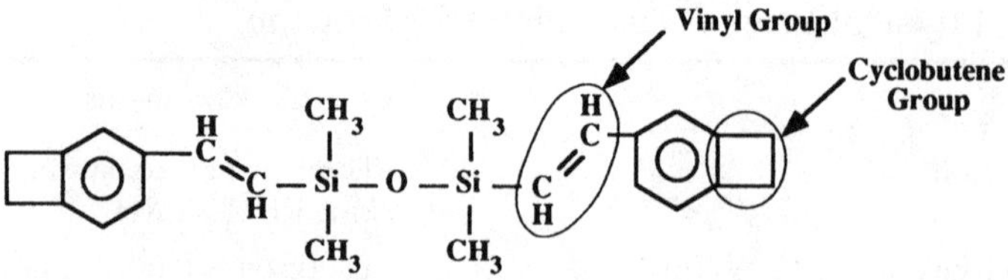

Figure 4-19. Divinyl siloxane of bis-benzocyclobutene monomer.

Table 4-26. Electrical and Physical Properties of DVS bis-BCB*

Property	Value
Flexural modulus, klb/in^2	475
Young's modulus, klb/in^2	340
Glass transition temperature, °C	>350
Temperature stability in N$_2$, °C	to 350
Breakdown voltage, V/cm	4.0×10^6
Volume resistivity, Ω·cm	9.0×10^{19}
Dielectric constant, 10 kHz to 10 GHz	2.7–2.6
Dissipation factor, 10 kHz to 10 GHz	0.0008–<0.002
Water absorption, 24-h boil, %	0.23
Coefficient of thermal expansion, 25 to 300°C, ppm/°C	65
Index of refraction, 589 nm	1.56

*DVS bis-BCB is supplied in mesitylene solution at 35 to 62% solids, depending on deposition-thickness requirements.

SOURCE: Ref. 46.

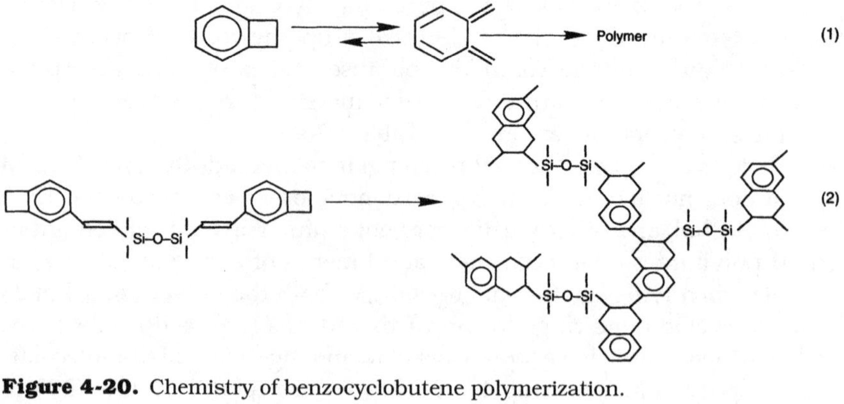

Figure 4-20. Chemistry of benzocyclobutene polymerization.

300°C) compared with polyimides (400 to 425°C)—an important feature in its use as a coating for heat-sensitive parts such as flat panel displays or assembled microcircuits. However, curing should be performed in an inert atmosphere such as dry nitrogen since the cured coating has a tendency to oxidize, as shown by infrared spectroscopy and by the yellowish discoloration that develops when the coating is heated in air.[48] Recently, a photocuring version of BCB was announced by Dow Chemical.[49] It is a negative-acting formulation based on DVS bis-BCB and a bisarylazide photo-cross-linking agent. Well-resolved conductor line widths and spacings of 16 μm for a 5.6-μm-thick film have been reported.[49]

In thin-film electronic applications, BCB resins are B-staged (partially polymerized), producing low-molecular-weight prepolymers that are readily soluble in several organic solvents. Solutions of various solids contents up to 62 percent are possible, allowing the viscosity to be adjusted to control thickness and planarization. Resin coatings are generally applied by spin-coating from mesitylene solutions.

Silicon Dioxide. Thin-film silicon dioxide (SiO_2) is being used as an inorganic dielectric by nCHIP in its MCM-D process. nCHIP has developed a SiO_2 deposition process that solves a long-standing problem of achieving stress-free, pinhole-free thin films over wide areas of substrate. A plasma CVD method involving the chemical reaction between silane and oxygen gases at temperatures of less than 400°C is used. The process produces films up to 20 μm thick that are in compression which offset the intrinsic tensile stresses resulting from metal conductor films.[50] Among some key advantages of using silicon dioxide dielectric rather than organic dielectrics are its higher thermal conductivity; harder surface, making it easier to wire-bond and rework dice; high thermal stability; and negligible water absorption and outgassing. A key limitation is its very poor planarization. The nCHIP company builds its multilayer structure on silicon wafers and uses sputtered aluminum, electroplated copper, or combinations of aluminum and copper for the conductor traces (see Chap. 3). Properties of silicon dioxide are given in Table 4-27.

Polyxylylenes (Parylenes). Polyxylylenes, commercially known as Parylenes (a trade name of Union Carbide), are unique among polymer coatings in that they are vapor-deposited without decomposing. Most polymer coatings cannot be vaporized by heating, even under high vacuum, because of their low decomposition temperatures. Parylene, however, deposits by an entirely different mechanism—the polymer is generated in situ on the surface at almost room temperature by the recombination of

Table 4-27. Properties of Silicon Dioxide

Dielectric constant, 1 MHz	3.5–4.0
Dissipation factor, 1 MHz	0.0001
Dielectric strength, V/mil	>10,000
Volume resistivity, $\Omega \cdot$cm	>10^{16}
CTE, ppm/°C	0.3–0.5
Thermal conductivity, W/(m$\cdot$K)	1.35
Density, g/cm^3	2.2
Refractive index	1.45
Planarization, %	0–2
Deposition temperature, °C	300–400
Melting point, °C	1710
Young's modulus, klb/in^2	10,000

gaseous di-radical monomers. The free-radical monomers are generated in a chamber adjoining the deposition chamber by the sublimation and thermal cracking of a white solid dimer at 680°C (Fig. 4-21). The basic polymer, Parylene N, is a pure hydrocarbon (N = no atoms other than C and H). Parylene C (C = chloro) is the monochlorinated version, and Parylene D (D = di-chloro) is the dichlorinated version. A Parylene F (F = fluorinated) analog was also synthesized by Union Carbide but has not been widely used. Today, Parylene C is the most popular type because of its faster rate of deposition, although Parylene N has somewhat better electrical and moisture resistance properties.

The key benefit of electronic modules derived from the vapor deposition of Parylene is the ultrathin integral coatings that are produced and the thorough conformal coverage of high-density, closely packed devices. The vapor penetrates around and beneath wire-bonded dice, then condenses to form the solid film. The film has also been demonstrated to strengthen weak wire bonds and to freeze any loose particles, preventing them from migrating later to cause short circuits. Parylene comes close to being an ideal coating for electronic component protection because of its high purity, excellent coverage (even in thicknesses of a few hundred angstroms), low moisture permeability, and excellent electrical properties. (Tables 4-28 and 4-29).

Although Parylenes have been evaluated as interlayer dielectrics for MCM-D interconnect substrates,[51] they have not yet been implemented in any multichip module product. This is probably due to the difficulty in depositing thick films of 10 to 25 µm required for controlled impedance, lack of planarization, and poor thermal stability in an air ambient. The electrical properties of Parylene C degrade at tempera-

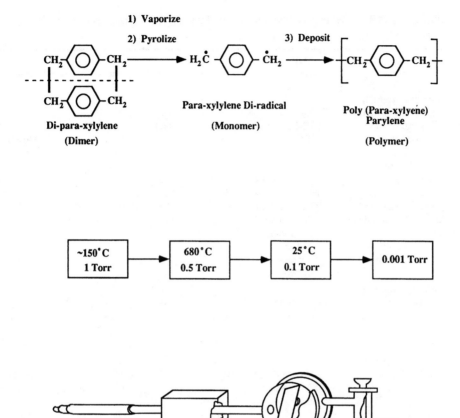

Figure 4-21. Parylene vapor deposition process.

tures as low as 150°C while Parylene N is quite stable (Fig. 4-22). As a passivation for assembled multichip modules, all three Parylene coatings are effective in shielding the module from moisture and preventing metal migration, corrosion, and electric leakage currents. As early as 1968, Licari and Lee through a NASA-funded program[52] tested metal-oxide semiconductor (MOS) devices before and after coating with Parylenes N, C, and D and after extended thermal and environmental exposures. They found no significant changes in MOS field effect transistor parameters such as I_{DSS}, BV_{SDS}, V_{GST}, and BV_{DSS}.[53]

Table 4-28. Electrical Properties of Parylenes Compared to Other Coating Types

Properties*	Method or conditions	Parylene C	Parylene N	Epoxies†	Silicones†	Urethanes†
Dielectric strength						
Short-time, V/mil						
1-mil films	ASTM D 149-64	5,600	7,000			
Corrected to $\frac{1}{8}$ in	ASTM D 149-64	590	700	400–500	550	450–500
Step-by-step						
1-mil films	ASTM D 149-64	4,700	6,000			
Corrected to $\frac{1}{8}$ in	ASTM D 149-64	550	550	380	550	450–500
Volume resistivity (23°C, 50% humidity), $\Omega \bullet$cm	ASTM D 257-61 (1-in^2 mercury electrodes)	8.8×10^{16}	1.4×10^{17}	10^{12}– 10^{17}	2×10^{15}	2×10^{11}– 10^{15}
Surface resistivity (23°C, 50% humidity), Ω	ASTM D 257-61 (1-in^2 mercury electrodes)	10^{14}	10^{13}			
Dielectric constant						
60 Hz	ASTM D 150-65T	3.15	2.65	3.5–5.0	2.75–3.05	4–7.5
10^3 Hz	(1-in^2 mercury electrodes)	3.10	2.65	3.5–4.5		4–7.5
10^6 Hz		2.95	2.65	3.3–4.0	2.6–2.7	6.5–7.1
Dissipation factor						
60 Hz	ASTM D 150-65T (1-in^2 mercury electrodes)	0.020	0.0002	0.002– 0.01	0.007– 0.001	0.015– 0.017
10^3 Hz		0.019	0.0002	0.002– 0.02	—	0.05– 0.06
10^6 Hz		0.013	0.0006	0.03– 0.05	0.001– 0.002	

*Properties measured on Parylene films 0.001 in thick.
†Properties and methods as reported in *Modern Plastics Encyclopedia*, issue for 1968, vol. 45, no. 1A, McGraw-Hill, New York, 1967.
SOURCE: *By permission of Union Carbide Corporation.*

Subsequently, Parylenes were found to be effective in conformally coating printed-wiring boards and immobilizing loose particles in hybrid microcircuits. Among the key drawbacks of Parylenes have been their difficulty of removal to rework a circuit and the difficulty and high cost of masking portions of a circuit from the high penetration of the vapors.

Fluoropolymers. Fluoropolymer dielectrics, the most prominent member of which is Teflon TFE, also known chemically as PTFE (polyte-

Table 4-29. Physical and Thermal Properties of Parylenes*

Property	Parylene N	Parylene C
Coefficient of friction	0.25	0.29
Abrasion index, mg loss per 100 cycles	2.6	4.4
Water absorption (2-mil films), %/24 h	0.01	0.06
Thermal conductivity, cal/s•m•°C	3×10^{-4}	
Coefficient of thermal expansion, in/(in•°C)	6.9×10^{-5}	3.5×10^{-5}
Nitrogen gas permeability*	15	0.6
Oxygen gas permeability	55	5
Carbon dioxide gas permeability	420	14
Moisture-vapor transmission, g•mil/(100 in²•24 h)	15	0.6

*This and other gas permeability values are in cm^3•mil/(100 in²•24 h•atm).

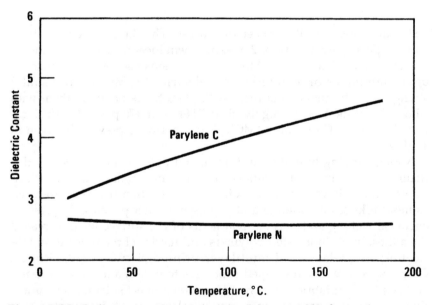

Figure 4-22. Dielectric constant versus temperature at 1 kHz for parylenes.

trafluoroethylene), are ideal dielectrics from an electrical and thermal stability standpoint. Being ultrapure highly symmetric, completely fluorinated polymers, they combine very low dielectric constants and dissipation factors with high thermal stability. Teflon-FEP (fluorinated ethylene propylene) is also a completely fluorinated polymer, but it has a slightly different structure. Its melting temperature is lower than that of

Table 4-30. Typical Properties of Fluorocarbon Coatings

Property	Teflon* TFE	Teflon FEP	TFE with 15% glass fiber
Melting point, °C	327	265	
Tensie strength, lb/in^2	3500–5000	3000	2800–3600
Elongation, %	300–600	300	325
Flexural modulus, klb/in^2	40–90	95	312
CTE, ppm/°C (23 to 60°C)	26	22–27	14–37
Thermal conductivity, W/(m•K)	0.24		0.37
Dielectric strength, V/mil	480	600	
Volume resistivity, Ω•cm	>10^{18}	>10^{18}	10^{13}
Dielectric constant (60 Hz to 2 GHz)	2.1	2.1	2.4–2.5
Dissipation factor (60 Hz to 2 GHz)	0.0003	0.0002–0.0012	0.0753–0.0029
Water absorption, %	<0.01	<0.01	0.015

*Trade name of Du Pont.

PTFE, and therefore it is easier to process. The dielectric constants of either type are slightly above 2, making them ideal for the very high-frequency circuits (Table 4-30). Fluorocarbon resins used as printed-circuit board laminates or as interlayer dielectrics for MCMs are usually strengthened by reinforcing them with glass fibers or filling them with silica. The properties of a highly filled PTFE resin (70 percent with silica) used by Rogers Corp. for an MCM-L process were previously given in Table 4-10.

A longstanding limitation of fluorinated polymers has been the difficulty in adhering metallization and other materials to them. In fact, fluorocarbons have such low coefficients of friction that they are used as mold releases in molding and processing other plastics. Proprietary chemical etching and plasma treatments of the surface have increased the adhesion of fluorinated polymers and rendered them useful in fabricating single-layer and multilayer printed-circuit boards. Rogers Corp. especially has developed processes to obtain adequate adhesion and is a leading fabricator of fine-line interconnect substrates by additive plating of copper using a sequential process. It is reported that diffusion layers between the copper and fluorocarbon are not necessary, as with copper and polyimide.

Epoxies. Because of their lower cost and lower processing temperatures, epoxies are more suitable than polyimides for the low-end consumer and industrial multichip module markets. Glass-reinforced epoxy laminates, especially FR-4 with photoetched copper conductors, are almost universally used for multilayer printed-circuit boards. These same laminates with finer geometries are now being used as interconnect substrates for MCM-L and chip-on-board circuits. High-purity, low-k epoxy resins are being used to fabricate the top signal and die bonding site layers of multilayer cofired ceramic and plastic laminate substrates. A few companies have successfully employed photocurable epoxies, previously developed and used as PCB solder maskants. These epoxies provide the fine-line features required for the signal layers and die bonding sites. Vias 35 to 50 μm in diameter and conductor lines 25 μm wide with 50-μm spacings have been achieved by using photosensitive epoxy solder maskants. Their dielectric constants are low, ranging from 2.3 to 4 at 1 MHz depending on the formulation, but their dissipation factors (0.01 to 0.035) are not as low as those of some other polymer dielectrics.

Epoxy resins may be spin-coated, screen-printed, sprayed, or flow-coated depending on their viscosities. They are compatible with copper deposited by sputtering or plating without requiring barrier coatings, have excellent adhesion, and cure at temperatures of 125 to 175°C. However, epoxies are not as thermally stable as polyimides and are more prone to outgassing.

4.2.2 Thick-Film Dielectrics

Thick-film dielectrics in either paste or "green" tape form may serve both as the interlayer dielectric and as the support substrate for MCM-Cs. Thick-film paste dielectrics consist of mixtures of ceramic powders such as alumina mixed with glass frit, oxides of various metals, organic binders, and solvents. The green tape used for cofired ceramics consists essentially of the same ingredients except that the glass content is much higher and a polymeric binder gives the tapes a rubbery consistency after casting and drying. Tape dielectric is preferred to paste dielectric in fabricating high-density multilayer substrates consisting of more than seven layers. Ceramic tape can be purchased in different thicknesses, and its ultimate thickness after shrinkage due to firing can be controlled. Hence structures up to 60 layers having tens of thousands of vias have been produced. The tape is easily handled, processed, and inspected as individual sheets prior to laminating and sintering. LTCC has gained prominence as an interconnection and packaging approach intermediate in cost and performance between MCM-L and MCM-D. A key advantage of LTCC over HTCC is its compatibility with low-resistivity conductors, primarily gold, silver, and copper. Its low sintering temperatures (850 to 950°C) and its ability to fire in air also permit integrating resistors, inductors, and capacitors with the multilayer conductor wiring structure. These passive devices consisting of metal oxides would be completely reduced under the high firing and reducing atmosphere conditions of HTCC. Refer to

Eq. 1	$XO + H_2 \rightarrow X + H_2O$
Eq. 2	$XO + C \rightarrow X + CO$
Eq. 3	$2C + O_2 \rightarrow 2CO$
Eq. 4	$XO + CO \rightarrow X + CO_2$
where XO may be $BATiO_3$, Fe_3O_i, RuO_2, etc	

Figure 4-23. Generic equations for reduction of metal oxides in an HTCC firing ambient.

the equations of Fig. 4-23. Most mixed oxides (XO) are reduced by hydrogen (Eq. 1) or by carbon (Eq. 2) which can be generated from the organic binder in the tape after firing. Carbon monoxide (CO), another reducing agent, can also be generated by the incomplete oxidation of carbon (Eq. 3) or by the reduction of metal oxides by carbon. In air firing, oxygen maintains these oxides in their normal oxidation state. However, in a reducing condition at high temperatures, these oxides are partially or totally reduced, as shown. Even firing in an inert nitrogen ambient is risky since carbon monoxide (CO) may be generated from the decomposition of the organic binder. The CO so formed is a strong reducing agent (Eq. 4).

LTCC green tapes are commercially available from several suppliers, including Du Pont, Ferro, and ElectroScience Laboratories. Properties of LTCC tape dielectrics are given in Table 4-31.

Another low-temperature cofired tape composition based on cordierite glass ceramic has been developed by IBM. It consists of 18 to 23 percent alumina, 50 to 55 percent silicon dioxide, 18 to 25 percent magnesium oxide, 0 to 3 percent phosphorus pentoxide, and 0 to 3 percent boron oxide.[54] The sintered material has a relatively low dielectric constant and a CTE closely matching that of silicon and is compatible with copper metallization. The properties of this sintered dielectric are reported to be[55]

Density:	2.62 g/cm³
Flexural strength:	210 MPa
Dielectric constant:	5.2
CTE:	3 ppm/°C

The cofired ceramic tape sinters below 1000°C and can be cofired with copper in an inert controlled ambient. The sintered copper has an electrical resistivity of 3.5 μΩ·cm. The electrical, thermal, mechanical, and chemical properties of other ceramic compositions were already discussed under substrate materials.

Table 4-31. Comparison of Selected LTCC Properties

Property	Du Pont 951 standard applications	Ferro A6 microwave applications	Under development
	Physical characteristics		
Flexural strength	340 MPa	130 MPa	
Density	3.1 g/cm^3	2.5 g/cm^3	<1.8 g/cm^3
Thermal stability			
Intermittent	650°C	850°C	
Continuous	250°C	300°C	
Dimensional tolerance	± 0.2–0.5%	± 0.2–0.5%	± 0.001 in
Camber	0.001 in/in	0.001 in/in	
CTE	5.5 ppm/°C	6.5 ppm/°C	~3 ppm/°C
Thermal conductivity	3 W/(m•K)	2 W/(m•K)	16–20 W/m•K)
	Design capabilities		
Line width	≥ 100μ (4 mils)	≥ 100μ (4 mils)	50μm (2 mils)
Pitch	≥ 200μ (8 mils)	≥ 200μ (8 mils)	100μm (4 mils)
Maximum layers	50	50	~100
Layer thickness	100μm (4 mils) 150μm (6 mils) 225μm (8.5 mils)	100μm (4 mils)	<75μm (3 mils)
Via diameter	≥ 100μm (4 mils)	≥ 100μm (4 mils)	50μm (2 mils)
Max. part size	125 mm (5 in)	125 mm (5 in)	>400 mm (16 in)
Line density	25 cm/(cm^2•layer)	25 cm/(cm^2•layer)	100 cm/(cm^2•layer)
	Electrical properties		
Dielectric constant	7.8	5.9	<3.0–>10,000
Dissipation factor	0.15%	<0.2%	
Insertion loss			
@ 1 GHz	0.2 dB/cm	0.08 dB/cm	
@ 10 GHz		0.18 dB/cm	
Line impedance	30–80Ω ± 5%	30–80Ω ± 3%	
Breakdown voltage	>4000 V/layer	>4000 V/layer	
Line resistivity	3–8 mΩ/square	1.5–5 mΩ/square	<0.3 mΩ/square
Line capacitance	<3.0 pF/cm		
Line inductance	~3 nH/cm		

4.3 Metals.

Metals serve a number of functions in MCMs; two main functions are as electrical conductors for interconnecting devices and thermal conductors for drawing heat away from devices. Other functions include pad sites for wire bonding or for tape automated bonding (TAB)

attachment of devices, pads for adhesive or solder attachment of dice
to the substrate, heat spreaders, resistor material, and ohmic contacts
for chip capacitors, resistors, and transistors. Desirable engineering
and manufacturing properties for MCM metallizations are listed in
Table 4-32. Metallizations may be classified as *primary* or *secondary*
types and as *thin film* or *thick film*. The primary types serve an electrical or a thermal function such as the conductor traces constituting the
signal lines, ground planes, voltage power planes, via interconnections, heat sinks, and thermal spreaders. Secondary metallizations are
auxiliary to the primary types and serve as diffusion barriers, adhesion promoters, corrosion and oxidation barriers, and metals that facilitate device attachments such as solders. Properties of both types are
given in Table 4-33.

4.3.1 Primary Thin-Film Metallizations

The most widely used primary metallizations for thin-film MCM-D
interconnect substrates are copper and aluminum. Copper has excellent electrical and thermal conductivity, is inexpensive, and is easy to
process by sputtering, vacuum evaporation, electroless plating, or
electrolytic plating. Copper has been the standard conductor in epoxy
and polyimide multilayer printed-circuit boards for many decades.
Among its drawbacks are its high chemical reactivity and corrosion
potential, although in practice these problems have been obviated by
plating the outer exposed layers with gold or coating them with solder. Another drawback—its diffusion into polyimide dielectric with
time at elevated temperature—has also been overcome by encapsulating the inner copper layers with diffusion barrier metals such as
chromium, nickel, or titanium or by employing dielectric materials
other than polyimides that are not susceptible to diffusion.

Although not as electrically or thermally conductive as copper, aluminum is also widely used because of the years of successful experience that semiconductor manufacturers have had in using aluminum
for IC fabrication. Aluminum is easily deposited by sputtering or by
evaporation and is compatible with polyimide and other dielectrics
without requiring barrier coatings. Aluminum is also inexpensive and
is suitable for monometallic wire bonding of devices using ultrasonic
bonded aluminum wire. All-aluminum monometallic interconnections
offer the maximum reliability under elevated-temperature conditions
and, because of aluminum's low atomic number, are more radiation-
resistant than gold-to-aluminum wire bonds.

Table 4-32. Desirable Properties for MCM Metallizations

Engineering properties	Manufacturing properties
High electrical conductivity	Low material cost
High thermal conductivity	Low equipment cost and maintenance
CTE closely matching those of the dielectric and substrate	Availability of material and equipment
High adhesion to dielectric	Reproducible thickness
Resistant to metal migration and electromigration	Easily patterned to form fine lines and spacings
Resistant to diffusion	
Resistant to oxidation and corrosion	
Easily bondable	
Easily solderable	

Table 4-33. Thin-Film Metallizations Used in Multichip Modules

Metal	Symbol	Electrical resistivity, $\Omega \cdot cm \times 10^{-6}$	CTE, ppm/°C	Thermal conductivity, W/(m·K)	Melting temperature, °C	Deposition methods
		Primary metallizations				
Copper	Cu	1.67	16.5–17.6	393–418	1083	A, B, E
Aluminum	Al	2.65	23	240	660	A, B
Gold	Au	2.35	14.2	297	1063	A, B, D, E
Silver	Ag	1.62	19.7	418	961	A, B, E
Nickel	Ni	6.9	13.3	92	1453	A, B, D, E
		Secondary metallizations				
Chromium	Cr	13–20	6.3	66	1890	A, B, D, E
Nickel	Ni	6.9	13.3	92	1453	A, B, D, E
Titanium	Ti	55	10.0	22	1675	A, B, C
Titanium/tungsten	Ti/W	5.5 (W)	4.5 (W)	178 (W)	3415 (W)	A
Palladium	Pd	10.8	11.0–12.0	70	1552	A, B, C, D, E
Platinum	Pt	10.6	9	73	1769	A, B, C, D
Tin/lead solder	Sn/Pb (60/40)	15–17	24.5	51	183–190	E. reflow

A = sputtered
B = vacuum-evaporated
C = electron-beam-evaporated
D = metalloorganic decomposition
E = electroplated

Gold is ideal as both a primary and a secondary metallization, meeting almost all MCM requirements. However, it is too costly to use gold to fabricate all the conductor layers of a multilayer substrate. Gold is therefore deposited only for the top circuit layer where it is required to protect the sublying layers from corrosion and to provide a surface compatible with die attachment and bonding, such as wire bonding and TAB attachment. In thin-film structures, gold is first sputtered as a thin layer of approximately 1000 to 2000 Å, then thickened by electrolytic plating to several micrometers and photoetched to pattern the top bonding pads. Gold has high electrical and thermal conductivities, and is extremely inert chemically.

Nickel has much lower electrical and thermal conductivities than copper, aluminum, or gold and thus is not generally used for the primary conductor traces. However, in some MCM designs, nickel is used as a via fill. Alcoa, e.g., uses an electroless process to plate nickel to fill in vias that have been etched in cured polyimide. After vias are etched in polyimide, nickel is plated to fill them in an electroless process. The primary conductor lines in this process, originally developed by AT&T, consist of electroplated copper.[56,57] Nickel and especially its alloys with chromium are used to produce precision thin-film resistors. For example, nichrome resistors are photodelineated and integrated with thin-film conductors.[58]

Silver possesses the highest electrical and thermal conductivities of all metals, but in its pure state silver is not used in circuits that have closely spaced conductors because of its propensity to migrate under moist conditions where traces of ionic contaminants and a small voltage bias are present. Even a few parts per million of chloride ions and a bias as low as 1 V have been demonstrated to result in silver migration and electric short-circuiting of lines spaced as far apart as 5 mils. Alloying the silver with palladium or platinum suppresses the migration, while coating the circuit with high-purity polymer or inorganic coating averts migration by preventing moisture from condensing on the surface. High-purity polymer coatings such as those of semiconductor-grade silicones and Parylenes have been found effective in preventing silver migration. Silver-filled epoxies and silver-terminated capacitors have been used reliably for over 30 years in hybrid microcircuits, which are efficiently cleaned and hermetically sealed in a dry nitrogen ambient.

4.3.2 Secondary Thin-Film Metallizations

It is, of course, desirable to use only primary metallizations in the manufacture of MCMs. However, in many cases, secondary metallizations must be used to promote adhesion of the primary metallization to the substrate or to the dielectric; to protect the primary metallization from corrosion, oxidation, or metal migration; or to serve as a diffusion barrier between material combinations. Chromium, nickel, titanium, titanium/tungsten, and palladium are being used to satisfy these requirements. Tin-lead solder and other solders are being used to provide a compatible surface for flip-chip solder attachment, TAB solder attachment, and surface protection.

Both chromium and nickel, either sputtered or plated, are widely used as barriers preventing the diffusion of copper into polyimide. The diffusion phenomenon and the barrier properties of chromium or nickel in preventing diffusion have been thoroughly studied and reported by many investigators.[35,59,60] Sputtered titanium/tungsten (10 percent Ti) is also effective as a diffusion barrier as well as an adhesion promoter.

Titanium is well known as an excellent adhesion layer in bridging gold with other surfaces such as ceramics and plastics. In a 10 to 15 percent alloy with tungsten, titanium provides a diffusion barrier between aluminum and gold.

4.3.3 Primary Thick-Film Metallizations

Thick-film metallizations are applied by screen-printing and firing conductor or resistor pastes onto a prefired ceramic substrate or onto a green ceramic tape. The pastes are heterogeneous mixtures of the metal, glass frit binder, organic binders, thixotropic compounds, and other additives. The fired conductors lack the optimum electrical properties inherent in thin-film vapor-deposited metals because of the residual glass, oxides, and other formulation ingredients that become part of the sintered mass. Thick-film conductor pastes of gold, silver, and alloys of gold or silver with palladium or platinum are commercially available and have been well characterized. Because of the large number of thick-film paste formulations on the market, their proprietary nature, and differences in composition, only ranges of material properties can be given (Table 4-34). The paste manufacturer should be consulted for exact parameters of specific formulations. The alloy films are better than the unalloyed metal films in reducing solder leaching and, in the case of silver, in suppressing silver migration. However, the addition of either palladium or platinum reduces the electrical conductivity of the films. The bulk resistivity of palladium or platinum is 4 times greater than that of gold and 6 to 7 times greater than that of silver.

Table 4-34. Primary Thick-Film Metallizations Used in Multichip Modules

Fired conductor paste	Symbol	Sheet resistance mΩ/square*	Line resolution, mils
Gold, fritless	Au	1.5–6	3–5
Gold, fritted	Au	1.8–7	4–5
Gold, mixed bonded	Au	2.0–7	3–4
Gold/palladium	Au/Pd	5.0–9	5–8
Gold/platinum	Au/Pt	30–80	5
Copper, fritless	Cu	1.0–1.5	5
Copper, fritted	Cu	1.5–2.5	5–7
Copper, mixed bonded	Cu	1.0–2	5
Silver	Ag	1.0–3	5.0–8
Silver/palladium	Ag/Pd	32–40	6–8
Silver/platinum	Ag/Pt	15–25	5–8
Silver/Pd/Pt	Ag/Pd/Pt	30–60	6–8
Tungsten	W	15	5

*Values vary with thickness; recommended thicknesses range from 12 to 18 μm, although some pastes can only be printed from 8 to 15 μm. Paste manufacturers should be consulted for exact values.

Tungsten, molybdenum, and molybdenum/tungsten pastes are primary metallizations for high-temperature cofired ceramic interconnect substrates. Although their electrical conductivities are not as high as those of gold, silver, or copper, they are the only metals that are compatible with the high firing temperatures required to sinter the ceramic.

4.3.4 Secondary Thick-Film Metallizations

Refractory metals such as tungsten, molybdenum, and molybdenum manganese used as primary metallizations in HTCC circuits are corrodible, nonsolderable, and difficult to bond to. It is also desirable to augment their relatively low electrical conductivities. Secondary metallizations are therefore used for the top layers. Electroless or electrolytic nickel plating followed by gold plating are customarily used and are very compatible with wire bonding, solder attachment, and brazing. The top gold surface also provides an excellent corrosion barrier.

Other secondary metallizations consist of screen-printing and firing a second paste over the first to provide compatibility of the top layer with wire bonding and TAB. The secondary conductor may be screen-printed selectively over the bonding sites. Solder coating of thick-film conductors is also used to attach surface-mounted components.

References

1. P. Kluge-Weiss and J. Gobrecht, "Directly Bonded Copper Metalization of AlN Substrates for Power Hybrids," *Proceedings of the Materials Research Society Symposium,* **40,** 1984.

2. K. H. Dalal, "Substrate Metalization Selection for High Power Circuits Based on Thermal Resistance and Temperature Cycling Reliability," *Proceedings of the Sixth International Applied Power Electronics Conference,* Dallas, TX, 1991.

3. D. C. Hopkins, S. H. Bhavnani, and K. H. Dalal, "Effect of Metalization Thickness on Thermal Conductance of First-Level Power Structure," *Microcircuits and Electronic Packaging,* **16**(3), 1993.

4. F. W. Martin, "Polymer Thick Film Extends Options for Hybrid and PCB Fabrication," *Circuits Manufacturing,* **17**(5), 1977.

5. F. W. Martin, "Low Firing Polymer Thick Film Enables Screen Printing of Resistors and Conductors on PC Boards," *Insulation/Circuits,* **23**(2), 1975.

6. K. M. Shambrook and P. A. Trask, "High-Density Multichip Interconnect (HDMI)," *Proceedings of the EIA/IEEE, Electronic Components Conference,* **656,** 1989.

7. J. Enloe et al., "Properties of AlN Package Material," *Proceedings of the National Electronics Packaging Conference,* Los Angeles, 1990.

8. Naval Command Control and Ocean Surveillance Center, San Diego, CA, contract no. N66001-88-C-0181, *VLSI Packaging Technology,* Final Report, April 1993.

9. E. Y. Luh, J. H. Enloe, A. Kovacs, and R. Luceroni, "Metallization of Aluminum Nitride Packages," *IEPS Journal,* **13**(2), 1991.

10. C. F. Coombs, ed., *Printed Circuits Handbook,* 3d ed., McGraw-Hill, New York, 1988.

11. V. J. Brzozowski and R. N. Horton, "Rigid and Flexible Printed Wiring Boards," chap. 8 in C. Harper, ed., *Electronic Packaging and Interconnection Handbook,* McGraw-Hill, New York, 1991.

12. G. L. Lucas, "Laminate Developments Enhance PCB Performance," *Electronic Packaging and Production,* April 1993.

13. D. J. Arthur, A. DeGreef-Safft, and G. S. Swei, "Advanced Fluoropolymer Dielectrics for MCM Packaging," *Proceedings of the International Conference on Multichip Modules,* Denver, CO, April 1992.

14. G. A. Bouska, "High Speed Low Dielectric Substrate Material," *Proceedings of the Second International SAMPE Electronics Conference,* June 1988.

15. G. W. Bogan et al., "Unique Polyaromatic Cyanate Esters for Low Dielectric Printed Circuit Boards," *Proceedings of the Second International SAMPE Electronics Conference,* June 1988.

16. C. Harper, ed., *Electronic Packaging and Interconnection Handbook,* McGraw-Hill, New York, 1991, chap. 8.

17. K. K. Aghajanian, "Processing and Properties of Silicon Carbide Reinforced Aluminum Metal Matrix Composites for Electronic Applications," *Proceedings of the ISHM*, 1991.

18. K. K. Aghajanian, "A New Infiltration Process for the Fabrication of Metal Matrix Composites," *SAMPE Quarterly*, **20**(4), 1989.

19. D. R. White, Metal matrix composites, U.S. patent 4828008, May 9, 1989.

20. D. R. White, S. Keck, I. Smith, and A. Silzars, "New High Ground in Hybrid Packaging," *Hybrid Circuit Technology*, December 1990.

21. A. L. Eustice et al., "Low-Temperature Cofirable Ceramics: A New Approach for Electronic Packaging," *36th Electronic Components Conference*, Seattle, WA, 1986.

22. J. I. Steinberg, S. J. Horowitz, and R. J. Bacher, "Low Temperature Cofired Tape Dielectric Material Systems for Multilayer Interconnections," *Solid State Technology*, January 1986.

23. D. Schroeder, "The Use of Low Temperature Cofired Ceramic for MCM Fabrication," *Proceedings of the International Conference on Multichip Modules*, Denver, CO, 1992.

24. L. W. Herron, R. N. Master, and R. R. Tummala, Methods of making multilayer glass-ceramic structures having internal distribution of copper-based conductors, U.S. patent 4234367, 1980.

25. R. R. Tummala and E. J. Rymaszewski, eds., *Microelectronics Packaging Handbook*, Van Nostrand Reinhold, New York, 1989.

26. R. R. Tummala and J. Knickerbocker, "Advanced Cofire Multichip Technology at IBM," *Proceedings of the IEPS*, San Diego, CA, September 1991.

27. F. Borchelt and G. Lu, "Use of CVD Diamond Substrates in Electronic Applications," *Sixth International SAMPE Electronics Conference*, Baltimore, MD, 1992.

28. J. A. Herb, M. G. Peter, and T. J. Fileds, "Diamond Films for Thermal Management Applications," *Proceedings of the ASM International Third Electronic Materials and Processes Conference*, San Francisco, 1990.

29. D. J. Pickrell and F. M. Kimock, "CVD Diamond Substrates: Current Status of the Technology," *Proceedings of the ISHM*, Dallas, November 1993.

30. *Photonics Spectra*, Laurin Publishing Co., Pittsfield, MA, November 1992.

31. ISHM Abstracts, Diamond and Diamond-Like Film Workshop, Breckenridge, CO, March 1991.

32. J. D. Craig, "Polyimide Coatings," *Electronic Materials Handbook*, vol. 1: *Packaging*, American Society for Metals, Metals Park, OH, 1989.

33. W. D. Weber, R. H. Hopla, A. J. Roza, and T. Maw, "Polyamic Esters," *Advanced Packaging*, summer 1993.

34. B. J. Bachman et al., "Evaluation of Polyimide as a Dielectric for Multichip Packaging," *Proceedings of the ISHM*, Baltimore, MD, 1989.

35. G. M. Adema, I. Turlik, P. L. Smith, and M. J. Berry, "Effects of Polymer/Metal Interaction in Thin-Film Multichip Module Applications," *Proceedings of the Electronic Components Conference*, 1990.

36. H. J. Neuhaus, "A High Resolution Anisotropic Wet Patterning Process Technology for MCM Production," *Proceedings of the International Conference on Multichip Modules*, Denver, CO, 1992.

37. J. Summers et al., "Wet Etching Polyimides for Multichip Module Applications," *Proceedings of the International Conference on Multichip Modules*, Denver, CO, 1992.

38. N. Bilow, A. Landis, et al., "Acetylene-Substituted Polyimide Adhesives," *SAMPE Journal*, January/February 1982.

39. N. Bilow, "Acetylene-Substituted Polyimides as Potential High Temperature Coatings," *ACS Symposium on Organic Coatings*, 177th American Chemical Society National Meeting, Honolulu, HI, 1979.

40. R. Rubner, *Siemens Forsch-u Entwicki-Ber.*, **5**, 235, 1976.

41. H. Ahne, H. Kruger, E. Pammer, and R. Rubner, in K. L. Mittal, ed., *Polyimides*, Plenum Press, New York, 1984.

42. K. K. Chakravorty and J. M. Cech, "Photosensitive Polyimide as a Dielectric in High Density Thin Film Copper Polyimide Interconnect Structures," *Journal of the Electrochemical Society*, **137**(3), 1990.

43. T. Ohsaki et al., "A Fine Line Multilayer Substrate with Photosensitive Polyimide Dielectric and Electroless Copper Plated Conductors," *Proceedings of the third IEEE International Electronics Manufacturing Technology*, 1987.

44. A. E. Nader et al., "Photodefinable Polyimides Designed for Use as Multilayer Dielectrics for Multichip Modules," *Proceedings of the International Conference on Multichip Modules*, Denver, CO, 1992.

45. E. Sweetman, "Characteristics and Performance of PHP-92," *Proceedings of the International Conference on Multichip Modules*, Denver, CO, 1992.

46. R. H. Heistand, "BCB: Planar and Simple," *Materials Engineering*, March 1992.

47. T. M. Stokich, W. M. Lee, and R. A. Peters, "Real Time FTIR Studies of the Reaction Kinetics for Polymerizing of Divinyl Siloxane Bis-Benzocyclobutene Monomers," *Proceedings of the Materials Research Society*, 1991.

48. D. Burdeaux, P. Townsend, J. Carr, and P. Garrou, *Journal of the Electronic Materials*, **19**: 1357, 1990.

49. E. W. Rutter et al., "A Photodefinable Benzocyclobutene Resin for Thin Film Microelectronic Applications," *Proceedings of the International Conference on Multichip Modules*, Denver, CO, 1992.

50. T. Horton and B. McWilliams, "MCM Driving Forces, Applications, and Future Directions," *Proceedings of the National Electronics Packaging Conference West*, Anaheim, CA, 1991.

51. N. Majid et al., "The Parylene-Aluminum Multilayer Interconnection System for Wafer Scale Integration," *Journal of Electronic Materials* **18**:301, 1989.

52. J. J. Licari, S. M. Lee, and I. Litant, "Reliability of Parylene Films," *Proceedings of the Met. Society, Tech. Conference Defects Electronic Materials and Devices*, Boston, 1970.

53. S. M. Lee, *Polymeric Films for Semiconductor Passivation*, Final Report, NASA contract NAS12-2011, 1969.

54. A. H. Kumar, P. W. McMillan, and R. R. Tummala, Glass-ceramic structures and sintered multilayer substrates thereof with circuit patterns of gold, silver, or copper, U.S. patent 4301324, 1981.

55. R. N. Master, L. W. Herron, and R. R. Tummala, "Cofiring Process for Glass-Ceramic/Copper Multilayer Ceramic Substrate," *Proceedings of the IEEE*, 1991.

56. L. W. Schaper, "Meeting System Requirements through Technology Tradeoffs in Multichip Modules," *Proceedings of the IEPS Conference*, 1990.

57. A. C. Adams et al., "High Density Interconnect for Advanced VLSI Packaging," Karl Suss Seminar, Publication 108, 1987.

58. J. J. Licari and L. Enlow, *Hybrid Microcircuit Technology Handbook*, Noyes, Park Ridge, NJ, 1988, chap. 3.

59. S. P. Kowalczyk, Y. H. Kim, G. F. Walker, and J. Kim, "Polyimide on Copper—The Role of Solvent on the Formation of Copper Precipitates," *Applied Physics Letter*, **52**(5), February 1988.

60. Y. H. Kim et al., "Adhesion and Interface Investigation of Polyimide on Metals," *Journal of Adhesion Science Technology*, **2**(2), 1988.

5
Thermal Management

Charles P. Minning

Thermal management is both the the art and science of transporting heat away from power-dissipating semiconductors in complex electronic assemblies such that the electrical performance and reliability of these assemblies are in compliance with customer-specified requirements. This art and science is a combination of design experience, analysis techniques, and materials selection for achieving the required semiconductor temperatures while also meeting electrical performance, size, weight, test, repair, and cost requirements. Aspects of thermal management include system architecture tradeoffs to minimize power dissipation, selection of compatible materials to minimize stresses induced by temperature gradients, selection of a cooling strategy that minimizes additional power and weight for coolant circulation, choice of coolant, and physical layout.

Thermal management of multichip modules (MCMs), which can be components of a larger assembly or stand-alone subsystems, is the subject of this chapter. Emphasis is placed on thermal management as an integral part of the MCM design process which extends from logic synthesis to design for final assembly and test. It will be shown that trends both in semiconductor technology and in electronic system design are such that the dissipation per unit volume of electronics has been increasing with time and will become a major factor in the design of compact three-dimensional electronic systems in the future (see Chap. 7). The mechanisms of power dissipation in semiconductor devices and the effects of temperature on device electrical performance and reliability are then discussed to develop the context for thermal management in the overall MCM design process. Typical MCM configurations and thermal interfaces in the external electronic system are then described, as are the heat-transfer mechanisms most commonly encountered in electronic cooling. Details of MCM thermal design, including the thermal resistance concept, heat spreading in MCM substrates, considerations for materials selection, and computational tools are then discussed. The chapter concludes with a discussion of future trends in MCM thermal management.

5.1 Background

The evolution of solid-state electronic circuits, from Shockley's invention of the transistor in 1951 and Kilby's invention of the integrated circuit[1] in 1958 to the complex microprocessors of the mid-1990s, has led to the development of military and commercial electronics applications unheard of 50 years ago. For example, computers have progressed from the vacuum-tube-based ENIAC (*electronic numerical integrator and computer*), which occupied an 80-foot square room and performed approximately 5000 instructions per second (0.005 MIPS) in the late 1940s,[2] to desktop computers that perform several orders of magnitude more quickly. The ENIAC dissipated approximately 140 kW of electric power and had a mean time between failure (MTBF) of 30 min. On the other hand, notebook computers of the mid-1990s dissipate less than 50 W and have MTBFs of months, if not years. The dramatic increases in computational performance and reliability and the decreases in size, weight, and power dissipation of computers were made possible by progress in semiconductor technology. However, the decrease in power dissipation does not mean that thermal management should be ignored in designing modern computers and advanced electronics for military applications. Indeed, in some advanced portable computers that employ the latest generation of microprocessors, such as Intel's Pentium chip, special liquid cooling techniques are required to keep the microprocessor chips from overheating.[3]

In Figure 5-1, it is seen that as individual circuit features, such as line widths and spacings, have decreased with time, system designers have integrated more functions into silicon dice. This has resulted not only in larger die sizes, but also in increased power dissipation per die.[4] The impact of this trend at the system level is shown in Fig. 5-2 for two airborne radars that use the same size *line replaceable units*

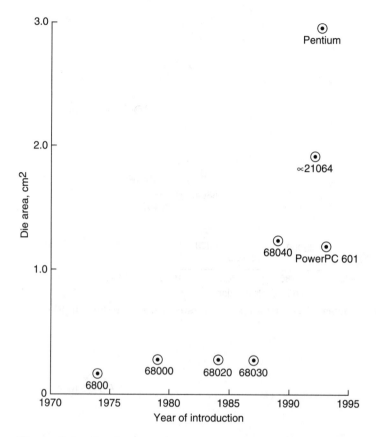

Figure 5-1a. Key features of microprocessors versus year of introduction to market. Die size versus year of introduction. The complexity, size and power dissipation of microprocessors have been increasing since the mid-1970s.[4-6] (*Courtesy of Motorola.*)

(LRUs) as the basic building blocks of their radar signal processors. Here it is shown that very large-scale integrated (VLSI) semiconductor dice packaged in six MCM-Ds (technology of the mid-1990s) soldered to printed-wiring boards (PWBs) on a standard avionics module (SAM) results in about a 92 percent decrease in volume, about an 82 percent decrease in weight, and about an 80 percent decrease in power dissipation relative to the same functions implemented in first-generation gate arrays in single-chip packages (technology of the early 1980s). Note, however, that the power dissipation per unit volume occupied by the electronics of the 1990s increased by a factor of 2.5 rel-

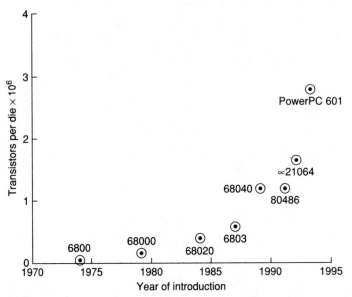

Figure 5-1b. Number of transistors per die versus year of introduction. (*Courtesy of Motorola.*)

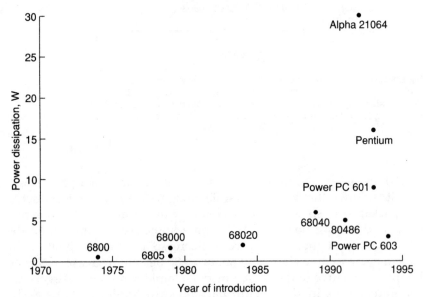

Figure 5-1c. Power dissipation versus year of introduction. (*Courtesy of Motorola.*)

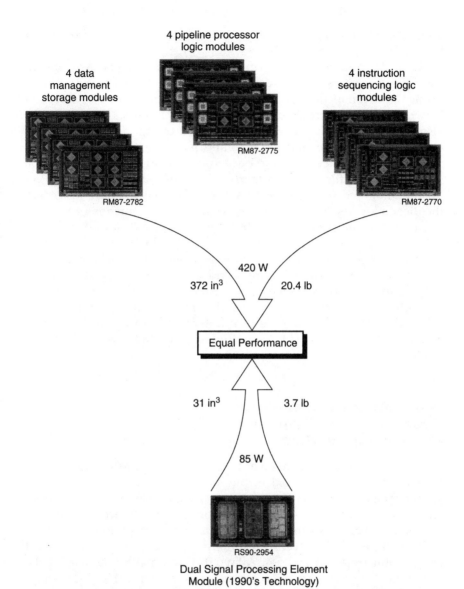

Figure 5-2. Size, weight and power reductions for signal processing elements of airborne radars. Advances in semiconductor and MCM technologies have made possible reductions of over 80 percent in these features. (*Courtesy Hughes Aircraft Co.*)

ative to the technology of the early 1980s. Design features of the radar LRUs are summarized in Table 5-1.

Multichip modules in the form of hybrid microcircuits have been in use for many years and are described in detail by Licari and Enlow.[7] From the thermal design point of view, there is much in common between hybrid microcircuits and multichip modules, but there are also some significant differences, as shown in Fig. 5-3. The common feature is that semiconductor dice and passive components such as resistors and capacitors are mounted on an interconnect substrate (thick-film, thin-film, or high-temperature cofired ceramic) which, in turn, is mounted in a hermetic package. The major difference is that the dice in MCMs are much larger, occupy a larger fraction of the substrate area (4 percent of the substrate area for the hybrid shown in Fig. 5-3a and 41 percent of the substrate area for the MCM shown in Fig. 5-3b), and dissipate more power. However, even though the power dissipations of the individual dice in hybrids were low, the power dissipation per unit of die area often ranged from 15 to 150 W/cm^2. Molytabs were often used to spread the power dissipated by the die over an area larger than the die before the heat entered the substrate. However, molytabs are not practical to use with large dice in the MCM-D shown in Fig. 5-3b.

5.2 The Nature and Consequences of Power Dissipation in MCMs

The quality of MCM thermal design is only as good as the accuracy to which the power dissipated in the individual chips and resistors is known. The mechanisms by which power is dissipated in diodes, bipolar transistors, and field-effect transistors and the effects of temperature on transistor performance are discussed in this section.

The complexity involved in estimating power dissipation in VLSI chips is illustrated in Fig. 5-4. A layout of the major functions of the Motorola Power PC 603 microprocessor is shown in Fig. 5-4a, and a three-dimensional bar chart showing power dissipation as a function of microprocessor operating mode and chip operating frequency is shown in Fig. 5-4b. Clearly, the accuracy of any thermal analysis has to correspond to the appropriate operating mode of the dice inside the MCM. In most thermal analyses, it is assumed that the power is evenly dissipated over the surface of the die, which is obviously contrary to the situation portrayed in Fig. 5-4. For silicon, this is usually not a problem, because its high thermal conductivity spreads the heat over

Table 5-1. Comparison of Design Features between Airborne Radars of the Mid-1980s and Mid-1990s

Design feature	Radar of mid-1980s	Radar of mid-1990s
Gate array technology	■ LSI logic LL9K series ■ 6K avg. usable gates ■ No BIST	■ LSI logic LCA 10K series ■ 50K avg. usable gates ■ BIST designed in
No. SAMs	12	1
Volume of SAMs	372 in^3	31 in^3
Weight of SAMs	20.4 lb	3.7 lb
Total power dissipation	420 W	85 W

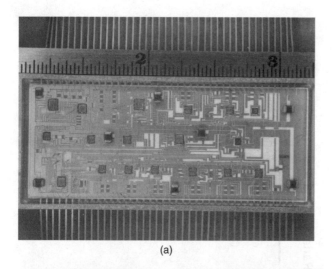

(a)

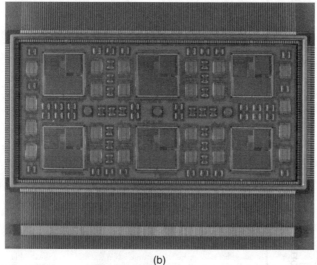

(b)

Figure 5-3. Comparison of (*a*) thin-film hybrid microcircuit of the 1970s and (*b*) a multichip module of the 1990s. The dice occupy a significantly larger portion of the interconnect substrate in an MCM. (*Courtesy Hughes Aircraft Co.*)

the surface of the chip. However, as indicated in Table 5-2, the much lower thermal conductivities of other semiconductors, such as gallium arsenide and indium phosphide, hinder the efficient spreading of heat in the die, and significant temperature gradients on the die may result unless extraordinary measures are taken to cool the die. The thermal conductivity of a semiconductor is also a function of temperature, as shown in Fig. 5-5.

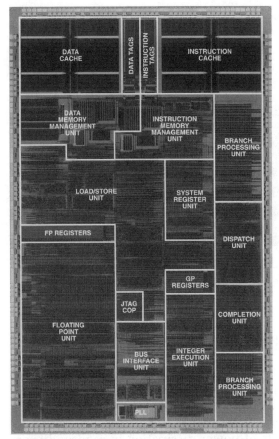

Figure 5-4a. Motorola Power PC 603 microprocessor. Location of power dissipating elements. (*Courtesy of Motorola.*)

5.2.1 Power Dissipation and Performance of Semiconductor Devices

Power dissipation in semiconductors results from a complex set of charge transport processes which are strongly influenced by factors such as materials, degree of doping, mode of operation (analog, digital, or a combination of both), transistor configuration, frequency, and temperature. The numerous transport processes and associated mater-

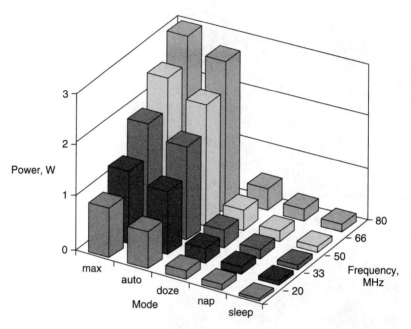

Figure 5-4b. Motorola Power PC 603 microprocessor. Power dissipation versus clock speed and operating modes. (*Courtesy of Motorola.*)

ial properties have been presented in the framework of nonequilibrium thermodynamics by Wachutka[9] in his review of power dissipation in and effects of temperature on semiconductor electrical performance.

The fundamental aspects of power dissipation in semiconductors and the consequential effects of temperature on semiconductor electrical performance are summarized in this section. Emphasis is placed on the performance of digital systems in the temperature range of − 55 to + 125°C. Operation of semiconductors at cryogenic temperatures has been described by Hanamura et al.[10] and operation at temperatures greater than 125°C by Hosticka et al.[11] The effect of temperature on the performance of analog systems is discussed by Cheng and Manos[12] and by Shoucair.[13]

The *pn* Junction Diode. *The* pn junction diode is a device common to all semiconductors, and an understanding of power dissipation and temperature effects on electrical performance of diodes will aid in understanding these phenomena in more complex semiconductor devices such as gate arrays and application-specific integrated circuits (ASICs).

A *pn* junction, shown conceptually in Fig. 5-6*a*, consists of two neighboring segments of semiconductor material such as silicon or

Table 5-2. Thermal Properties of Semiconductor Materials at 300 K

Material	Thermal conductivity, $W/(m \cdot K)$	Density, g/cm^3	Specific heat, $W \cdot s/(g \cdot K)$	Coefficient of thermal expansion, $10^{-6}/K$
Silicon	145	2.33	0.70	2.5
Gallium arsenide	80	5.32	0.35	5.9
Germanium	65	5.33	0.31	5.8
Indium phosphide	68	4.78	0.311	4.56

SOURCE: Ref. 8.

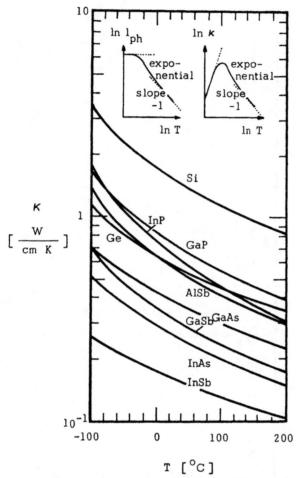

Figure 5-5. Thermal conductivity of semiconductor materials versus temperature. (*Source: Ref. 8.*)

germanium, one segment consisting of *n*-type material and the other consisting of *p*-type material. The *n*-type material is prepared by adding (i.e., doping) impurity atoms, such as antimony, phosphorus, and arsenic, to the semiconductor; the impurity atoms, in effect, add free electrons to the conduction band of the semiconductor. The *p*-type material is prepared by adding impurity atoms, such as boron, aluminum, and gallium, to the semiconductor. The impurity atoms in this case introduce "holes" in the valence band of the semiconductor (Ref. 14, pp. 10–12). In the *n*-type material, charge is carried primarily by

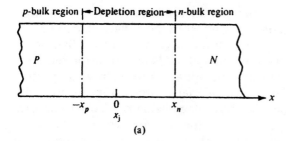

(a)

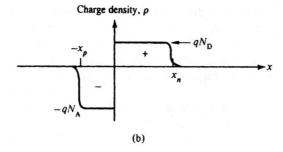

(b)

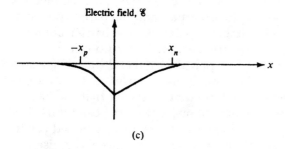

(c)

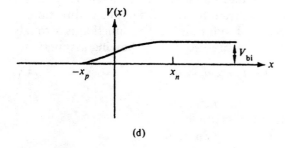

(d)

Figure 5-6. Cross section of a *pn* junction diode. (*Source: Ref. 15.*)

electrons (i.e., "majority" carriers) in the conduction band, but a small percentage of the charge is also carried by positively charged holes (i.e., "minority" carriers) moving in the direction opposite that of the electrons. Holes exist in n-type material as the result of the temperature-induced electron-hole pair generation and recombination process (Ref. 14, pp. 10–12). In the p-type material, charge is carried primarily by holes, with a much smaller percentage being carried by minority carrier electrons. The presence of electrons in the p-type material, as in the n-type material, is the result of the electron-hole pair generation and recombination process.

The p and n segments of the diode are electrically neutral except for a very narrow *depletion region* at the junction. Electrons from the n side diffuse across the junction and fill some of the holes on the p side. This movement of electrons results in the buildup of negative charge on the p side of the junction and a buildup of positive charge on the n side, as shown in Fig. 5-6b. As shown in Figs. 5-6c and d, respectively, the charge separation results in an electric field that counters further charge separation and in the resultant *barrier* or *built-in potential* at the junction.[14–16] The barrier potential is a function of impurity (dopant) concentrations and temperature. Typical values at 300 K are 0.3 V for germanium and 0.7 V for silicon (Ref. 14, p. 16).

If voltage is applied across the diode such that the p side is made more positive and the n side is made more negative, then the diode is *forward-biased*, the depletion regions shrinks, and the barrier potential is reduced. As the applied voltage is increased, the barrier potential eventually disappears, and electric current freely flows across the junction. When external voltage is applied such that the p side is made more negative and the n side is made more positive, the diode is *reverse-biased* and the depletion region widens such that the barrier potential increases to resist the flow of majority carriers and thus electric current across the junction. However, in the reverse-biased condition, minority carriers may still cross the junction, thus giving rise to a very small reverse saturation current. As the reverse-biased voltage is increased, the current flow due to minority carriers eventually causes the diode to break down and current flows in the opposite direction.

The current-voltage relationship for an ideal pn junction is strongly temperature-dependent and is expressed in the following manner:

$$I = I_0(e^{qV_a/(kT)} - 1) \tag{5-1}$$

where q = electronic charge of an electron = 1.6×10^{-19} C
 V_a = applied potential across diode, V
 k = Boltzmann's constant = 1.38×10^{-23} J/K
 I_0 = Reverse saturation current, A

Detailed derivations of Eq. (5-1) are presented by Neudeck[15] and Hodges and Jackson.[16] The reverse saturation current I_0 is a function of the area of the junction and temperature. The temperature coefficient for the reverse saturation current is given by (Ref. 16, p. 160)

$$\frac{1}{I_0} \frac{dI_0}{dT} = \frac{3}{T} + \frac{1}{T} \frac{V_b}{V_T} \tag{5-2}$$

where V_b = band gap voltage

$$V_T = \frac{kT}{q} = \text{thermal voltage}$$

The band gap voltage is a function of the semiconductor material and varies slightly with temperature; at room temperature (300 K) values of V_b are 1.1 V for silicon, 1.4 V for gallium arsenide, 0.7 V for germanium, and 1.3 V for indium phosphide (Ref 8, p. 33). For silicon diodes, the reverse saturation current doubles about every 8°C (Ref. 16, p. 160).

The temperature coefficient of the diode current for a fixed forward-biased voltage is given by (Ref. 16, p. 160)

$$\frac{1}{I} \frac{dI}{dT}\bigg|_{V_a=\text{const}} = \frac{1}{I_0} \frac{dI_0}{dT} - \frac{1}{T} \frac{V_a}{V_T} \tag{5-3}$$

For a forward-biased voltage of 700 mV at 300 K, the diode current increases about 6 percent per degree Celsius and doubles about every 12 °C.

The temperature dependence of the diode voltage at constant current is given by (Ref. 16, p. 161)

$$\frac{dV}{dT}\bigg|_{I=\text{const}} = \frac{V}{T} - V_T \left(\frac{1}{I_0} \frac{dI_0}{dT} \right) \tag{5-4}$$

For a forward current of about 1 mA in a silicon diode, the measured change in forward-biased voltage at 300 K is about $-2\,\text{mV/°C}$.

The effects of temperature on the current versus voltage characteristics of a diode are illustrated in Fig. 5-7. These temperature characteristics are usually of minor importance to the digital integrated-circuit (IC) designer, because signal voltages are much greater than the expected variation in diode voltage over the 0 to 70°C temperature range of interest for commercial applications and the -55 to $+125$°C temperature range for military applications. However, one of the cur-

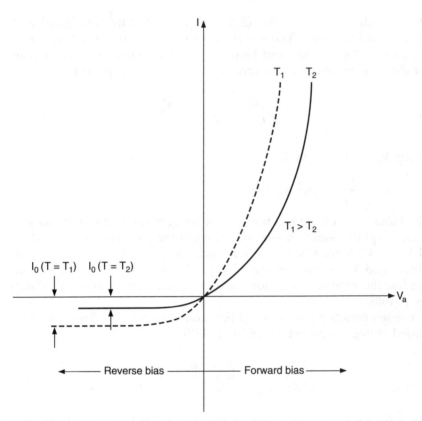

Figure 5-7. Effects of temperature on forward-biased voltage and on reverse saturation current of a *pn* junction diode.

rent goals in microprocessor design is to minimize power dissipation. For field-effect transistors, one way to reduce power dissipation is to reduce power supply voltage and, therefore, signal voltage from 5 to 3.3 V and lower in some applications. The effects of temperature on diode characteristics are therefore expected to become more important and require that closer attention be paid to system thermal design. Although the temperature effects on diode performance may be of marginal importance for digital ICs at present, this is not the case for analog ICs. The diode temperature coefficients given in Eqs. (5-2) to (5-4) are very important and become one of the subtle aspects of analog IC design (Ref. 16, p. 161).

The diode temperature characteristics are also convenient for measuring the temperature of an integrated circuit. The usual method is to measure the forward-biased voltage (at constant current) of diodes designated for temperature sensing on the surface of the IC[17-20] or to use diode chips to measure surface temperatures on hybrid microcircuit (or MCM) substrates.[21] Many examples of this method are presented in the thermal management literature, including the use of specially designed test chips for the verification of thermal models and determination of package thermal resistance.[22-25]

Transistors. A bipolar transistor consists of a section of one semiconductor type (either n or p type) sandwiched between two sections of the other type. A *pnp* transistor consists of an n-type semiconductor sandwiched between two sections of p-type semiconductor. Likewise, an *npn* transistor consists of a p-type semiconductor sandwiched between two sections of n-type semiconductor. Each section of the transistor is provided with a terminal for electrical connection to other circuit elements, such as resistors, capacitors, or other transistors. For either transistor type, the center section is referred to as the *base,* and the other sections are referred to as the *collector* and the *emitter.* Two *pn* junctions exist within the transistor (the collector-base junction and the emitter-base junction), and each junction has the diode characteristics previously described.

In practice, the center section of the transistor is made much narrower than the outer sections, and the outer sections are more heavily doped than the center section. In this situation, the depletion regions at each junction extend into the base far more than in either the emitter or the collector. For normal operation, the emitter-base (EB) junction is forward-biased (i.e., the electrical potential of the base is positive with respect to the emitter), and the collector-base (CB) junction is reverse-biased (i.e., the electrical potential of the base is negative with respect to the collector). In *npn* transistors, the forward bias of the EB junction reduces the junction barrier potential, thus causing electrons to flow from the emitter to the base. Since the base is lightly doped, almost all the current flowing across the EB junction consists of electrons flowing from the emitter to the base. Most of these electrons also flow across the collector-base junction. Thus, the forward-biased voltage on the EB junction controls the emitter and collector currents. The *pnp* transistor behaves in the same manner except that the majority carriers are holes rather than electrons (Ref. 14, pp. 63–69). Since the potential difference across the EB junction is much less than the potential difference across the CB junction, most of the power dissipation in a bipolar junction transistor occurs at the CB junction (Ref. 14, p. 181).

For the stable operation of a bipolar transistor, a resistor R_L is placed in series with the collector. The voltage applied across the outer terminal of this resistor (i.e., the terminal on the side of the resistor opposite the collector) and the emitter is referred to as the *supply voltage* V_{CC}. The voltage between the emitter and the collector V_{CE} is given by

$$V_{CE} = V_{CC} - I_C R_L \qquad (5\text{-}5)$$

At thermal equilibrium, the power dissipation is given by

$$P = I_C V_{CE} \qquad (5\text{-}6)$$

Bipolar transistor operation can be seriously affected by temperature. The two most sensitive parameters are the base-emitter voltage V_{BE} and the collector-base reverse saturation current I_{CBo}. The effects of temperature on these parameters are the same as for *pn* junction diodes. As the junction temperature increases, I_{CBo} increases which, in turn, causes I_C to increase, thus increasing the power dissipation. This process can result in a shift in the dc operating point of the transistor, or, worst case, in a thermal runaway condition where I_C keeps increasing until the CB junction overheats and burns out. Transistor circuit design techniques for providing temperature stability are described by Bell.[14]

A field-effect transistor (FET) is a voltage-operated device. These devices do not require any input (either signal or bias) current and dissipate substantially less power than bipolar transistors. There are two major types of field-effect transistors: junction FETs (JFETs) and metal-oxide semiconductor FETs (MOSFETs).

A JFET consists of one type of semiconductor material (for instance, *n* type), called the *channel*, bounded by two smaller sections of the opposite type of semiconductor material (in this case, *p* type) attached to its sides to form two *pn* junctions. The two ends of the channel are referred to as the *source* and *drain*, and the smaller sections at the sides of the channel are connected and referred to as the *gate*. The side sections are more heavily doped than the channel. When the gate terminal is not connected and a potential is applied along the channel (the drain is positive with respect to the source), a current flows through the channel. When the gate terminal is biased negative with respect to the source, the *pn* junctions are reverse-biased. The depletion regions extend into the channel. These regions are essentially insulators since they are depleted of charge carriers. This results in a narrower channel for current flow, a higher channel resistance, and a reduced current. If the bias voltage on the gate is made more negative, the depletion regions extend farther into the channel until they meet, and the current in the channel is shut off.

A MOSFET consists of two heavily doped *n*-type regions in a *p*-type substrate. These *n*-type regions serve as the source and drain regions of the transistor. A thin layer of silicon dioxide insulator is grown on the top surface of the substrate between the source and drain regions. Metal is deposited on the oxide layer to form the gate terminal. Metal is also deposited on the source and drain regions as well as on the backside of the substrate. The source and drain regions form *pn* junctions, which during normal operation are kept in a reverse-biased condition. As with a JFET, the drain is operated at a positive voltage relative to the source. In addition, both the substrate and the source are connected to ground during normal operation. When no bias voltage is applied to the gate, the source and drain form back-to-back diodes that prevent current flow when voltage is applied between the source and drain. If the source, drain, and substrate are connected to ground and a positive voltage (with respect to the source) is applied to the gate, electrons accumulate in the substrate region under the gate, forming, in effect, an induced *n*-type region if the gate voltage is high enough. The value of the gate voltage when this condition occurs is called the *threshold voltage* V_t. By applying voltage V_{DS} between the drain and source (with the drain positive with respect to the source), current can flow through the substrate in the region under the gate. The current is proportional to both the excess gate voltage $V_{GS} - V_t$ and the voltage between the drain and source V_{DS}.

The physics of power dissipation in FETs is rather complicated and is not discussed in detail here. However, operation of FETs can be viewed as the alternate charging and discharging of a capacitor, and in this situation, a generic expression for power dissipation in FETs is given by

$$P = \frac{1}{2} f V_{DD}^2 \, C \qquad (5\text{-}7)$$

where f = clock frequency
$\quad V_{DD}$ = power supply voltage
$\quad C$ = capacitance

The transistor switching speed and the leakage currents are the two FET operating parameters that are most affected by temperature.

5.3 Heat-Transfer Mechanisms

Heat is the flow of energy due to the existence of a temperature difference between two geographically separated points. The three basic mechanisms—conduction, convection, and radiation—by which this energy transfer takes place are described in this section. Once these mechanisms have been described, they are related to the concept of thermal resistance, a convenient means of understanding MCM thermal design.

5.3.1 Conduction

Heat conduction is energy transfer by molecular vibration in solids or stagnant fluids. In a multichip module, typical solids are semiconductor chips and interconnect substrates. An example of a stagnant fluid would be entrapped gas in a die-bond void between chip and interconnect substrate.

The conservation of energy in conduction-dominated systems is expressed in vector form as follows (Ref. 26, p. 44):

$$\rho c \, \frac{dT}{dt} + \nabla \cdot \mathbf{q} = u'''$$

(5-8)

where ρ = density, gm/cm^3
c = specific heat, W$\cdot$s/(g$\cdot$°C)
u''' = local heat generation per unit volume, W/cm^3
$\mathbf{q}$ = heat flux vector, W/cm^2

In microelectronics applications, the local heat generation term u''' takes into account resistance heating in electric conductors and is considered a negative quantity (Ref. 26, p. 31). In cartesian coordinates, the heat flux vector is given by

$$\mathbf{q} = \mathbf{i}q_x + \mathbf{j}q_y + \mathbf{k}q_z$$

(5-9)

where $\mathbf{i}$, $\mathbf{j}$, and $\mathbf{k}$ are unit vectors in the x, y, and z directions and q_x, q_y, and q_z are the x, y, and z components, respectively, of the heat flux vector. The second term on the left-hand side of Eq. (5-8) is given by

$$\nabla \cdot \mathbf{q} = \frac{\partial q_x}{\partial x} + \frac{\partial q_y}{\partial y} + \frac{\partial q_z}{\partial z}$$

(5-10)

For conduction in an isotropic medium, the heat flux vector $\mathbf{q}$ is given by the vector form of Fourier's law:

$$\mathbf{q} = -k\nabla T$$

(5-11)

where $q_x = -k \, \dfrac{\partial T}{\partial x}$, etc.

k = thermal conductivity, W/(cm$\cdot$K)

Thermal conductivities of typical semiconductor materials at 300 K are listed in Table 5-2, and the temperature dependence of k for some of these materials is given in Fig. 5-5. Thermal conductivities of other materials commonly used in microelectronics and MCMs are listed in Table 5-3.

Substitution of Eq. (5-10) into Eq. (5-8) yields the equation for time-dependent conduction in an isotropic medium with temperature-dependent thermal conductivity:

$$\rho c \, \frac{dT}{dt} = \nabla \cdot (k \nabla T) + u''' \tag{5-12}$$

This is the appropriate governing equation for determination of the temperature distribution in high-speed semiconductors, such as gallium arsenide and indium phosphide, where k is small, u''' is high, and the resulting temperature gradients on the surface of the chip can be very high. Equation (5-12) is nonlinear (due to the temperature dependence of the thermal conductivity) and usually requires numerical methods for determination of temperature distributions for the com-

Table 5-3. Thermal Conductivities of Materials Commonly Used in MCMs

Material	Thermal conductivity, W/(m•K)
Metals	
Copper	393
Gold	297
Tungsten	160
Molybdenum	146
Kovar	17
Aluminum	247
Ceramics	
Alumina (92%)	18
Alumina (96%)	20–35
Alumina (99%)	37
Alumina (LTCC, 50%)	2.4–3.0
Silicon carbide	270
Aluminum nitride	140–230
Beryllia	240
Silicon dioxide	7.
Diamond	1500–2000
Plastics	
Polyimide	0.2

plex geometries characteristic of semiconductor chips.[27] Closed-form analytical methods of solution for simple geometries are also available and are described by Arpaci.[26]

For steady-state applications, Eq. (5-12) reduces to the Poisson equation

$$\frac{\partial^2 T}{\partial x^2} + \frac{\partial^2 T}{\partial y^2} + \frac{\partial^2 T}{\partial z^2} = -\frac{u'''}{k} \tag{5-13}$$

The solution to this equation for heat spreading in MCM substrates is discussed in Sec. 5.4.1.

Mechanical interfaces are a common occurrence in microelectronic systems. The bond line between a chip and an interconnect substrate, between a hermetic MCM package and printed-wiring board, and between an SEM-E clamped to an integrated rack in an aircraft avionics bay are all examples of such interfaces. These interfaces should be dealt with cautiously, because poorly designed interfaces result in large thermal resistances (usually unexpected), high semiconductor temperatures, poor electrical performance, and premature electrical failure.

A typical joint or mechanical interface is shown in Fig. 5-8. The same material may be used on either side of the joint, or the materials may be entirely different. On a microscopic scale, the two surfaces of the joint appear rough, and only part of the overall surface area of the joint is actually in contact. Heat conduction takes place across the contacting high points and through fluid (gas or liquid) that may be trapped in the gaps between the high points. Heat transfer by radiation across the gaps may also occur if there is a large temperature difference across the interface.

If the interface is a bond line, where a material such as thermal grease, thermally conductive epoxy, or a metal eutectic fills the voids, the bond line thickness is much greater than the height of the roughness asperites, and the thermal effect of the bond line on heat transfer can be represented by Eq. (5-14). The thermal resistance of a void-free bond line can be determined from the bond line thickness, the thermal conductivity of the bonding material, and the total area of the joint. Bond line voids are a major concern in the thermal design of MCMs.

It is convenient to view the thermal effect of an unbonded joint as a fictitious temperature drop ΔT_i, which is defined by extrapolating to the centerline the temperature that exists in the materials on either side of the joint. The *interface thermal conductance* is defined as

$$h_i = \frac{Q}{A_{\text{total}} \Delta T_i} \tag{5-14}$$

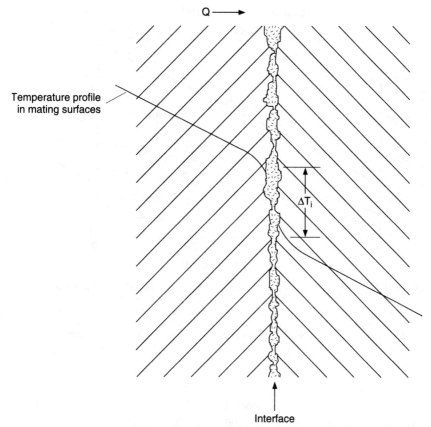

Figure 5-8. Microscopic view of mating surfaces of a mechanical interface or joint.

The interface thermal conductance depends on many variables, among them the contact pressure, surface finish and materials of the mating surfaces, hardness and stiffness of the mating surface materials, and presence of filler materials such as gas or liquid in the interstitial voids at the interface. For applications involving large surfaces, thin foils made of soft metals such as indium are placed between the mating surfaces to improve the thermal conductance. Typical values for interface thermal conductance are illustrated in Fig. 5-9, which shows the influences of contact pressure and surface finish, and in Fig. 5-10 which clearly shows the influence of altitude (i.e., the presence of air in the interstitial voids of the joint).

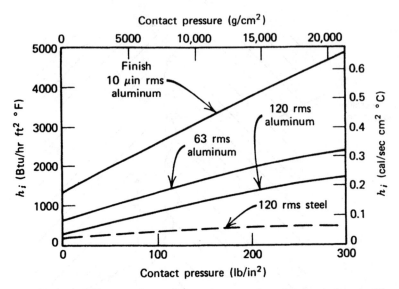

Figure 5-9. Influence of contact pressure, surface finish, and materials on interfacial thermal conductance between mating plates in air. The interfacial conductance increases significantly with contact pressure and surface smoothness for soft materials such as aluminum. Contact pressure has little effect on the conductance for hard materials such as steel.[28]

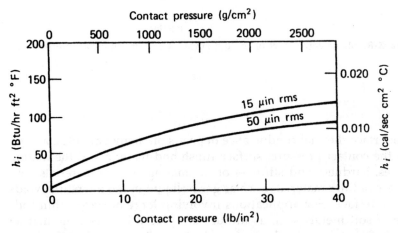

Figure 5-10. Interfacial thermal conductance between two aluminum (2024-T6) plates in a vacuum. Values of the thermal conductance at 10^{-4} torr are significantly lower when the interface exists in a vacuum and there is no gas to fill the interstitial voids at the interface.[28]

In dealing with interface thermal conductances, thermal designers should be aware that many of the data reported in the technical literature are for small-area samples in a controlled laboratory environment. In the real world of electronic design, these interface areas are often quite large, and the structures that compose the interface warp, which leads to nonuniform contact pressure over the interface area. The actual thermal conductance of the interface can be an order of magnitude less than that published in the literature. For instance, the average interface thermal conductance for a 5052 aluminum plate measuring $10.16 \times 15.24 \times 0.317$ cm bolted at the four corners with four no. 10-32 screws to a 2.54-cm-thick cold plate is only 0.125 W/(cm²•°C) in air at 1 atm and only 0.028 W/(cm²•°C) in a vacuum. The surface finish of these plates was about 32 µin rms (Ref. 28, p. 68).

5.3.2 Convection

Convection is the transfer of heat to or from a surface by a moving fluid. The fluid may move in an orderly manner (laminar flow) or in a chaotic manner (turbulent flow). Fluid motion may be produced by a fan or pump (forced convection) or may result from buoyancy effects due to the presence of temperature gradients within the fluid (natural convection).

The rate of convective heat transfer is given by Newton's law of cooling

$$Q = hA \, \Delta T \qquad (5\text{-}15)$$

where Q = rate of heat transfer, in watts
 A = surface through which heat is transferred, m²
 ΔT = temperature difference between surface and some point in fluid, °C
 h = convective heat transfer coefficient, W/(m²•°C)

The convective heat-transfer coefficient is a function of the flow geometry, fluid properties, temperature, and flow regime (i.e., laminar or turbulent). Theoretical methods for analyzing convective heat-transfer problems and predicting h are presented in depth by Arpaci and Larsen.[29] General overviews of both theory and experimental data are presented by Kays and Crawford[30] and Kreith.[31] Kraus and Bar-Cohen[32] present a summary of convective heat transfer for electronic equipment applications.

Convective heat-transfer correlations are usually presented in the form of the dimensionless Nusselt number Nu as a function of other dimensionless parameters that characterize the geometry and the nature of the flow:

$$\text{Nu} = \frac{hL}{k} = f(\text{Re,Pr,Gr}) \qquad (5\text{-}16)$$

where L = characteristic dimension of the flow, m
       k = thermal conductivity of fluid, W/(m•K)

The dimensionless parameters Re, Pr, and Gr are defined as follows:

$$\text{Re} = \text{Reynolds number} = \frac{VL\rho}{\mu} \qquad (5\text{-}17a)$$

$$\text{Gr} = \text{Grashof number} = \frac{g\beta\rho^2\,L^3\,\Delta T}{\mu^2} \qquad (5\text{-}17b)$$

$$\text{Pr} = \text{Prandtl number} = \frac{\mu c_p}{k} \qquad (5\text{-}17c)$$

The parameters and fluid properties used in the above relations are as follows:

V = fluid velocity, m/s

ρ = fluid density, g/cm^3

c_p = fluid specific heat, W•s/(g•°C)

β = fluid thermal expansion coefficient, (°C)$^{-1}$

g = gravity = 9.8 m/s^2

The Reynolds, Grashof, and Prandtl numbers are similarity parameters that characterize the momentum and thermal energy transfer processes taking place in the application of interest. The Reynolds number represents the ratio of inertial forces to viscous forces, the Grashof number represents the ratio of buoyancy forces to inertial forces, and the Prandtl number represents the ratio of momentum diffusion to thermal diffusion.

Thermal properties of some typical coolants are listed in Table 5-4.

The value of the ratio Gr/Re^2 determines whether the flow is in the natural- or forced-convection regime. When this ratio is much greater than 1, natural convection predominates; and when this ratio is much less than 1, forced convection is the dominant flow regime. For values

Table 5-4. Thermal Properties of Typical Coolants

Property	Air (1 atm, 27°C)	Helium (1 atm, 27°C)	Water (30°C)	PAO* (30°C)	FC-70 (30°C)	FC-77 (25°C)
ρ, g/cm^3	1.1766×10^{-3}	1.626×10^{-4}	0.996	0.784	1.97	1.78
μ, g/(cm·s)	1.853×10^{-4}	1.99×10^{-4}	7.977×10^{-3}	5.344×10^{-2}	0.187	1.6×10^{-2}
C_p, W·s/(g·K)	1.005	5.193	4.179	2.218	1.072	1.045
k, W/(m·K)	2.614×10^{-2}	0.155	0.615	0.151	0.0713	0.065
β, (°C)$^{-1}$	3.33×10^{-2}	3.33×10^{-3}	3.02×10^{-4}	N/A	10^{-3}	1.4×10^{-3}
Pr	0.711	0.667	5.42	78.5	281.2	25.7

*Polyalphaolefin.

of Gr/Re^2 near unity, both natural- and forced-convection effects are present, the flow situation is very complex, and experimental prototypes may be required for hardware thermal design.

The value of the Reynolds number determines whether the flow is laminar or turbulent. For forced convection in ducts, the flow is laminar when $Re < 2300$ and fully turbulent when $Re > 10^4$, where the characteristic length L in Eq. (5-17a) is given by

$$L = \frac{4A}{P} \tag{5-18}$$

where A = flow cross-sectional area, m^2
P = wetted perimeter of duct, m

For external flows, the transition from laminar to turbulent flow begins at about $Re \approx 5 \times 10^5$, where the characteristic length is taken as the distance downstream from the leading edge where the fluid first contacts the surface.

The Nusselt number for forced convection is typically of the form

$$Nu = c\,Re^m\,Pr^n \tag{5-19}$$

Some typical relations for external flow past a flat plate are given by Eqs. (5-20a) through (5-20d) for the local Nusselt number Nu_x, where x is the distance downstream from the leading edge, and for the average Nusselt number Nu_L or a plate of length L.
Laminar flow:

$$Nu_x = 0.332\,Re_x^{0.5}\,Pr^{0.33} \tag{5-20a}$$

$$\overline{Nu_L} = 0.664\,Re_L^{0.5}\,Pr^{0.33} \tag{5-20b}$$

Turbulent flow:

$$Nu_x = 0.0296\,Re_x^{0.8}\,Pr^{0.33} \tag{5-20c}$$

$$\overline{Nu_L} = 0.037\,Re_L^{0.8}\,Pr^{0.33} \tag{5-20d}$$

Typical equations for fully developed flow in ducts (i.e., where the Nusselt number is independent of the distance from the entrance to the duct) are given in Table 5-5.

For natural convection, heat-transfer correlations are generally of the form

$$Nu = c(Gr\,Pr)^n \tag{5-21}$$

Table 5-5. Heat Transfer Correlations for Fully Developed Flow in Ducts

Duct shape	Aspect ratio*	Nusselt number (constant heat flux)	Nusselt number (constant temperature)
Round tube	—	4.364	3.66
Rectangular	1.0	3.61	2.98
	2.0	4.12	3.39
	3.0	4.79	3.96
	4.0	5.33	4.44
	8.0	6.49	5.6
	∞	8.235	7.54

*Aspect ratio = duct width/duct height.
SOURCE: Ref. 31.

Table 5-6. Values of c and n for Natural Convection from Vertical and Horizontal Plates and Cylinders

Geometry	Gr Pr	c	n
Vertical plate	10^4–10^9	0.59	0.25
	10^9–10^{12}	0.13	0.33
Horizontal plate (hot surface up)	10^4–10^7	0.54	0.25
	10^7–10^{11}	0.15	0.33
Vertical cylinder	10^4–10^7	0.54	0.25
	10^7–10^{11}	0.15	0.33
Horizontal cylinder	10^3–10^9	0.53	0.25

Values of c and n for the average Nusselt number for both vertical and horizontal flat plates are given in Table 5-6. The flow is laminar when the Grashof number (based on the distance from the leading edge of a vertical plate) is greater than about 4×10^8.

Some typical values of h for free and forced convection are listed in Fig. 5-11.

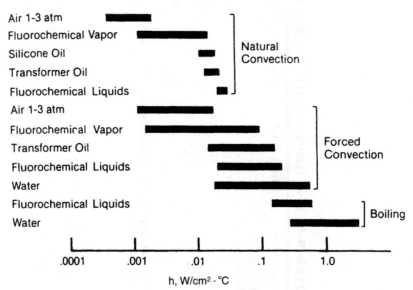

Figure 5-11. Values of convective heat-transfer coefficient for different coolants.[34]

5.3.3 Radiation

The third mode of heat transfer occurs through the emission of electro-magnetic energy from a surface and the transmission through and reflection and absorption of this energy by other surfaces at different temperatures. Radiation heat transfer can occur in solids (i.e., molten glass, lenses, and windows), in gases, and in a vacuum.

The maximum radiant energy that can be emitted from a surface is given by the Stefan-Boltzmann equation

$$Q_b = \sigma A T^4 \qquad (5\text{-}22)$$

where σ = Stefan-Boltzmann constant
$\quad\quad = 5.6688 \times 10^{-12} \ W/(cm^2 \cdot K)$
$\quad Q_b$ = radiant energy emitted by a blackbody

This relation is for an ideal blackbody, which not only emits the maxi-mum amount of radiant energy at a given temperature but also allows all incident radiation to pass into it (no reflection) and absorbs all inci-dent radiation (no energy transmitted through the body). The net rate of radiant energy transfer from a surface i in an enclosure containing N black surfaces is given by

$$\frac{Q_i}{A_i} = \sum_{j=1}^{N} \sigma F_{i-j}(T_i^4 - T_j^4) \qquad (5\text{-}23)$$

The quantity F_{i-j} represents the fraction of energy emitted by surface i that is incident on surface j and is variously referred to as the *shape factor*, *configuration factor*, and *angle factor* in the heat-transfer literature. These "shape factors" depend on the shape of the surface and the geo-metric relationship between the surface and the other surfaces of the enclosure.

The following relationships hold for the shape factors:

$$A_i F_{i-j} = A_j F_{j-i} \qquad (5\text{-}24)$$

$$\sum_{j=1}^{N} F_{i-j} = 1 \qquad (5\text{-}25)$$

For a surface that views itself, such as the inner surface of a round tube, the shape factor F_{i-i} is nonzero. An extensive catalog of shape fac-tors can be found in Siegel and Howell.[33]

The blackbody is an approximation of radiating bodies encountered in practice. The ratio of the radiant energy emitted by a real body to

Table 5-7. Emissivities of Some Materials Used in Multichip Modules

Material	Emissivity, ε
Black paint	0.96–0.98
Nickel-plated surface (electroless)	0.06–0.17
Gold (electrodeposited)	0.02
Aluminum	
Sandblasted	0.41
Black anodized	0.86
Machine-polished	0.03–0.06
Alumina	0.33
Copper (polished)	0.023–0.052
Kapton film on aluminum (25°C)	
25 μm thick	0.55
50 μm thick	0.68
75 μm thick	0.78
100 μm thick	0.8

that emitted by a blackbody at the same temperature is called the *emissivity* ε. Emissivities must be measured experimentally, and emissivities for materials used in multichip modules are listed in Table 5-7.

Radiant energy incident on a real surface is partially reflected, absorbed, and transmitted. Mathematically, this is expressed as follows:

$$\alpha + \rho + \tau = 1 \qquad (5\text{-}26)$$

where α = absorptance
ρ = reflectance
τ = transmittance

Kirchhoff's law, a very useful relation for radiation heat transfer, states that

$$\alpha = \varepsilon \qquad (5\text{-}27)$$

This relation is strictly true for thermodynamic equilibrium where there is no net exchange of radiant energy between surfaces. However, experimental evidence indicates that this relation is valid for most applications where there is radiant energy exchange between surfaces (Ref. 33, p. 66). In radiation heat-transfer applications of interest in electronic packaging, most materials are opaque, that is, $\tau = 0$, and Eq. (5-26) reduces to

$$\alpha + \rho = 1 \qquad (5\text{-}28)$$

Substitution of Eq. (5-27) into Eq. (5-28) then yields the relationship between reflectance and emissivity:

$$\rho = 1 - \varepsilon \qquad (5\text{-}29)$$

All real applications of radiation heat transfer in electronic packaging involve energy exchange between reflecting surfaces. In these situations, the net energy exchange between diffuse surfaces i and j is given by

$$\frac{Q_{i-j}}{A_i} = \sigma \mathscr{F}_{i-j}(T_i^4 - T_j^4) \qquad (5\text{-}30)$$

where $\mathscr{F}_{i-j}$ = transfer factor. The following relations hold for the transfer factor:

$$A_i \mathscr{F}_{i-j} = A_j \mathscr{F}_{j-i} \qquad (5\text{-}31)$$

$$\sum_{j=1}^{N} \mathscr{F}_{i-j} = \varepsilon_i \qquad (5\text{-}32)$$

The transfer factor, which is often referred to as "script F" in the heat-transfer literature, is an algebraically complicated function of the geometry and surface properties of the enclosure under consideration. These factors are determined through a set of equations which can be represented in matrix form as follows (Ref. 33, p. 286):

$$\mathscr{F} = \mathbf{D}^{-1}\mathbf{M} + \varepsilon \qquad (5\text{-}33)$$

Several examples for calculating the transfer factors can be found in Kraus and Bar-Cohen[32] and Siegel and Howell.[33] The terms in Eq. (5-33) are given in matrix form below (note that $\mathbf{D}^{-1}$ is the inverse of matrix $\mathbf{D}$):

$$\mathscr{F} = \begin{bmatrix} \mathscr{F}_{11} & \mathscr{F}_{12} & \cdots & \mathscr{F}_{1N} \\ \mathscr{F}_{21} & \mathscr{F}_{22} & \cdots & \mathscr{F}_{2N} \\ \cdots & \cdots & \cdots & \cdots \\ \mathscr{F}_{N1} & \mathscr{F}_{N2} & \cdots & \mathscr{F}_{NN} \end{bmatrix} \qquad (5\text{-}34a)$$

$$D = \begin{bmatrix} \dfrac{1}{\varepsilon_1} & -\rho_2 \dfrac{\mathcal{F}_{12}}{\varepsilon_2} & \cdots & -\rho_N \dfrac{\mathcal{F}_{1N}}{\varepsilon_N} \\[2em] -\rho_1 \dfrac{\mathcal{F}_{21}}{\varepsilon_1} & \dfrac{1}{\varepsilon_2} & \cdots & -\rho_N \dfrac{\mathcal{F}_{2N}}{\varepsilon_N} \\[2em] \cdots & \cdots & \cdots & \cdots \\[1em] -\rho_1 \dfrac{\mathcal{F}_{N1}}{\varepsilon_1} & -\rho_2 \dfrac{\mathcal{F}_{N2}}{\varepsilon_2} & \cdots & \dfrac{1}{\varepsilon_N} \end{bmatrix}$$

(5-34b)

$$M = \begin{bmatrix} -1 & \mathcal{F}_{12} & \cdots & \mathcal{F}_{1N} \\ \mathcal{F}_{21} & -1 & \cdots & \mathcal{F}_{2N} \\ \cdots & \cdots & \cdots & \cdots \\ \mathcal{F}_{N1} & \mathcal{F}_{N2} & \cdots & -1 \end{bmatrix}$$

(5-34c)

$$\varepsilon = \begin{bmatrix} \varepsilon_1 & 0 & \cdots & 0 \\ 0 & \varepsilon_2 & \cdots & 0 \\ \cdots & \cdots & \cdots & \cdots \\ 0 & 0 & \cdots & \varepsilon_N \end{bmatrix}$$

(5-34d)

5.3.4 The Thermal Resistance Concept

It is common practice to view a thermal flow path as a series of resistances R between the power-dissipating chips and the heat sink. The term *thermal resistance* is analogous to electrical resistance in that the heat flow is directly proportional to the temperature drop in the direction of the heat flow path. In other words,

$$Q = \frac{\Delta T}{R} \tag{5-35}$$

where Q = power dissipation, W
$\quad\quad \Delta T$ = temperature difference between endpoints of resistance, °C
$\quad\quad R$ = thermal resistance, °C/W

For simple conduction through a constant-area flow path of length t and thermal conductivity k, the resistance R is given by

$$R = \frac{t}{kA} \tag{5-36}$$

For convective heat transfer from surface A, the thermal resistance is given by

$$R = \frac{1}{hA} \tag{5-37}$$

The temperature difference associated with Eq. (5-35) is $T_{\text{surface}} - T_{\text{fluid}}$. Since most correlations for h are determined experimentally, care must be taken in using the correct value for T_{fluid} in complex flow geometries.

An example of a complex thermal circuit for a multichip module in the form of a pin grid array is shown in Fig. 5-12.

Another parameter often mentioned in the literature on electronic thermal management is the junction-to-case thermal resistance θ_{jc}. This resistance is defined as follows:

$$\theta_{jc} = \frac{T_j - T_{\text{case}}}{Q_{\text{chip}}} \tag{5-38}$$

where T_{case} is the case temperature directly under the chip. This parameter has been used in industry for a long time, and standard procedures such as Mil-Std-883C, method 1012, and SEMI Standard 1321 have been prepared to provide a framework for consistent and repetitive results and as guidelines for the presentation of results. The thermal resistance θ_{jc} has been used extensively for single-chip packages and is particularly convenient for determining junction temperatures on complex PWBs, such as those shown in Fig. 5-2 for the radar technology of the 1980s containing many such packages. For hermetically sealed packages where conduction is the only heat-transfer mechanism inside the package, θ_{jc} is supposed to be an intrinsic property of the

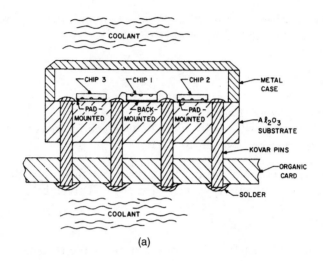

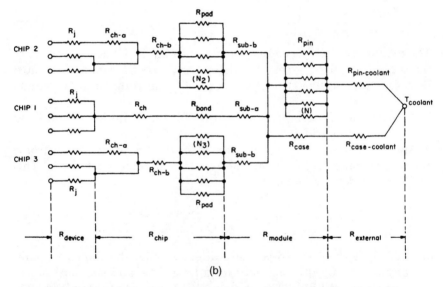

Figure 5-12. Typical thermal resistance circuit for a multichip module. (*a*) Cross-sectional view of MCM and its thermal environment; (*b*) thermal resistance network. (*Source: J.H. Seely and R.C. Chu, Heat Transfer in Microelectronic Equipment, Marcel Dekker, 1972.*)

chip-package combination and independent of the package location or environment in the next-higher-level assembly. Also θ_{jc} has been used as a figure of merit for evaluating competing thermal design concepts. It thus seems natural to use θ_{jc} to evaluate thermal designs of competing MCM technologies.

However, there are several drawbacks to using θ_{jc} in thermal models and as a figure of merit. First, values of θ_{jc} are hard to obtain if they are not listed in a vendor's specification; this is especially true if a third-party vendor is used to furnish custom chips such as gate arrays and ASICs in custom packages. Second, as pointed out by Dutta,[34] the "intrinsic" assumption of θ_{jc} appears questionable, because changing the package environment may significantly alter the heat flow paths and temperature distribution within the package. Since MCMs contain several chips, several θ_{jc}'s are required, which diminishes the utility of the concept. In all likelihood, the mounting surface temperature of large-area MCMs will be nonisothermal as well as different when mounted on flow-through air-cooled SAMs or on edge-cooled SEM-Es. Thus, although θ_{jc} has been used extensively in the past, we will not use this concept in discussing MCM thermal designs.

5.4 MCM Thermal Design

The objective of MCM thermal design is to develop a low-resistance flow path by which heat generated by the ICs and other active devices can flow to the heat sink.

5.4.1 Heat Spreading in Substrates

Power dissipated in a chip enters the substrate through an adhesive bond line for bottom-bonded chips or through solder bumps for flip chips. Since the chip-to-substrate contact area is much smaller than the total area of the substrate, the heat will diffuse by conduction away from the area immediately under the chip toward cooler regions of the substrate. The resulting temperature distribution in the substrate—and therefore the substrate's contribution to the total thermal resistance between the power-dissipating junctions on the chip and the cooling medium for the electronic assembly—depends on how the substrate is attached to the next higher level of assembly. For the purposes of this chapter, it is assumed that chips are back-bonded to one side of the substrate and that the opposite side of the substrate is bonded to the package.

Calculation of the temperature rise across the substrate is mathematically complex. An approximate method will be discussed first to

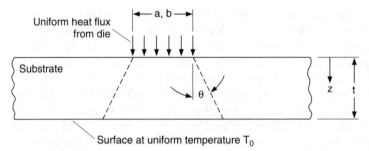

Figure 5-13. Geometry for "approximate" model for heat spreading in MCM substrate.

establish a "feel" for the magnitude of the temperature rise across the substrate. Exact mathematical solutions will then be discussed, and the results of the approximate and exact solutions compared.

In the approximate "heat-spreading" model, it is assumed that the power dissipation Q in a rectangular die of length a and width b diffuses toward the bottom of the substrate, which is assumed to be at uniform temperature T_0, in a volume defined by angle θ, as shown in Fig. 5-13.

From Eq. (5-11), Q is given by

$$Q = qA(z) = -kA(z)\frac{dT}{dz} \tag{5-39}$$

For a given z, the heat flows through area $A(z)$ defined by

$$A(z) = (a + 2z\tan\theta)(b + 2z\tan\theta) \tag{5-40}$$

The differential temperature drop across a differential volume of thickness t is found by substituting Eq. (5-40) into Eq. (5-39) and rearranging to give

$$dT = \frac{(-Q/k)\,dz}{(a + 2z\tan\theta)(b + 2z\tan\theta)} \tag{5-41}$$

Integration of Eq. (5-41) between $z = 0$ and $z = t$ then yields the following relations for the substrate thermal resistance for a rectangular die:

$$R_{\text{substr}} = \frac{T(z=0) - T_0}{Q} = \frac{1}{2k(a-b)\tan\theta}\ln\left[\left(\frac{a}{b}\right)\left(\frac{b + 2t\tan\theta}{a + 2t\tan\theta}\right)\right] \tag{5-42}$$

For a square die, the substrate thermal resistance is given by

$$R_{substr} = \frac{2t}{ka(a + 2t \tan \theta)} \tag{5-43}$$

In most applications, MCM substrates are bonded inside a hermetic package, and the assumption of constant temperature at $z = t$ is unrealistic since additional heat spreading takes place in the package. Heat spreading in both substrate and package is illustrated in Fig. 5-14; in this example, it is assumed that minimal heat spreading occurs in the thin adhesive bond line between substrate and case. Equations (5-42) and (5-43) are also applicable to the case if the values of a and b are increased by $2t_{substr} \tan \theta$ and if t is taken as the case thickness.

A key assumption in the derivation of Eqs. (5-42) and (5-43) is that the die spacing is large enough that the influence of power dissipation in neighboring dice is negligible. The validity of this assumption is questionable for modern MCMs where dice cover a significant percentage of the substrate surface. Assuming that power dissipation from each die spreads through a volume defined by the spreading angle θ, the spacing S_0 at which a power-dissipating neighbor begins to influence the die temperature is given by Eq. (5-44) and illustrated in Fig. 5-15a:

$$S_0 = 2(t_{substr} + t_{case}) \tan \theta \tag{5-44}$$

When die spacing is less than S_0, the available volume for heat flow is reduced, as indicated in Fig. 5-15b.

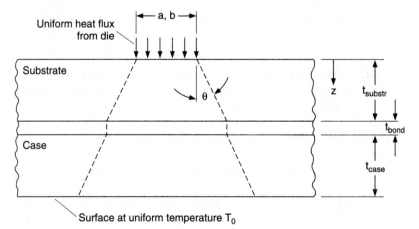

Figure 5-14. Geometry for "approximate" model for heat spreading in an MCM substrate bonded in a hermetic case.

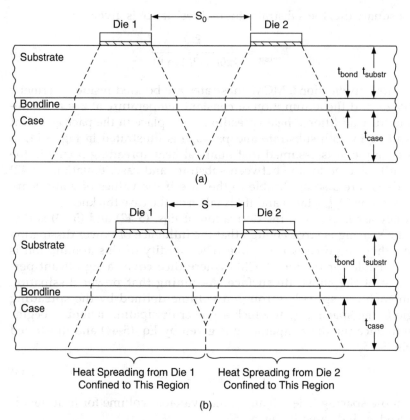

Figure 5-15. Geometry for "approximate" model for heat spreading. (a) Spacing at which power-dissipating neighbor begins to influence die temperature; (b) die spacing $S < S_0$ leads to less area for heat spreading from neighboring die.

To illustrate the relevance of this aspect of MCM thermal design, consider the large gate arrays and their neighbor memory dice shown in Fig. 5-3b. Here the spacing between a gate array and memory die is about 90 mils ($S = 0.229$ cm), and the thicknesses of the substrate and package are 25 mils (0.064 cm) and 40 mils (0.102 cm), respectively. Assuming $\theta = 45°$, the chips should be spaced 130 mils ($S_0 = 0.330$ cm) apart to avoid affecting each other thermally. However, system size and weight requirements dictate that $S < S_0$.

The "approximate" heat-spreading model with $\theta = 45°$ has been used extensively for hybrid microcircuits such as that shown in Fig. 5-3a, but there is no theoretical basis for this choice of θ. It is not clear that θ is independent of the thermal conductivity. In addition, application of this model is tedious for an MCM that contains many closely spaced chips. Detailed mathematical analyses of heat flow in MCM substrates have been performed by Maly and Piotrowski,[35] David,[36] and Ellison.[37]

Maly and Piotrowski[35] developed an algorithm for optimizing the placement of temperature-sensitive and power-dissipating components to maximize the reliability of power hybrids. Maly and Piotrowski provide an equation for determining the temperature at any location (x,y) on the surface of an MCM substrate due to the presence of a heat source at a different location (x_j,y_j). It was assumed that the substrate is bonded to an isothermal cold plate.

David[36] solved the Laplace equation [that is, Eq. (5-13) with $u''' = 0$] by means of Fourier analysis to determine the temperature distribution in a multilayer MCM in which the substrate, adhesive bond layers, and package are of length L_1 and width W. In this analysis, it is assumed that (1) the power dissipation is uniform over the surface of each die, (2) all materials are isotropic, (3) neither radiation nor convection heat transfer occurs at the top surface of the substrate where the dice are mounted, (4) there is no heat conduction out of the substrate through wire bonds to the package, and (5) the MCM is bonded to a constant-temperature cold plate. The solution to this complicated problem is discussed in detail, and an example is presented for a hybrid containing 11 power-dissipating dice mounted on a two-layer thick-film/alumina substrate; the substrate is mounted in a Kovar package. The author also compared the thermal resistance for a single square die determined by this analysis with that predicted by Eq. (5-39) and found that agreement was best for $\theta = 32.5°$.

Ellison[37] presents an exact three-dimensional mathematical analysis of an MCM containing up to four layers in which the effects of heat transfer through wire bonds, convection cooling from both top and bottom surfaces of the substrate, and anisotropic thermal conductivities are taken into account. This analysis is the theoretical basis for the TAMS (thermal analyzer of multilayer structures) computer program, which is discussed in Sec. 5.5. The author used the TAMS program to determine chip temperatures on square alumina (glazed and unglazed) substrates cooled by natural convection in air. Typical results, reported as substrate thermal resistance (defined as maximum substrate temperature at a chip location divided by the total power dissipation on the substrate) versus number of chips for different values of total thermal resistance of the leads, are shown in Fig. 5-16. For this example, a significant portion of the total power dissipation leaves the substrate through the leads.

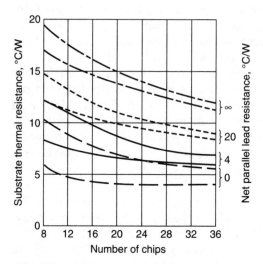

Figure 5-16. Substrate thermal resistance versus number of 50 × 50 mil chips on a 2 × 2 in square alumina substrate. Lower and upper curves of each set are for unglazed and glazed substrates, respectively.[37]

5.5 Tools for MCM Thermal Design

There are a number of computer programs available for MCM thermal analysis. These programs fall into four categories: (1) programs tailored specifically for electronic thermal management applications, (2) network thermal analysis programs that make use of discrete or lumped-parameter elements, (3) thermal analysis programs based on finite elements, and (4) thermal analysis routines embedded in suites of tools for computer-aided design of MCMs.

Licari and Enlow[7] provide a listing for a program for evaluation of the thermal resistance due to heat spreading in MCM substrates and intermediate material layers between the substrate and a constant-temperature heat sink. This program is based on the "approximate" model for heat spreading described in Sec. 5.4.1. The program, which is written in BASIC for the Tektronix 4050 computer, is capable of handling up to 20 separate material layers and permits user specification of angle θ (see Fig. 5-14).

TAMS (thermal analyzer of multilayer structures) is another thermal analysis program specifically tailored for electronic packaging applications. This program combines classical Fourier series analysis for heat spreading in substrates with lumped-parameter representations of heat losses through packaged leads and through the top and bottom surfaces of the package. Theoretical background for this program can be found in Refs. 37 to 40. The program can handle up to four layers, anisotropic thermal conductivity of the layers, and heat sources on the surfaces of as well as buried within the substrate. TAMS is written in FORTRAN IV, and a full program listing is found in Ref. 40.

The general network thermal analyzers are probably the most widely used programs for thermal analysis for microelectronics. These programs treat complex thermal problems as a network of thermal resistors (as shown, e.g., in Fig. 5-12) and thermal capacitors if transient conditions are part of the problem. Examples are TNETFA (*transient network thermal analyzer*) and SINDA (*systems improved numerical differential analyzer*). SINDA is available through COSMIC (University of Georgia, Athens), and a PC-based version of SINDA is available through Network Analysis Associates (Fountain Valley, CA). TNETFA is written in FORTRAN IV, and a full program listing is found in Ref. 40.

The resistors and capacitors in a thermal network are treated as lumped-parameter elements, as shown in Fig. 5-17. The temperature distribution in the network is determined for a finite number of volume elements called *nodes*. There are three types of nodes used in the computational scheme: diffusion nodes, arithmetic nodes, and boundary nodes.

Diffusion nodes are those nodes that possess thermal mass (i.e., thermal capacitance) and are capable of storing thermal energy in those situations where the power dissipations and/or the boundary temperatures change with time. The power dissipation in a volume element is assumed to occur at the center of the element, as indicated for node 7 in Fig. 5-17. The temperature is calculated at the center of the node, and it is assumed that the entire mass of the node is at that temperature. In other words, the temperature calculated in this manner is an average temperature for the node. For the example as shown in Fig. 5-17, solid nodes 1, 4, 7, 9 and fluid nodes 11 and 12 would be considered diffusion nodes. The thermal capacitance of a diffusion node is the product of the mass of the node and the specific heat of the material that makes up the node. In other words, for a diffusion node i, the thermal capacitance C_i is given by

$$C_i = \rho_i \, (\text{vol})_i c_{pi} \qquad (5\text{-}45)$$

T_{B1}, T_{B2} = fixed temperature boundary nodes
T_{Fi}, = inlet fluid temperature

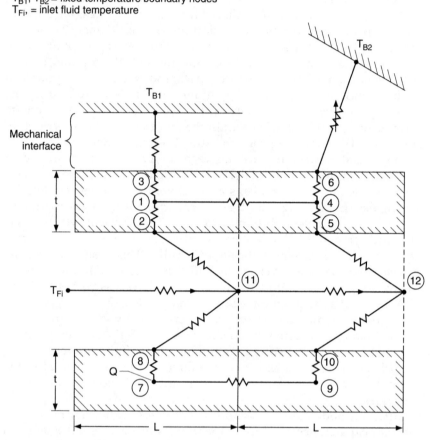

Figure 5-17. Thermal network showing different types of thermal resistances and computational nodes for determining temperature.

Arithmetic nodes have no thermal mass and hence no thermal capacitance. These nodes are usually associated with surfaces at mechanical interfaces, solid-fluid interfaces, and surfaces exchanging thermal radiation. In Fig. 5-17, nodes 2, 3, 5, 6, 8, and 10 are considered arithmetic nodes. Node 3 is connected to a boundary node through a mechanical interface, node 6 exchanges radiant energy with another boundary node, and nodes 2, 5, 8, and 10 are associated with convection to a fluid flowing in a duct.

Boundary nodes have no thermal capacitance, and the temperatures (either steady-state or time-varying depending on the nature of the problem) of these nodes are specified by the thermal analyst prior to computation of the temperature distribution. In Fig. 5-17, the boundary nodes are shown as surfaces with temperatures T_{B1} and T_{B2} and with inlet fluid temperature T_{Fi}.

The interconnections between nodes i and j are usually expressed in terms of the thermal conductance G_{ij}. The conductance of a thermal path is the reciprocal of the thermal resistance, which is discussed at length in Sec. 5.3.4. There are three types of thermal conductors: linear, nonlinear, and one-way.

Linear conductors are of the form

$$Q_{i,j} = G_{i,j}(T_i - T_j) \tag{5-46}$$

where $Q_{i,j}$ is the heat transfer between nodes i and j. These terms are used for conduction in a solid, conduction across a mechanical interface, and convective heat transfer between a surface and a fluid. Assuming that the thickness of the network (i.e., the dimension normal to the page) shown in Fig. 5-17 is w, the conductance between nodes 1 and 4 is given by

$$G_{1,4} = k\,\frac{wt}{L} \tag{5-47}$$

The conductance across the mechanical interface is given by

$$G_{3,B1} = h_{\text{interface}}wL \tag{5-48}$$

where $h_{\text{interface}}$ is the interfacial thermal conductance. The conductance between a surface and a flowing fluid, such as that between nodes 2 and 11 in Fig. 5-17, is given by an equation of the same form as Eq. (5-48) but with the appropriate value of the convective heat-transfer coefficient instead of the artificial thermal conductance.

The nonlinear conductance is used for radiant energy exchange between surfaces and is indicated in the thermal network by a resistor crossed with an arrow. These conductances are of the form

$$Q_{i,j} = G_{i,j}(T_i^4 - T_j^4) \tag{5-49}$$

where $G_{i,j}$ is given by

$$G_{i,j} = \sigma A_i \mathcal{F}_{i,j} \tag{5-50}$$

and σ is the Stefan-Boltzmann constant. Calculation of the transfer factors $\mathcal{F}_{i,j}$ is discussed in Sec. 5.3.3. In the example, the nonlinear conductance associated with radiant energy exchange between node 6 and boundary $B2$ is given by

$$G_{6,B2} = \sigma(wL)\mathscr{F}_{6,B2} \tag{5-51}$$

One-way connectors are used to model fluid flow in heat exchangers such as cold plates for cooling electronic boxes. These conductors allow energy to be transferred such that the fluid carries away the energy to the next downstream node. These connectors are indicated in the network as the resistors with arrows pointing in the direction of flow, as shown between the fluid inlet and node 11 and between nodes 11 and 12. One-way connectors reflect the change in enthalpy of the fluid stream while flowing from node i to node j and are of the form

$$G = \dot{m}c_p \tag{5-52}$$

where $\dot{m}$ is the mass flow rate between nodes and c_p is the specific heat of the fluid.

Fluid flow in electronic equipment is difficult to model, and often the flow distribution and the pressure drop in the fluid stream must be determined simultaneously to obtain an accurate picture of the fluid flow around power-dissipating components. There are options in both the TNETFA and SINDA programs for modeling the flow distribution and pressure drop. For extremely complex situations where there is close coupling between heat transfer and fluid flow and for situations that are not amenable to analysis with network thermal analyzers, computational fluid dynamics (CFD) programs such as FIDAP (Fluid Dynamics International, Evanston, IL), FLOTRAN (Compuflow, Inc., Charlottesville, VA), and PHOENICS (CHAM of North America, Huntsville, AL) are available. FIDAP and FLOTRAN use finite element techniques to solve for flow patterns, pressure distributions, and convective heat-transfer coefficients. PHOENICS is a general-purpose CFD program used for simulation of fluid flow, heat and mass transfer, and chemical reactions.

Once the values for the conductors, power dissipations, and boundary temperatures are specified, the network thermal analyzer performs an energy balance for the arithmetic and diffusion nodes and iterates this procedure until the maximum temperature change for any node in the network after an iteration falls below a value specified by the thermal analyst. In transient problems, the computations are performed by using finite time increments, which must be carefully selected to avoid excessive computational time (too small a time step) and to avoid instability in the solution routine (too large a time step). An acceptable value for the time increment can be found by using the following relation for each (nonboundary) node in the network

$$\text{Time increment for node } i = \frac{C_i}{\sum_j G_{ij}} \qquad (5\text{-}53)$$

and then selecting the smallest value for the time increment in the solution routine. Note that the index j in Eq. (5-53) pertains only to those nodes connected to node i.

Thermal analysis capabilities are now included in computer-aided design (CAD) tool suites used in MCM design. These tool suites are part of a larger design framework, such as the Mentor Graphics FALCON framework, which includes logic synthesis, electrical performance simulation, schematic capture, design verification, and generation of manufacturing aids. Mentor Graphics' MCM STATION is a suite of tools in the FALCON framework used to design multichip modules. The AUTOTHERM program is a finite element thermal analysis routine embedded in MCM STATION. The analysis capabilities of these embedded thermal analysis routines are generally restricted to problems with linear conductors. The major advantage of embedding thermal analysis in a larger design framework is the ease of maintaining a consistent set of data among the design steps and between the final design and the manufacturing aids, test vectors, and final product. This approach to MCM design also facilitates back annotation and design changes anywhere in the design sequence. Back annotation refers to a design change to improve thermal performance that will be reflected anywhere in the design where this change may influence electrical performance, routing, parasitics, etc.

References

1. J. S. Kilby, "Invention of the Integrated Circuit," *IEEE Transactions of Electronic Devices,* **ED-23:**648–654, 1976.

2. W. H. Desmonde, *Computers and Their Uses,* Prentice-Hall, Englewood Cliffs, NJ, 1969.

3. "Dolch Displays Pentium Portable," *Electronic Engineering Times,* November 15, 1993, p. 110.

4. M. Mahalingam, "Thermal Management in Semiconductor Device Packaging," *Proceedings of the IEEE,* **73:**1396–1404, 1985.

5. L. Geppert, "Not Your Father's CPU," *IEEE Spectrum,* **30:**20–23, 1993.

6. R. Profeta, personal communication, Chip Supply, Inc., Orlando, FL, January 13, 1994.

7. J. J. Licari and L. R. Enlow, *Hybrid Microcircuit Technology Handbook: Materials, Processes, Design, Testing and Production,* Noyes, Park Ridge, NJ, 1988.

8. H. F. Wolf, *Semiconductors*, Wiley-Interscience, New York, 1971.

9. G. K. Wachutka, "Rigorous Thermodynamic Treatment of Heat Generation and Conduction in Semiconductor Device Modeling," *IEEE Transactions on Computer-Aided Design*, **9**:1141–1149, 1990.

10. H. S. Hanamura, A. Aoki, T. Masuhara, O. Minato, Y. Sakai, and T. Hayashida, "Operation of Bulk CMOS Devices at Very Low Temperatures," *IEEE Journal of Solid State Circuits*, **21**:484–490, 1986.

11. B. J. Hosticka, K.-G. Dalsab, D. Krey, and G. Zimmer, "Behavior of Analog MOS Integrated Circuits at High Temperatures," *IEEE Journal of Solid-State Circuits*, **20**:871–874, 1985.

12. S. Cheng and P. Manos, "Effects of Operating Temperature on Electrical Parameters in an Analog Process," *Circuits and Devices*, July 1989, pp. 31–38.

13. F. S. Shoucair, "Design Considerations in High Temperature Analog CMOS Integrated Circuits," *IEEE Transactions of Components, Hybrids, and Manufacturing Technology*, **CHMT-9**:242–251, 1986.

14. D. Bell, *Electronic Devices and Circuits*, 3d ed., Prentice-Hall, Englewood Cliffs, NJ, 1986.

15. G. W. Neudeck, *The PN Junction Diode*, 2d ed., vol. 2 in Modular Series on Solid State Devices, Addison-Wesley, Reading, MA, 1989.

16. D. A. Hodges and H. G. Jackson, *Analysis and Design of Digital Integrated Circuits*, 2d ed., McGraw-Hill, New York, 1988.

17. G. K. Baxter, "A Recommendation of Thermal Measurement Techniques for IC Chips and Packages," *Proceedings of the 15th International Reliability Physics Symposium*, IEEE, Las Vegas, April 1977, pp. 204–211.

18. F. F. Oettinger, "Thermal Evaluation of VLSI Packages Using Test Chips—A Critical Review," *Solid State Technology*, **27**:169–179, February 1984.

19. B. S. Siegal, "Factors Affecting Semiconductor Device Thermal Resistance Measurements," *Proceedings of the 4th IEEE Semiconductor Thermal and Temperature Measurement Symposium*, San Diego, CA, February 1988, pp. 12–18.

20. D. A. Shope, W. J. Fahey, J. L. Prince, and Z. J. Staszak, "Experimental Thermal Characterization of VLSI Packages," *Proceedings of the 4th IEEE Semiconductor Thermal and Temperature Measurement Symposium*, San Diego, CA, February 1988, pp. 19–24.

21. P. Gregory, "Thermal Characteristics of a Hybrid Microcircuit," *Microelectronics Journal*, **13**:12–15, 1982.

22. J. N. Sweet, D. W. Peterson, D. Chu, B. L. Bainbridge, R. A. Gassman, and C. A. Reber, "Analysis and Measurement of Thermal Resistance in a Three-Dimensional Silicon Multichip Module Populated with Assembly Test Chips," *Proceedings of the 9th IEEE Semiconductor Thermal Measurement and Management Symposium*, Austin, TX, February 1993, pp. 1–7.

23. D. L. Blackburn, "A Review of Thermal Characterization of Power Transistors," *Proceedings of the 4th IEEE Semiconductor Thermal and Temperature Measurement Symposium*, San Diego, CA, February 1988, pp. 1–7.

24. S. C. O. Mathuna, T. Fromont, W. Koschnick, and L. O'Connor, "Test Chips, Test Systems and Thermal Test Data for Multichip Modules in the ESPRIT-APACHIP Project," *Proceedings of the 9th IEEE Semiconductor Thermal and Temperature Measurement Symposium*, Austin, TX, February 1993, pp. 117–126.

25. R. Pendse and B. J. Shanker, "A Study of Thermal Performance of Packages Using a New Test Die," *Proceedings of the 4th IEEE Semiconductor Thermal and Temperature Measurement Symposium*, San Diego, CA, February 1988, pp. 50–54.

26. V. S. Arpaci, *Conduction Heat Transfer*, Addison-Wesley, Reading, MA, 1966.

27. K. Poulton, K. L. Knudsen, J. J. Corcoran, K.-C. Wang, R. L. Pierson, R. B. Nubling, and M.-C. F. Chang, "Thermal Design and Simulation of Bipolar Integrated Circuits," *IEEE Journal of Solid State Circuits*, **27**:1379–1387, 1992.

28. D. S. Steinberg, *Cooling Techniques for Electronic Equipment*, Wiley, New York, 1980.

29. V. S. Arpaci and P. S. Larsen, *Convection Heat Transfer*, Prentice-Hall, Englewood Cliffs, NJ, 1984.

30. W. M. Kays and M. E. Crawford, *Convective Heat Transfer*, 3d ed., McGraw-Hill, New York, 1993.

31. F. Kreith, *Principles of Heat Transfer*, International Textbook, Scranton, PA, 1958.

32. A. D. Kraus and A. Bar-Cohen, *Thermal Analysis and Control of Electronic Equipment*, Hemisphere Publishing, Washington, 1983.

33. R. Siegel and J. R. Howell, *Thermal Radiation Heat Transfer*, 3d ed., Hemisphere Publishing, Washington, 1992.

34. F. E. Altoz, "Thermal Management," chap. 2 in C.A. Harper, ed., *Electronic Packaging and Interconnection Handbook*, McGraw-Hill, New York, 1991.

35. W. Maly and A. P. Piotrowski, "Heat Exchange Optimization Technique for High-Power Hybrid IC's," *IEEE Transactions of Components, Hybrids, and Manufacturing Technology*, **CHMT-2**:226–231, 1979.

36. R. F. David, "Computerized Thermal Analysis of Hybrid Circuits," *Proceedings of the 27th Electronic Components Conference*, IEEE, Arlington, VA, May 1977, pp. 324–332.

37. G. N. Ellison, "The Effect of Some Composite Structures on the Thermal Resistance of Substrates and Integrated Circuit Chips," *IEEE Transactions of Electron Devices*, **ED-20**:233–238, 1973.

38. G. N. Ellison, "A Thermal Analysis Computer Program Applicable to a Variety of Microelectronic Devices," *Proceedings of the International Conference on Hybrid Microelectronics,* International Society for Hybrid Microelectronics, Minneapolis, MN, 1978.

39. G. N. Ellison, "Extensions of the Closed Form Method for Substrate Thermal Analyzers to Include Thermal Resistances from Source-to-Substrate and Source-to-Ambient," *IEEE Transactions of Components, Hybrids, and Manufacturing Technology,* **15:**658–666, 1992.

40. G. N. Ellison, *Thermal Computations for Electronic Equipment,* Robert E. Krieger Publishing Co., Malabar, FL, 1989.

6

Assembly Processes

Assembly processes for multichip modules (MCMs) are not too different from those that have been used for over 30 years to assemble hybrid microcircuits. The attachment of substrates and bare dice with electrically conductive or electrically insulative epoxy adhesives and the interconnection of the dice by wire bonding, though greatly optimized and automated through the years, are still the dominant processes. However, multichip modules are driving many of the existing technologies to their limits, resulting in orders-of-magnitude increases in densities and performance. For example, flip-chip dice are now being produced with hundreds, even thousands, of fine-pitch bumps in area arrays. Interconnecting these bumped dice by solder reflowing or by using anisotropic electrically conductive adhesives is emerging as the preferred interconnect method for high-density, low-cost, high performance modules.

6.1 Die and Substrate Attachment

The materials and assembly processes for attaching dice and substrates in multichip modules are generally the same as those used in hybrid microcircuits. As with hybrids, in over 90 percent of multichip modules, low-cost, easily reworked epoxy adhesives are used. Solder or alloy attachment is primarily used for high-power circuits or for circuits that must meet class K, space-grade outgassing requirements. A comparison of the advantages and limitations of epoxy and alloy attachments is given in Table 6-1. Polyimide adhesives, either metal-filled or unfilled, are much more thermally stable than epoxy adhesives (400 to 500°C versus 175 to 250°C, respectively) and as such are

Table 6-1. Comparison between Epoxy and Eutectic or Alloy Attachment

	Epoxy		Eutectic or alloy	
Advantages	Limitations	Advantages	Limitations	
Low processing temperatures (150 to 180°C cure)	Risk of excessive outgassing (H$_2$O, NH$_3$, etc.)	High thermal conductivity, best approach for high-power circuits	High processing temperatures; 300°C can degrade wire bonds and devices	
Low-cost batch processing (screen printing, stamping, or automated dispensing) available in paste, tape, or preform	Risk of excessive contaminants (Cl$^-$, F$^-$, Na$^+$)		Limited rework due to higher solder melt temperatures	
	Risk of particle generation	Available as preforms		
Ease of rework, softens under temperature and pressure	Possible interface resistance increase		Risk of metal particle generation or splatter	
Available in both electrically conductive and insulative versions	Cleanliness control for best adhesion	No outgassing circuits meet class K, low H$_2$O requirement ≤3000 ppm	Only available in electrically conductive versions	
	Poor to moderate thermal conductivity			

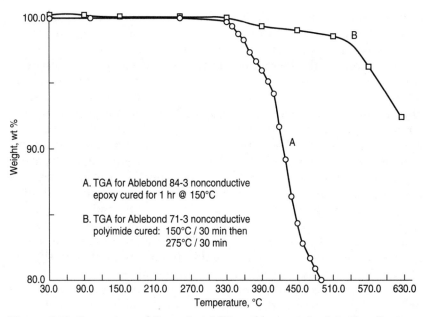

Figure 6-1. Comparison of thermal stabilities of epoxy and polyimide adhesives. (*Courtesy Ablestick Laboratories.*)

used for higher-temperature applications. Figure 6-1 contains thermo-gravimetric analysis (TGA) curves showing the improved thermal performance of the polyimide adhesives at temperatures above 250°C. However, epoxies satisfy the requirements for most hybrid and multi-chip module applications since epoxies are seldom exposed to temperatures higher than 150°C.

Epoxy adhesives, when first introduced for hybrid assembly (1960s), were considered risky and unreliable. In fact, numerous failures occurred that were attributed to the poor grades of epoxies commercially available then. Considerable progress has been made since then, first in establishing requirements and specifications, then in formulating improved adhesives to meet the specifications, and finally in qualifying the adhesives in functional circuits. Today, almost all hybrid microcircuits and multichip modules are assembled with epoxy adhesives that have been proved highly reliable.[1-7] NASA, foreseeing the need to control the outgassing, ionic impurity content, and general quality of adhesives, under several contracts to Rockwell International, prepared and issued in 1982 the first specification, MSFC-Spec-592: *Specification for the Selection and Use of Organic Adhesives in Hybrid Microcircuits*. This specification, modified to some extent, was adopted by other government agencies and ultimately became the basis for the current method 5011 of Mil-Std-883C. Many commercial epoxy formulations and other adhesives that meet or exceed the requirements of Mil-Std-883C and have a long history of reliable use in hermetically sealed modules are now available.

Two types of adhesives, electrically conductive and electrically non-conductive (insulative), formulated as either pastes or films, are in wide use. Conductive silver-filled epoxies are extensively used to attach transistors, integrated circuits (ICs), and chip capacitors to substrates and to attach substrates to packages. Gold- and copper-filled epoxies are also available but are used only in special situations. Metal-filled conductive adhesives are used to produce ohmic contacts to transistor, diode, capacitor, and resistor chips. However, if carefully applied, the same conductive adhesive can be used to attach all other devices and the substrate. Even though electrical conductivity is not required in those cases, the metal filler provides improved thermal conductivity. Dispensing a single adhesive is also highly desirable from a production standpoint since this avoids the extra process steps of handling, dispensing, and curing a second adhesive.

The electrical conductivities of silver- or gold-filled epoxies are orders of magnitude higher than those of the unfilled versions, yet still orders of magnitude lower than those of the pure metals. Even though the epoxies may be highly filled with metal (70 to 80 percent), their conductivity still does not approach that of the metal because the metal particles become encapsulated by a thin insulation coating of the resin, preventing optimum electrical conduction. Volume resistivities for silver-filled epoxies typically range from 10^{-4} to 10^{-5} $\Omega \bullet$cm. Volume resistivities for the insulative epoxies are high, ranging from 10^{12} to 10^{16} $\Omega \bullet$cm, although these values may degrade appreciably for some adhesives when exposed to high humidity.

Metal-filled adhesives have higher thermal conductivities than unfilled versions, which is important in dissipating heat from the die and keeping junction temperatures low (Tables 6-2 and 6-3). In most cases, the adhesive serves as the main conduit for transferring heat from the die to the substrate and from the substrate to the package. The thermal conductivities of most electrically conductive epoxies range from 1 to 3 W/(m$\bullet$K) while the unfilled versions (electrically nonconductive) range from 0.2 to 0.3 W/(m$\bullet$K). Method 5011 of Mil-Std-883C specifies 2 W/(m$\bullet$K) as the minimum value for electrically conductive adhesives and 0.2 W/(m$\bullet$K) as the minimum for nonconductive adhesives. Some recent formulations filled with very highly thermally conductive materials such as aluminum nitride, beryllia, boron nitride (cubic), or diamond are reported to have thermal conductivities as high as 11.6 W/(m$\bullet$K), for example, diamond-filled epoxies.

Table 6-2. Properties of Electrically Conductive Adhesives for Die and Substrate Attachment*

Resin	Filler	Form	Volume resistivity, $\Omega \cdot cm$	Thermal conductivity, W/(m•K)	$T(g)$, °C	CTE[†] (ppm/°C)
Epoxy	Silver	Paste	1×10^{-4} – 7×10^{-5}	1.87–3.0	127	46
Epoxy	Gold	Paste	8×10^{-4}	1.7–2.0	69–107	50–57
Epoxy	Copper	Paste	2×10^{-3}	1.38	59	46
Epoxy	Silver	Film	2×10^{-4}	1.0–2.1	81	66
Polyimide	Silver	Paste	1×10^{-4}		280	41
Thermoplastic	Silver	Paste or film	10^{-4}–10^{-5}	3.0		
Thermoplastic	Gold	Paste	10^{-4}	2.0		

*These values are typical ranges indicative of the generic classes and were compiled from data sheets from several companies, including Ablestick Laboratories, Epoxy Technology, Amicon, Emerson & Cuming, A.I. Technology, and Staystick. Specific formulations and other data may be obtained from these companies.

[†]Values are below $T(g)$; values will be higher above $T(g)$.

Table 6-3. Properties of Electrically Insulative Adhesives for Die and Substrate Attachment*

Resin	Filler	Form	Volume resistivity, $\Omega \cdot cm$	Thermal conductivity, W/(m•K)	$T(g)$, °C	CTE[†] (ppm/°C)
Epoxy	None	Paste	4×10^{14} – 3×10^{15}	0.2 –0.7	130	35
Epoxy	Alumina	Paste	1–4×10^{12}	0.52–1.4 2	85	41
Epoxy	Diamond	Paste	10^{14}	11.6	−25	120
Epoxy	Diamond	Film	$>10^{14}$	11.6	−25	110
Epoxy	Glass fabric	Film	7×10^{12}– 8×10^{14}	0.27–1.12	102–140	50–83
Epoxy	AlN	Paste	$>10^{14}$	3.6	−25	120
Polyimide	Silica	Paste			253	
Thermoplastic	Ceramic	Paste	10^{15}	0.27–1.0	>150	
Thermoplastic		Film	$>10^{14}$			
Thermoplastic	AlN	Film	10^{13}	1.0		

*These values are typical ranges indicative of the generic classes and were compiled from data sheets from several companies, including Ablestick Laboratories, Epoxy Technology, Amicon, Emerson & Cuming, A.I. Technology, and Staystick. Specific formulations and other data may be obtained from these companies.

[†]Values are below $T(g)$; values will be higher above $T(g)$.

Automatic drop-dispensing, screen-printing, and stamping paste adhesives are popular production methods for attaching ICs and other chip devices. Film adhesives (preforms), however, are more practical for attaching substrates or very large dice (500 mils square or larger). Figure 6-2 shows a close-up of an automated pick-and-place machine for attaching large dice to an MCM-D interconnect substrate.

Some modifications to adhesive formulations have had to be made to accommodate MCMs, e.g., the addition of flow control agents so that automatic dispensing or stamping could be used. Flow must be controlled and minimized when dice that are closely spaced (some at 40 mils or less) are attached and where bonding pads have a fine pitch (4 to 6 mils). Excessive epoxy flow contaminates adjacent bonding sites on the substrate and requires additional cleaning steps for removal. Control of flow and thickness is even more important in attaching large dice. Uneven adhesive thicknesses result in tilted dice, which interfere with automatic wire bonding and TAB bonding. Furthermore, thickness variations in the substrate attachment adhesive can significantly reduce the conduction of heat from the substrate to the package base. For this reason, film adhesives in which the thickness is uniform and flow-out is controlled by precutting the film to shape are used almost exclusively for substrate attachment.

Stress is still another adhesive parameter that is more critical to the reliability of MCMs than to hybrid microcircuits. Increased stresses are generated from thermal expansion mismatches when large silicon dice or large silicon substrates are attached to alumina ceramic or plastic laminate substrates. Exact matching of the coefficients of thermal expansion (CTEs) of the mating materials that have high moduli of elasticity is not easy. Using adhesives that can relieve and dissipate

Figure 6-2. Automatic pick-and-place machine for positioning and attaching large dice to an interconnect substrate.

stresses is a more viable approach. Epoxy adhesives having low moduli of elasticity (flexible) and low glass transition temperatures are reported to be stress-free and more reliable for MCM assembly.[8,9]

Last, the ability and ease of reworking MCMs are more an issue with MCMs than with hybrids. Benign rework procedures are essential for the high-end MCMs because of the high costs associated with the custom dice, the interconnect substrate, and the assembled modules (see Chap. 9).

6.2 Cleaning

Cleaning is performed at various stages in the fabrication, assembly, testing, and rework of multichip modules. However, cleaning is often overdone and may do more harm than good, especially if it is not thorough. Localized contaminants may merely be spread around, or residues from the evaporation of the cleaning solvent may be left behind. A good practice is to ensure minimal handling, thus reducing the number of cleaning steps. However, there are three assembly steps in which cleaning is essential: (1) after substrate and die attachment, (2) prior to vacuum baking, and (3) after rework.

Cleaning should be performed after die and substrate attachment regardless of whether organic adhesives or solders are used. Curing epoxy adhesives results in outgassing of organic volatiles that can condense on vicinal wire-bonding pads. Migration ("bleeding") of one of the epoxy constituents (resin or hardener) onto bonding pads is also a common occurrence. Fluxes and flux residues from solder attachment are also known to migrate to adjacent areas. Flux residues and almost all other contaminants affect both initial wire bondability and long-term interconnect reliability.

Prior to hermetic sealing or plastic encapsulation, it is important to ensure that the assembly is free of ionic, organic, and particulate contaminants which may have been introduced during handling, assembly, and electrical testing. Vacuum baking is normally performed after final cleaning and prior to sealing, to ensure the removal of solvents, absorbed moisture, and volatile constituents from epoxies and other organic materials used in assembly.

For high-density MCMs, several rework steps may be necessary prior to or after sealing. Rework requires extra handling and often aggressive procedures which generate debris. Delidding and removal of dice require reconditioning of the surfaces and cleaning prior to die replacement and resealing. Besides the in-line specified cleaning steps, operators, at their discretion, often insert extra cleaning steps such as flushing the part with methylethyl ketone or isopropyl alcohol.

Generally, the same established solvents and cleaning methods as used for hybrid circuits can be used for MCMs. Plasma cleaning, vapor degreasing, solvent spray cleaning, solvent immersion, and ultrasonic cleaning are some of the widely used methods. Cleaning in an oxygen-argon plasma is particularly effective in oxidizing and removing organic residues generated from the attachment of the dice and substrates with epoxy adhesives. Although it is a very effective cleaning process, plasma can degrade the electrical parameters of some devices; hence some initial testing should be performed to determine the susceptibility of dice to the plasma conditions. Flux and ionic residues are much more difficult to remove and generally require a combination of aqueous and nonaqueous solvents assisted by mechanical scrubbing. A 50-50 mixture of isopropyl alcohol and naphtha or Freon azeotropes have been used for many years to remove flux residues. Ultrasonic cleaning of assembled circuits has been found to be risky, causing damage to fine wire and wire bonds. Ultrasonics, however, can be used safely and effectively to clean bare substrates, packages, and lids prior to assembly.

In vapor degreasing and spray cleaning, the chlorofluorocarbon (CFC) solvents such as the Freons have been the mainstay solvents for both the semiconductor and hybrid circuit industries. Unfortunately, CFCs are being restricted and phased out by the federal government because of their ozone depletion effects. This presents a serious challenge for the entire semiconductor and microelectronics industry to select alternate solvents that are as efficient, nonflammable, nontoxic, and inert as the CFCs. Aqueous cleaning, coupled with scrubbing or ultrasonics, though effective in removing flux and chloride contaminants from printed-wiring boards, has been discouraged for cleaning multichip modules due to the fragile nature of the wires, wire bonds, chip devices, and thin-film metallization. Entrapment, absorption, and tenacious retention of water by epoxies, polyimides, and ceramic surfaces can also result in corrosion, high electric leakage currents, metal migration, and failure to pass the water content requirement specified in Mil-Std-883. Replacements for CFC solvents such as alcohols, alcohol-water mixtures, glycols, organic esters, citric acid–water solutions and CO_2 jet spraying are being investigated.

6.3 Electrical Connections

In spite of numerous attempts over the years to replace wire bonds, and in spite of all the predictions that wire bonding would be displaced by more advanced processes, wire bonding remains the dominant process for electrically interconnecting dice to substrates and substrates to packages. Nevertheless, alternatives to wire bonding such as tape automated bonding (TAB), ribbon bonding, flip-chip bonding, overlay batch metallization (a General Electric process), and anisotropic adhesive bonding have many advantages in facilitating the pretesting and assembly of dice for multichip modules. Processes that are over 30 years old such as IBM's C4 flip-chip bonding are resurfacing as improvements over wire bonding in producing short interconnect paths and high packing densities. Table 6-4 pictorially shows the unique features of and differences among these processes. A brief discussion of each follows.

6.3.1 Wire Bonding

One-mil-diameter gold wire thermosonically bonded by automatic equipment is the most widely used method for both hybrid circuits and MCMs. Ultrasonically bonded 1-mil-diameter aluminum wire is used to a lesser extent but has advantages in being a room-temperature process and in providing a monometallic all-aluminum interconnection at the die level. Thermosonic, ultrasonic, and thermocompression wire bonding are well-established processes that have been treated extensively in the literature and need not be reiterated here.[10-12] However, in going from hybrid circuit bonding to multichip module bonding, there are some extra controls and care that must be taken. In MCMs the number of wire bonds required per module is several thousand instead of several hundred, as in hybrids, and the wire bond pad pitch has been reduced to 6 mils and even to 4 mils. In fact, some application-specific integrated circuit (ASIC) dice now require over 500 wire bonds to interconnect each die. Both the increase in the number of bonds and their proximity to each other and to the edge of the die require more accurate wire bonding equipment, improved process controls, and greater human care in order to obtain high first-pass yields. To avoid wire short-circuiting to the edge of a die, controlled wire-loop profiles are programmed so that they traverse the die surface and clear the edge of the die[13] (Fig. 6-3). The probability of damaging adjacent wires during rework is, of course, much greater in highly dense modules.

Although there exists an extensive history of reliability for wire-bonded dice, the well-known factors that can degrade wire bonds must be proved to be absent for each new design. This is especially true for MCMs because of the wide variety of materials and processes that can be, and are being, used. Ionic contaminants from substrate processing, outgassing from organic dielectrics, entrapped plating and etching solutions, and photoresist residues can degrade wire bonds, especially when the action is accelerated by elevated temperature and

Table 6-4. Comparison of Die Interconnect Methods

	WIRE BOND	TAB (FACE-UP)
	IC	IC
Substrate metallization	Not critical: thin- or thick-film gold	Not critical: Au or solder
Extra IC processing	Not required	Au or solder bumped I/O (wafer level)
Interconnect	Gold or Al wire	Au weld on solder Cu tape lead (Au plated/welded or soldered)
Bonding process	Thermocompression, Thermosonic, ultrasonic	Chip-specific tape TAB ILB/OLB gang or single-point
Capital equipment	Manual or automatic wire bonder	Pick and place, TAB form and excise ILB and OLB
Die attachment	Conductive/nonconductive adhesive or metallurgical	Conductive/nonconductive adhesive
I/O pitch (min.) I/O location	6 mils proved Staggered	4–8 mils Perimeter
Reworkability	Yes	OLB—yes
Chip-to-chip spacing	50–70 mils	100 mils standard, 70 mils feasible
Chip pretestability and burn-in	N/A	Possible
Reliability issues	Reliability proven Nondestructive Pull testing	Visual inspection possible Bond integrity being addressed
Electrical	Speed and power limited	Low inductance, resistance
Heat dissipation	Best for metallurgical attachment	Good - through adhesive and Cu leads
Technological maturity	Very mature	Demonstrated by many, most applications single-chip, high-volume
Key issues	Device pretesting at speed difficult	IC bumping required Tape, hand tooling required for each IC type Tape quality/delivery delays Burn-in

TAB (FACE DOWN)	FLIP CHIP	MICROBUMPS
ATTACH BUMPED TAPE HESIVE I/Os LEADS IC	AIR GAP SOLDER I/Os IC (Si)	Au ADHESIVE MICROBUMPS IC
critical; Au or solder	Low CTE (Si on AlN required for large chips (>170 mils) Solder bond pads (size critical)	CTE less critical Planarity critical
r solder bumped I/O - er level)	Solder bumped I/O pads (wafer level)	Au microbumped I/Os (3 µm thick) (Wafer level)
veld or solder Cu tape leads plate/welded or soldered)	Solder	Metal contact (pressure), adhesive bond
-specific tape, ILB/OLB: gang or single-	Solder reflow (batch) Heat, inert atmosphere	UV cure under high pressure (6000–12,000 kg/cm^2
and place, TAB form and e nd OLB	Face-down aligner/bonder	Custom aligner bonder Critical accuracy (±2 µm) Planarity, high pressure, UV cure
conductive adhesive or dhesive	Nonconductive adhesive or no adhesive	UV cured acrylic resin
nils neter	8 mils Perimeter or area	10 µm (0.4 mil) Perimeter (or area with trans- parent substrate)
—yes	Difficult (solder volume control oxidation without flux)	Unknown
30 mils feasible	10–30 mils feasible (rework limited)	Unknown
ble	Not available	Not available
l inspection possible integrity being addressed	Inspection difficult CTE matched substrate required to reduce solder fatigue Difficult to clean beneath die	Inspection extremely difficult Limited temp. tests look OK; long-term tests and RGA unknown
-speed performance may ected by adhesive	Best performance	High-speed performance may be affected by adhesive
—through adhesive and ads	Poor through solder I/Os (depends on total I/O area)	Very good through microbumps and very thin adhesive (3 µm)
onstrated with air gap r chip: 8-mil pitch, sol- Au OLB	Demonstrated in production: IBM Delco, Motorola (8-mil pitch, min.)	Demonstrated 10-µm pitch (Matsushita)
mping required hand tooling required for IC type quality/delivery delays in	Device pretesting, heat dissipa- tion, inspectability, rework IC bumping required	Device pretest, incompatibili- ty with nonplanar substrates. Nonstandard IC microbump- ing, special processing equip- ment (high pressure, leveling, UV cure)

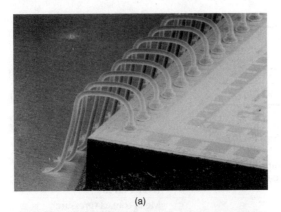

(a)

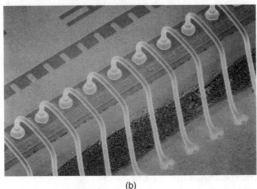

(b)

Figure 6-3. Top views of close-proximity wire bonds connecting die to substrate—second bonds are less than 380 μm from the edge of the die. (*Courtesy Kulicke and Soffa, Inc.*)

exposure to moisture. Intermetallic growth occurring in gold-to-aluminum bonds which degrades their strength and increases electrical resistance is a thoroughly characterized and well-documented phenomenon.[11–13] Initial bondability and yield, especially for thermocompression bonding and to a lesser extent for thermosonic and ultrasonic bonding, are also reduced because of organic residues on the bonding pads. Ultrasonic bonding is more forgiving of organic residues since the inherent scrubbing action of the ultrasonic energy displaces organic residues during bonding. A major source of organic residues on bonding pads results from the outgassing of epoxy attachment adhesives during curing and recondensation on adjacent pads. Bleed-out occurring during epoxy curing is yet another source of contaminants. Bleed-out is the separation and lateral flow of either the epoxy resin or the hardener during curing. With closely spaced ICs, excessive bleed-out quickly contaminates the bonding pads. Fortunately, plasma cleaning has been found to be an effective method of removing both bleed-out and condensed outgassed residues.[14] Oxygen or an oxygen/argon plasma can remove even invisible traces of organic residues, but one must first ascertain that the devices are not damaged by the plasma conditions. Most companies plasma-clean their circuits after epoxy curing and prior to wire bonding, even though the best epoxies, those having low outgassing and low bleed-out, are used.

The average wire bond pull strengths of dice thermosonically bonded with 1-mil-diameter gold wire to MCM substrates (polyimide/gold top pads) are excellent (9 to 10 g). Even after aging at 150°C for 1000 h there is essentially no change in the wire bond pull strength (Table 6-5).

6.3.2 Tape Automated Bonding (TAB)

TAB has been an established interconnection process for over 30 years and has proved to be very cost-effective in the high-production assembly of single-chip devices in single packages. It has not been found cost-effective for hybrid circuits or multichip modules because of the difficulty of procuring all devices in TAB form. However, for the high-density high-performance MCMs, TAB is generating renewed interest due to its better electrical performance (lower and more controlled inductance of bonds) and the ability to pretest and burn in devices. ASICs and high-density gate arrays that are required in many multichip modules are quite expensive; costs range from several hundred to several thousand dollars per die. Pretesting the dice, which is possible

Table 6-5. Wire Bond Pull Strength Results

	Hours at 150°C			
	0	257	500	1000
Alumina HDMI Substrate				
X (avg), g	9.14	9.85	9.69	9.60
σ, g	0.53	0.75	0.57	0.65
Maximum, g	10.80	12.10	11.80	11.60
Minimum, g	7.10	8.20	8.30	6.70
No. of bonds	201	201	201	203
Silicon HDMI Substrate				
X (avg.), g	10.38	10.38	10.11	9.78
σ, g	0.76	0.73	0.71	0.73
Maximum, g	12.30	12.70	13.00	11.70
Minimum, g	8.50	8.70	8.40	8.10
No. of bonds	201	201	200	202

with dice on TAB, and starting the assembly with a known good die (KGD) are therefore imperative in avoiding the high cost of reworking and replacing faulty dice.

In TAB, integrated circuits having bumped I/O pads are positioned in apertures of a polyimide film (Kapton, trade name of Du Pont) tape and then inner-lead-bonded (ILB) to corresponding photodelineated metal pads on the tape. The tape has sprocket holes along its edges similar to movie film whose size may be 16, 35, or 70 mm. The dice are automatically moved, positioned, and gang-bonded to the tape as the tape moves on reels.

Having bumps on the die is essential to the TAB process. Bumping is a metallurgical process that modifies the aluminum pad metallization to gold or solder by first vapor-depositing an adhesion and barrier metal such as titanium, chromium, or palladium. The bumps, generally 1 to 2 mils high, provide intimate contacts between the pads on the die and corresponding pads on the tape. The bumps also raise the tape slightly above the die, preventing short circuiting at the edges. Bumps are best processed on the dice at the wafer stage where batch processes can be used. Bumps are best processed by the semiconductor device manufacturer. Hence, for a user, the most expedient approach is to purchase the dice already inner-lead-bonded on the tape. Since not all dice are available in TAB form, users have been faced with applying the bumps themselves or sending the wafers to a specialty house for bumping, then procuring the tape, and inner-lead-bonding. This process, of course, is expensive, involves long turnaround times, and jeopardizes yield. The design and procurement of TAB are also expensive, especially where small quantities of a large variety of die types are involved. Design and procurement are long-lead-time items since each type and size of die require a die-specific tape. Such limitations, however, should be weighed against the savings resulting from the assurance of using pretested known good dice.

Once the dice have been inner-lead-bonded to the tape, they may be automatically tested and burned in. This ability to pretest the dice is a major benefit of multichip modules since it increases the first-time-through yield and avoids costly rework at the assembled, higher-value-added stage. Unlike wire bonding or flip-chip bonding, TAB lends itself to burning in and pretesting the dice prior to assembly.

The last steps in TAB processing are to excise and separate the individual dice from the tape, to form the outer leads, and to bond the leads to the interconnect substrate.[15] Outer-lead bonding (OLB) may be done by solder reflowing (effected by heat or laser), thermocompression gang bonding, or automated single-point thermosonic bonding. In flip TAB, the dice are mounted face-down, which results in even greater packaging density than normal face-up TAB (Fig. 6-4). With flip

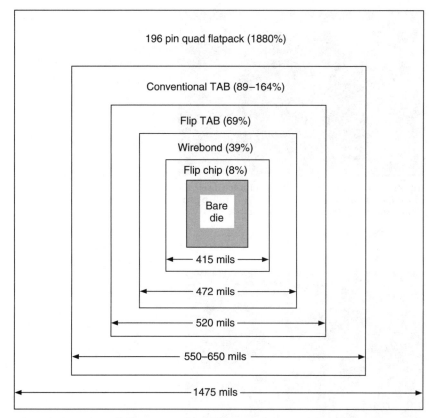

Figure 6-4. IC packaging density for various interconnection methods. (Numbers in parentheses are area enlargements over the bare die. The die in this example is 400 mils square. This drawing is not to scale.)

TAB, chip-to-chip spacings as small as 40 mils have been achieved. A 2.8 × 2.8 in, 400-MFLOP signal processor in which 28 chips were assembled by using flip TAB was previously shown in Fig. 1-10. The dice for this module were pretested and burned in on the TAB prior to assembly and were reported to give first-pass yields of 95 percent at the MCM stage. Figure 6-5 shows an inner-lead-bonded test chip on 4-mil-pitch pads. Figure 6-6 shows a TAB die outer-lead-bonded to a metal lead-frame assembly. TAB can accommodate bond pitches as small as 4 mils, but 2- to 3-mil pitches are projected for the near future.[16]

TAB connections provide better thermal and electrical performance than wire bonds because of the greater cross-sectional area of their leads (3 to 4 mils wide compared to 1-mil diameter for wire)(see Table 6-6).[17] TAB strengths are also higher (30 to 50 g) than wire bond pull strengths (9 to 12 g). TAB outer-lead bond pull strengths measured for

Figure 6-5. VLSIC test chip assembled on TAB 400 mils square with 360 leads on 4-mil pitch. (*Courtesy International Micro Industries, Inc.*)

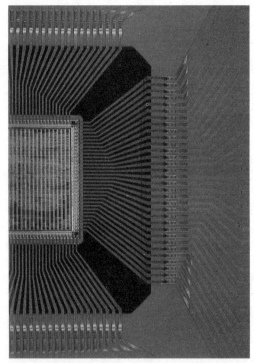

Figure 6-6. A 128 I/O TAB lead pattern and metal lead-frame assembly, gold-to-gold single-point thermocompression outer-lead-bonded using the IMI 1310 bonder. (*Courtesy International Micro Industries, Inc.*)

Table 6-6. Comparison of Wire Bond and TAB Electrical and
Thermal Properties

	Wire bond*		TAB†
	Aluminum	Gold	Copper
Lead resistance, Ω	0.142	0.122	0.017
Lead-to-lead capacitance, 6-mil spacing	0.025	0.025	0.006
Lead inductance, nH	2.621	2.621	2.10
Lead conduction, °C/(m•W)	79.6	51.6	8.3
Lead convection, free, °C/(m•W)	336.5	336.5	149.5

*Wire bond 0.001 in diameter, 0.10 in long.
†TAB lead 0.001 by 0.004 in, 0.10 in long.
SOURCE: Ref. 17.

six large gate array dice, each having 204 leads, all passed a minimum
of 25 g after Mil-Std-883 screen testing. There was also no degradation
of the underlying circuitry such as the signal lines or the polyimide
dielectric (Table 6-7). The greater robustness and the lower height profile have made TAB connections especially suited to ultrathin electronic packages such as wallet-size calculators, organizers, liquid-crystal
displays, and a host of other high-volume consumer products.[18] The
sturdier bonds and lower profiles of TAB dice make them compatible
with plastic encapsulants, glob-topped coatings, or plastic molding
compounds. TAB devices are better able to withstand the stresses

Table 6-7. TAB Pull Strength Results

Test	Method*	Pull strength
Thermal shock	1011, condition C	>25 g
Temperature cycle	1010, condition C (− 65 to 150°C, 100 cycles)	>25 g
Thermal aging	1008, condition C (1000 h @ 150°C)	>25 g
Constant acceleration	2001, condition D (10,000g)	>25 g
Constant acceleration	2001, condition D (20,000g)	>25 g Die detached from substrate and broke. TAB remained intact.

*Per Mil Std. 883.

caused by shrinkage of the plastic resin during curing and differences in the coefficients of thermal expansion after curing.

6.3.3 Flip-Chip Bonding

In flip-chip bonding, solder bumps are batch-produced on the I/O pads of the chip, and then the chip is flipped over (face down) and mated with corresponding bumps (pads) on the interconnect substrate. On heating the solder to its melting temperature, the bumps collapse and fuse with the corresponding metal bonding pads on the substrate.[19] The surface tension of the molten solder prevents the chip from collapsing during solder reflow and also prevents solder from bridging between solder bump connections. Known also as the C4 process (controlled collapse chip connection), flip-chip bonding was developed by IBM and has been used for over 30 years to assemble IBM's mainframe computer modules. Originally, only a few bumps arranged at the perimeter of the die were required; But recently, with the advent of high-I/O-count dice (>500) and the requirement for very short electrical paths for high-speed circuits, processes that generate closely spaced bumps in an area array have been developed.[20] In the original IBM thermal control modules, each die had an area array of 121 solder bumps. In recent designs, arrays of 762 solder bumps per die have been achieved.[21]

Prior to creating solder bumps on the die, the aluminum metallization on the I/O pads must be modified to a metallurgy compatible with the solder. This is best done at the wafer stage by sequentially evaporating thin films of chromium (1000 Å), copper (10,000 Å), and palladium (1000 Å). Chromium serves as an initial adhesion and barrier layer for the aluminum. Copper provides a solder-wettable surface while palladium functions as an oxidation barrier. The I/O pads on the chips are then delineated by patterning and etching through a photoresist mask, which may be either a liquid or a solid dry film photoresist. Once the chip pads have been defined, solder bumps 2 to 5 mils high are formed by evaporating solder alloys through metal masks or by electroplating. A wide variety of solder alloys may be employed, although lead-tin and indium alloys are most popular.

As with TAB, flip-chip bonding is generating renewed interest for high-speed multichip modules. Flip-chip interconnects offer the shortest signal electrical paths from the die to the substrate. The inductance and capacitance of the solder joints are lower than those of wire or TAB leads. To accommodate a large number of I/Os, dice with bumps in an area array are preferred to those on the perimeter, not only to achieve higher density but also to dissipate stresses more evenly over a larger area and to provide improved thermal transfer. By employing several additional process steps, the perimeter I/O pads of a die can be redistributed to form an area array. Again, this is best done at the wafer stage where low-cost batch photolithographic processes can be used.[22]

As with any technology, flip-chip bonding has some limitations. Several issues must be addressed, especially for dice that have a large number of bumps in an area array:

1. Difficulty in procuring off-the-shelf dice that already have bumps and are ready for assembly. Unless large quantities of dice are procured, semiconductor manufacturers are reluctant to add bumps to their dice or redesign for area arrays. Users are then faced with having to develop in-house processes or contract for outside services to prepare the wafers and insert bumps on the dice.

2. Inability to inspect the integrity of the internal connections once assembled.

3. Potential for entrapment of contaminants (ions, fluxes, etc.) beneath the dice and difficulty in cleaning out residues. Design guidelines call for the solder bumps to be several mils high after reflow to allow sufficient space for cleaning solvents to penetrate. Electrically nonconductive adhesives may be used as an underfill to prevent entrapment of ionic contaminants and to enhance thermal transfer, though at the expense of added processing steps and increased difficulty in rework.

4. High probability of solder bridging between closely spaced bumps.

5. Relatively poor thermal transfer, although the solder bumps augment heat transfer to the substrate and the greater the number of bumps, the better the thermal transfer. In fact, some firms purposely fabricate extra bumps (not electrically functional) to increase thermal conduction. The IBM thermal conduction module, however, relies primarily on removing heat from the backs of the flipped dice by connecting them to a heat plate and heat exchanger through copper pistons and springs.

6. Difficulty in removing failed dice and reworking. The greater the number of solder bumps, the more difficult it is to separate them. Several practical rework methods, however, have been developed and are discussed in Chap. 9.

A unique extension of flip-chip bonding has been reported[23] and patented by Matsushita.[24] Microsize bumps formed on the die are mated with corresponding pads on the substrate by pressing the die through a liquid resin and then curing the resin under pressure. By applying pressure, the resin is squeezed out from between the bumps and pads. The resin may be an ultraviolet light-curing acrylic suitable for large glass panel displays or a heat-curing resin such as epoxy suitable for opaque substrates. The connections are direct pressure contacts (not solder reflowed) and are permanently held in compression by the cured resin that has been squeezed out and surrounds the bumps (Fig. 6-7). Bumps as small as $5 \times 5 \times 3$ µm on 10-µm pitch are reported. The contact resistance is less than 7 mΩ per connection, but the planarity of the surfaces is important in achieving 100 percent good connections.

6.3.4 Anisotropic Adhesive Interconnections

Adhesives that can be rendered electrically conductive in the z-direction while remaining insulating in the xy direction are referred to as *anisotropic* or *directional* adhesives. They are a new breed of adhesives that may be used in lieu of solder to interconnect fine-pitch flip-chip or TAB dice to high-density interconnect substrates. Anisotropic adhesives are also used as interconnects in edge connectors and flexible cable. Among its widest applications, anisotropic adhesives are used to connect TAB chips to the ITO (indium tin oxide) electrodes of flat panel displays.

Anisotropic adhesives are available in either paste or film. They are formulated by homogeneously dispersing metal particles of controlled shape and size (typically 6 to 10 µm) in a polymer resin so that when pressure is applied between two mating surfaces, the particles act as

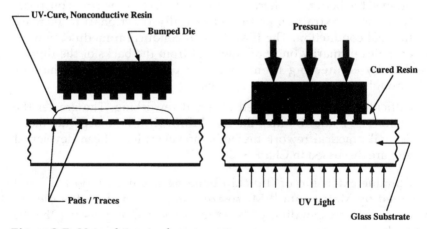

Figure 6-7. Matsushita microbump process.

shims, initially forming a pressure electrical contact. On curing the adhesive under pressure (500 to 1000 lb/in^2), the connections become permanent while the surrounding adhesive, though still filled with metal particles, remains nonconductive since in those areas the particles are still completely surrounded by resin (Fig. 6-8). The polymer matrix may be composed of either thermosetting resins, such as epoxies, or thermoplastic, such as polyvinyl butyral, polyetheretherketone (PEEK), or polyetherimide (ULTEM, trade name of General Electric). A key manufacturing advantage of the thermoplastic versions is that they soften and flow quickly on heating to their melting temperature and thus can be more easily removed when a faulty die is replaced. A wide selection of conductive particles can be used including carbon, nickel, copper, silver, gold, and metal alloys. It has been reported that using metal-plated thermoplastic particles as the filler reduces stresses and produces more reliable connections.[25] In other formulations involving epoxy resins, the hardener is stabilized by encapsulating it with an insulating resin according to a patent issued to Hitachi.[26] Under pressure, the thin plastic sheath breaks, releasing the hardener and causing it to react with and cure the epoxy resin. A connection pitch of 4 mils has been achieved although others report pitches as low as 1.6 mils.

Anisotropic adhesives have become especially popular in assembling low-cost compact consumer products, noteworthy examples of which include liquid-crystal displays, wallet-thin calculators, watches,

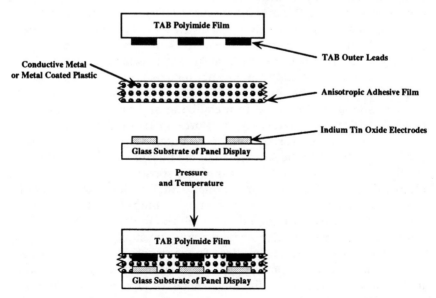

Figure 6-8. Principle of anisotropic adhesive connections.

personal organizers, and "smart" cards. Even for the high-end electronics, anisotropic adhesives have an advantage over solder connections in being processed at much lower temperatures, typically 110 to 150°C instead of the customary 280 to 320°C for solder alloys. These lower processing temperatures reduce stresses, which can cause substrates and dice to crack. Using anisotropic adhesives instead of solder also avoids the need for fluxes and the halogenated solvents required to remove flux residues. Further interest in adhesive connections arises from the Environmental Protection Agency's drive to eliminate lead solders and other lead-containing materials.

In spite of these benefits, considerable work is still needed to ensure the reliability of anisotropic adhesives, especially the long-term stability of the electrical connections (contact resistance) under temperature, bias, and humidity exposures. Anisotropic adhesives generally have higher electrical resistance than their metallurgical counterparts. The organic binder can impart resistances as high as 100 mΩ per bond. Leakage currents and metal migration between closely spaced connections are also concerns that require further investigation.

6.3.5 Overlay Batch Interconnections

A unique wireless interconnection approach has been developed by General Electric for its multichip modules. The process is called the *overlay process* or *high-density interconnect* (HDI) (see Chap. 3). After the dice are attached in a closely packed array on a support substrate, a polyimide film is laminated to the tops of the dice, and vias are formed at the pad sites of the dice by a programmed excimer laser. The surface is then metallized and photoetched, which fills in the vias and interconnects the dice from the top as a batch process.[27,28] The process, similar to that used in metallizing and interconnecting functions within an IC, produces very short signal paths and batch interconnections between ICs. A further advantage of the General Electric process derives from the separation of the thermal levels from the electrical levels. In most MCM processes, the substrate serves as both the electrical routing and the main thermal path, which is really a compromise and presents a problem in designing high-density modules. In the General Electric process, however, all the routing and interconnections are done over the dice while the support substrate conducts heat away from the bottom of the dice. The support substrate or frame is typically alumina, but for very high thermal dissipation, the substrate may be aluminum nitride, diamond, metal, or any other high-thermal-conductivity material.

6.3.6   Testable Ribbon Bonding

Testable ribbon bonding (TRB, a trademark of Hughes Aircraft) was developed and patented by Hughes Aircraft as a soft-tooled alternative to TAB.[29,30] The advantages of TRB over TAB include faster turnaround time from testing to packaging and the lower cost realized by eliminating the device-specific TAB tooling required for tape fabrication, inner-lead bonding, excising, and forming. Unlike TAB, bumped dice are not required, thus eliminating the many metal deposition, photoetching, and electroplating steps involved in bumping dice. Any off-the-shelf chip can be used. Testing and burning in of the dice prior to assembly, to ensure good dice, are performed by preassembling and testing the dice on a disposable carrier.

The TRB process involves ribbon-bonding the dice to a disposable carrier by using an automatic ribbon bonder having a custom chip-alignment stage. An outer vacuum chuck of the stage holds the carrier in place while the central pedestal with independent x, y, z, and θ motion aligns and holds the chip through an aperture in the carrier. The top plate of the vacuum chuck accommodates a wide range of chip sizes. The ribbon leads are formed and bonded in a single operation, and upon release of the vacuum, the chip is maintained in place by the leads. Three-mil-wide, 1-mil-thick gold ribbon is used to bond chip I/Os having 6-mil pitch. Finer, 2-mil-wide ribbon was used to demonstrate bonding of 4-mil-pitch dice. The test carrier has fine-pitch inner-lead bonding pads for bonding the gold ribbon to the chip and fan-out conductors for the outer pads, which are connected to standard test sockets. Carriers can accommodate a single chip or an array of chips.

Once the chips have been electrically tested and burned in on the carriers, the chips are removed by excising the leads using a YAG laser. The sequence of steps for TRB is shown in Fig. 6-9. Chips are assembled into the MCM with conventional die-attachment adhesives and processes. By using conventional pick-and-place equipment, epoxy adhesive is dispensed, the chip is removed from a Gel-Pack by a vacuum pickup, and its leads are aligned with the MCM bonding pads and then attached. After curing of the epoxy, the die is outer-lead-bonded to the substrate by an automatic thermosonic wire bonder equipped with a wedge tool.

As with TAB, the TRB chips may be bonded either face up or face down (flip TRB). Flip TRB provides higher packaging density and potentially better electrical performance. To prepare the chip for face-down assembly, it is bonded with its top surface coplanar with the carrier surface. The leads, after excising, form short arched beams. A comparison of chips assembled face up and face down is shown in Fig. 6-10.

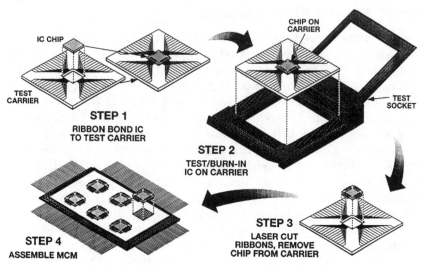

Figure 6-9. Testable ribbon bonding (TRB) process flow.

Minimum chip-to-chip spacings of 70 mils were achieved in the face-up configuration and 45 mils in the face-down configuration. By comparison, TAB requires chips to be spaced approximately 200 mils apart.

6.4 Environmental Protection and Packaging

There are two basic approaches to environmental protection of MCMs: hermetic sealing in separate-cavity packages or nonhermetic encapsulation with organic or inorganic materials. The type and extent of protection that must be provided depend on the reliability and life expectancy of the module. Very high-density, high-value MCMs, such as used in military, space, or medical electronics, should be vacuum-baked and then hermetically sealed in an inert ambient. A hermetic package protects expensive ICs from moisture, contaminants, and handling. In some microprocessors, e.g., the cost of the dice alone may be several thousand dollars. The interconnect substrate for such high-end modules may add an extra $1000. If devices fail after sealing, rework can be performed by delidding (machining off the metal lid), removing the malfunctioning die, cleaning the bonding site, and rebonding a new die. In many applications, up to three deliddings and rework are permitted. Even more serious are circuit failures that occur at the system level, as in manned spacecraft or implanted cardiac pacemakers. Hermetic sealing ensures long-term reliability in these critical applications.

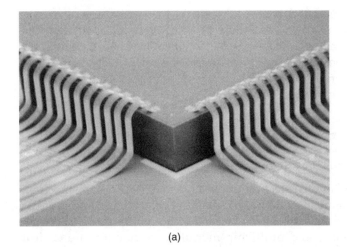

(a)

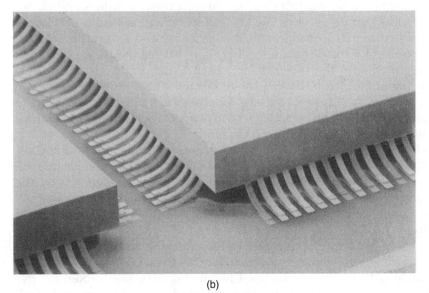

(b)

Figure 6-10. Testable ribbon bonded die, 228 I/Os on 6.6-mil pitch. (*a*) "Face-up" TRB (70-mils chip-to-chip spacing); (*b*) flip TRB (45-mils chip-to-chip spacing). (*Courtesy Hughes Aircraft Microelectronics Div.*)

For the low-end market applications such as commercial pagers, cellular phones, calculators, laptop computers, and radios, *chip-on-board* (COB) packaging is a low-cost widely used approach. Basic to the COB approach is the encapsulation or coating of the bare chips after assembly on the interconnect substrate. A separate package is not used, and an entire level of interconnections is thus avoided. The circuit may be completely encapsulated with high-purity, low-stress epoxy or silicone, or some of or all the dice may be selectively encapsulated, the latter being referred to as *glob-topped packaging*. Once encapsulated, for all practical purposes, the modules cannot be reworked. Hence COB packaging is generally used for electronics that are considered expendable, as are many consumer products.

6.4.1 Importance of Material Selection for Hybrid and Multichip Module Packages

The key elements in the selection of package materials are the mechanical integrity, chemical and thermal stability, electrical compatibility, and thermal conductivity. These factors, although applicable to both hybrid microcircuits and multichip modules, assume a more critical role for multichip modules because of the greater densities and higher electrical performances of MCMs. In fact, where packages previously provided merely an environmental enclosure and a form-fit structure important for handling and testing, with MCMs the package is often integrated with the substrate and participates in the electrical functioning of the circuit. Package materials for MCMs must therefore also be selected based on electrical properties such as low dielectric constant, low dissipation factor, low loss over a wide frequency range, low crosstalk, and controlled impedance.

In designing and fabricating packages for hybrid circuits or MCMs, diverse materials such as ceramics, metals, glass, and even plastics are often incorporated. The reliability of such packages depends on the compatibility and mechanical integrity of these materials of construction under environmental extremes. The mainstay material for hybrid circuit packages for the past 30 years has been Kovar, plated with nickel or gold and having glass-to-metal feedthroughs. Specially formulated glasses having a coefficient of thermal expansion (CTE) matching that of Kovar are used for the hermetic feedthroughs. After the hybrid is assembled in the metal package, it is vacuum-baked; then, in a dry nitrogen ambient, it is hermetically sealed with a gold- or nickel-plated Kovar lid. Although these materials and processes are quite mature, there is a limit on the size of the package and on the number of feedthroughs that can be fabricated. Generally metal packages having glass-to-metal feedthroughs are limited to sizes less than 1.5×1.5 in and to feedthrough pitches greater than 50 mils.

With the advent of high-density MCMs, packages 2×2 in and larger having more than 300 I/Os (in some cases approaching 1000 I/Os) are required. And I/O lead pitches of 25 mils and less, which could not be reliably fabricated with glass feedthroughs, are required. For such high-density MCM packages, both HTCC (high-temperature cofired ceramic) and LTCC (low-temperature cofired ceramic) having leads integral with the ceramic are used. In cofired ceramic, alumina containing various amounts of glass and binders are cast or extruded into "green" tape, then cut, drilled for vertical interconnections, screen-printed with conductor pastes, laminated, and fired. Both cofired ceramic processes are capable of integrating power, ground, and signal layers in a monolithic structure, thus integrating the substrate with the package. However, LTCC has an advantage in being able to use high-conductivity conductors such as gold, silver, or copper and in further integrating passive components such as resistors and capacitors. HTCC is restricted to the use of high-resistivity metal conductors, primarily tungsten. Because of these and other advantages, previously discussed in Chaps. 3 and 4, LTCC is emerging as the ceramic of choice for high-performance multichip modules.

6.4.2 Hermetic Packaging

In hermetic packaging, a metal or ceramic cavity package encloses the assembled circuit; then a metal lid is seam-welded, laser-welded, or soldered on. Ceramic lids may also be sealed directly to ceramic package cases with a low-melting-point glass. Hermeticity is defined not in the absolute sense of the word but in meeting a finite helium leak rate, as measured by a mass spectrometer according to procedures in Mil-Std-883, method 1014.

Metal Packages.   Both metal and ceramic packages have been extensively employed to handle, test, and seal off circuits from the environment, ever since the inception of ICs and hybrid circuits. Kovar metal packages having glass-to-metal feedthroughs are widely used to hermetically seal hybrid circuits having low I/O counts. As custom IC dice are being made denser and larger, they require a higher number of I/Os (some over 500) on very fine pitch (4 to 6 mils). In such cases, the classical Kovar metal packages have proved to be limited in the I/O density that can be reliably produced. The proximity of glass feedthroughs that can be constructed in a package is generally limited to 50-mil pitch or greater. As the number and density of glass feedthroughs increases, so does the probability of the glass fracturing and packages losing hermeticity. Packages may have feedthroughs and leads emanating radially from two sides or from all four sides or axially from the bottom of the package. If the pins emanate from the bottom, they may be configured along the perimeter or in an area pin grid array. Kovar packages may be either gold-plated or nickel-plated and sealed with similarly plated Kovar lids.

With the advent of MCMs, a need arose for large packages having a high density of leads or pins (200 to 1000 leads per package with a lead pitch of 25 mils or less). Glass-feedthrough metal packages could not satisfy these requirements. Only the cofired ceramic integral-lead packages combine the hermetic quality, structural integrity, and I/O densities required.

Ceramic Integral-Lead Packages. Ceramic (primarily alumina) packages having closely spaced integral leads can be produced either by the HTCC or LTCC process. Lead frames may be brazed to the cofired metallization pads on two sides or on all four sides of the package. The I/Os may also be designed to emanate from the bottom in either a perimeter or a pad array format. Pins may be brazed to the pads to produce a plug-in package. Ceramic packages can also be designed and produced with a very low height profile by creating recesses in the ceramic tape prior to firing. Dice can be mounted in the recesses so that the dice are flush with the top surface. LTCC packages having a total height of only 0.070 in. have been fabricated. This is to be compared with the more conventional packages whose heights range from 0.1 to 0.3 in. LTCC packages are gaining prominence among multichip module manufacturers because of the ease and rapidity with which packages alone and packages integrated with substrates can be designed and fabricated. Further benefits of LTCC include the ability to cofire with low-resistivity gold, silver, or copper conductors at relatively low temperatures; low insertion losses important for radio-frequency microwave circuits; and CTEs that can be tailored to closely match either silicon or gallium arsenide devices.

Although most ceramic packages in use today are produced from cofired alumina, advances have been made in designing and producing very high-density hermetic packages from aluminum nitride. Aluminum nitride (AlN) has 30 to 40 times the thermal conductivity of LTCC and a much closer CTE match to silicon. AlN packages 2×4 in having 368 leads on 30-mil pitch and 4×4 in having 612 leads on 25-mil pitch have been developed[32,33] (Fig. 6-11). These packages have Kovar or molybdenum seal rings brazed with silver or copper-silver to the post-metallized and nickel-plated cofired base. Lead frames are subsequently brazed to the outer I/O pads. Smaller packages designed with an array of pins brazed to cofired metal pads on the bottom of the package have also been produced.[34] Aluminum nitride packages must be cofired with tungsten or other refractory metal pastes because of the high sintering temperatures required (approximately 1900°C in a forming gas or nitrogen ambient). Aluminum nitride packages provide approximately a 30-fold increase in thermal conductivity over the LTCC alumina counterparts and have a CTE closely matching that of silicon.

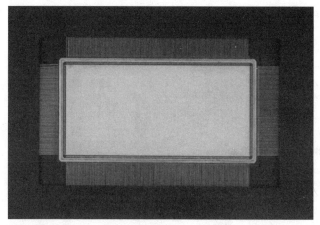

Figure 6-11. Aluminum nitride package 2 × 4 in having 368 leads on 30-mil pitch (*Courtesy Coors Electronic Package Co.*).

6.4.3 Nonhermetic Sealing: Chip-on-Board and Plastic Packaging

Nonhermetic sealing generally refers to a noncavity package in which the assembled module is protected by an organic polymer, an inorganic passivation coating, a glob-topped plastic encapsulant over selected dice, or an encapsulant or transfer molding compound covering the entire circuit. The key feature of COB packaging is the encapsulation or passivation of bare chips assembled on an interconnect substrate, thus eliminating the separate-cavity package (Fig. 6-12). COB packaging results in substantial cost savings and reductions in weight, volume, and height of electronic modules. Ultrathin wallet-size electronic calculators and other electronic memory circuits can be produced by recessing the dice in cavities of the substrate, further reducing the height of the package (Fig. 6-13). Although the interconnect substrate may be any one of the three main types (MCM-D, -C, or -L) generally COB packaging is associated with the laminate type (MCM-L). Chips assembled on FR-4 epoxy-glass laminates, then glob-topped or encapsulated with low-stress, high-purity plastic resins, are very popular for the lowest-cost consumer electronics.[36] The glob-topped encapsulants may be high-purity epoxies, silicones, epoxy siloxanes, or polyimide siloxanes.

Because of the potentially higher cost savings and greater densities that can be realized from plastic packaging compared to hermetic packaging, for many years the military has been interested in evaluating and ascertaining the reliability of plastic-packaged modules for

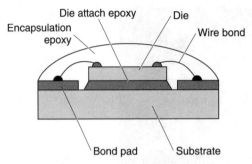

Figure 6-12. COB packaging eliminates one level of interconnections.

Figure 6-13. Chip in recessed cavity reduces COB height.

weapon and space systems. Plastic packaging was originally rejected as unreliable in protecting single-chip or multichip circuits from long-term moisture and temperature extremes, but advances in chip passivation and in high-purity, stress-free encapsulants have prompted the Air Force to reinvestigate this area under the so-called Reliability without Hermeticity (RwoH) Program.[37] For many years, single-chip IC packages have been transfer-molded or encapsulated with epoxies or silicones and have performed reliably in commercial products. However, both military and space environments and government specifications require screen tests that are much more severe than those used for consumer products. In addition, multichip modules contain many closely packed expensive dice interconnected by thousands of wire bonds, increasing the risk of failure when they are encapsulated and stress-tested. Reworking encapsulated dice is difficult. Malfunctioning dice cannot be removed without destroying other good dice or damaging the interconnect substrate. To use COB packaging successfully for multichip modules, extra attention must be paid to materials selection and process controls so that yields are high and rework is unnecessary. These are some examples:

- Ensure that the surface is clean and dry prior to encapsulation since small amounts of ionic contaminants become sites for moisture condensation and entrapment, destroying the adhesion and integrity of the encapsulant.

- Ensure the purity of the encapsulant since impurities such as sodium or potassium ions cause leakage currents, metal migration, and reduction in electrical insulation resistance, especially under moist conditions. Chloride ions are also deleterious; even in low parts per million, they react chemically and catalytically with aluminum, eventually etching it away.

- Control the degree of cure of the encapsulant since undercuring degrades its electrical and physical properties while overcuring results in brittleness and increased stresses.

- Select encapsulants whose coefficients of thermal expansion and moduli of elasticity are compatible with the MCM materials. Large differences in thermal expansion among materials can induce stresses and breakage of wires, wire bonds, silicon dice, and thin-film conductor metallization.[38] The rate of expansion of plastics increases sharply above the glass transition temperature T_g. Hence it is desirable to use encapsulants that have a T_g higher than the highest temperature that the module will be exposed to during processing, testing, or actual use.

Further limiting the use of plastic encapsulants in high-reliability applications is the lack of a rapid nondestructive test to ensure reliability. Unlike hermetically sealed cavity packages, the helium fine-leak and gross-leak tests cannot be used, nor can the moisture and residual gas analysis (RGA). Nondestructive, cost-effective tests for encapsulated modules do not exist yet. However, evaluation and qualification of passivation coatings, encapsulants, and cleaning and curing processes using very sensitive triple-track resistor chips have found them to be effective, at least for initial selection and qualification of a resin and process. One widely studied test chip is the ATCO1 chip, designed and developed by Sandia National Laboratories.[39] The chip consists of a variety of structures intended for corrosion testing under humidity-bias conditions. Included are several triple-track resistors, each consisting of thin-film interdigitated aluminum lines of various widths and spacings, ladder structures, straight-line structures, and van der

Pauw sheet resistance structures. The chip is approximately 0.250 in square and has a minimum aluminum feature size of 1.25 μm. The chips, encapsulated or protected by various coatings, are exposed to accelerated tests, and attempts are made to correlate changes in resistor values, interline insulation resistance, corrosion, or total failures (opens and short circuits) with life expectancy. Besides temperature cycling, humidity and temperature cycling, salt spray exposure, and bias testing under 85/85 conditions (85 percent relative humidity and 85°C),[40] the rather severe HAST (highly accelerated stress test) is widely used. HAST is somewhat like a pressure cooker test except that the parts are electrically biased.[41] As the test proceeds, the time to failure of each coated chip is measured; the better coatings prolong the time to failure. The failure mechanism is often attributed to corrosion or metal migration due to the presence of mobile ions or other residues within the encapsulant or on the surface at the interface. Table 6-8 lists some of the advantages and limitations of COB packaging.

Table 6-8. Advantages and Limitations of COB Packaging

Advantages	Limitations
Low material cost; no separate package	Requires very clean, dry surface preparation and perhaps an adhesion promoter
Low tooling cost	Not easily reworked; generally a throw-away
Wide choice of substrates from low-cost plastics to high-thermal-conductivity Si or AlN	Limited testing
Usable with any die	Generally limited to 8 dice or 1000 wire bonds
Reduced volume, weight, and height compared to surface mount technology (SMT) or hermetic cavity packages	Limited power dissipation if plastic substrates used
Accommodates high-I/O-count dice	
Eliminates one level of interconnections	Stress effects and compatibility with ICs and surfaces must be established
	Long-term stability of plastic encapsulant under operating/environmental conditions must be verified.

References

1. B. L. Weigand and S. V. Caruso, "Development of a Qualification Standard for Adhesives Used in Hybrid Microcircuits," *Proceedings ISHM,* 1983, Philadelphia, PA.

2. J. J. Licari, K. L. Perkins, and S. V. Caruso, "Evaluation of Electrically Insulative Adhesives for Use in Hybrid Microcircuit Fabrication," *IEEE Transactions of Parts, Hybrids, and Packaging,* **PHP-9,** 1973.

3. C. Mitchell and H. Berg, "Use of Conductive Epoxies for Die Attach," *Proceedings ISHM,* 1976, Vancouver, B.C., Canada.

4. J. C. Bolger and C. T. Mooney, "Volatile Organic and Extractable Ionic Contaminants in Epoxy Die Attach Adhesives," *Proceedings NEPCON,* Anaheim, CA, 1983.

5. P. J. Planting, "An Approach for Evaluating Epoxy Adhesives for Use in Microelectronic Assembly," *IEEE Transactions of Parts, Hybrids, and Packaging,* **PHP-11,** 1975.

6. D. M. Shenfield and C. M. Zyetz, "An Investigation of the Effect of 150°C Storage on the Electrical Resistance of Conductive Adhesive Attachment of 2N222A Transistors," *Proceedings ISHM,* Philadelphia, PA, 1983.

7. J. J. Licari, B. L. Weigand, and C. A. Soykin, *Development of a Qualification Standard for Adhesives Used in Hybrid Microcircuits,* NASA CR-161978, Marshal Space Flight Center, AL, 1981.

8. K. Chung et al., "MCM Die Attachment Using Low-Stress, Thermally Conductive Epoxies," *Surface Mount Technology,* May 1991.

9. K. Chung et al., "Tack-free Flexible Film Adhesives," *Hybrid Circuit Technology,* May 1990.

10. J. J. Licari and L. Enlow, *Hybrid Microcircuit Technology Handbook,* chap. 7, "Assembly Processes," Noyes, Park Ridge, NJ, 1988.

11. G. Harman, *Wire Bond Reliability and Yield,* ISHM Monograph, 1989.

12. C. Harper, ed., *Electronic Packaging and Interconnection Handbook,* McGraw-Hill, New York, 1991, chaps. 6 and 7.

13. L. Day, "Wire Bondability," *Advanced Packaging,* winter issue, 1993, IHS Publishing, Libertyville, IL.

14. M. L. White, "The Removal of Die Bond Epoxy Bleed Material by Oxygen Plasma," *Proceedings of the IEEE 32d Electronic Components Conference,* 1982.

15. C. Harper, ed., *Electronic Packaging and Interconnection Handbook,* McGraw-Hill, New York, 1991, chap. 10.

16. P. Hoffman, "TAB Implementation and Trends," *Solid State Technology,* **31,** June 1988.

17. P. Burggraaf, "TAB for High I/O and High Speed," *Semiconductor International,* June 1988.

18. R. Iscoff, "TAB: Forever a Niche Technology?" *Semiconductor International,* January 1993.

19. N. Koopman, "Solder Joining Technology," *Proceedings Material Research Society Symposium,* **154,** 1989.

20. P. A. Totta, "Flip Chip Interconnects in Advanced Products," *Proceedings IEPS,* San Diego, CA, September 1991.

21. *R. R. Tummala and J. Knickerbocker, "Advanced Cofired Multichip Technology at IBM," Proceedings IEPS, San Diego, CA, September 1991.*

22. E. E. Davidson, *IBM Journal of Research and Development,* **36**(5): 881 ff., September 1992.

23. K. Hatada and H. Fujimoto, "A New LSI Bonding Technology, Micron Bump Bonding Technology," *39th Electronics Components Conference,* 1989.

24. K. Hatada, Method of connecting a semiconductor device to a wiring board, U.S. Patent 4749120, June 7, 1988.

25. Y. Atsumi, Connecting structure for an electrical part, U.S. Patent 5001302, March 19, 1991.

26. I. Tsukagoshi et al., Composition for circuit connection, method for connection using the same, and connected structure of semiconductor chips, U.S. Patent 5001542, March 19, 1991.

27. C. W. Eichelberger, R. J. Wojnarowski, R. O. Carlson, and L. M. Levinson, "HDI Interconnects for Electronic Packaging," *SPIE Symposium: Innovative Science and Technology,* paper 877-15, January 1988.

28. C. A. Neugebauer, R. A. Fillian, R. O. Carlson, and T. R. Haller, "High Performance Interconnections between VLSI Chips," *Solid State Technology,* June 1988.

29. H. Congleton and R. E. Root, "TRB—A Soft-Tooled Approach to Device Pretest And Interconnection," *Proceedings Second International Conference Multichip Modules,* Denver, CO, April 1993.

30. H. Congleton et al., Testable ribbon bonding method and wedge bonding tool for microcircuit device fabrication, U.S. Patent 5007576, April 16, 1991.

31. R. Pond and W. Vitriol, "Custom Packaging in a Thick Film House Using Low Temperature Cofired Multilayer Ceramic Technology," *Proceedings ISHM, Dallas, September 1984.*

32. J. Enloe, J. Lau, and L. Dolhert, "Design Concepts in AlN Packaging," *Proceedings ISHM,* Chicago, IL, 1991.

33. L. Dolhert, A. Kovacs, W. Minehan, and E. Luh, "Performance and Reliability of Cofired AlN for Multichip Applications," *Proceedings NEPCON West,* Anaheim, CA, February 1992.

34. N. Iwase et al., "Aluminum Nitride Multilayer Pin Grid Array Package," *Proceedings of the 37th Electronics Components Conference,* 1987.

35. N. Iwase, et al., "Aluminum Nitride Multilayer Package Technologies for High Speed ECL and High Pin Count BiCMOS Devices," *Proceedings ISHM,* San Francisco, CA, 1992.

36. E. H. Newcombe, "COB Increases SMT Density," *Electronic Packaging and Production*, December 1992.

37. U.S. Air Force contract no. F33615-91-C-5710, "Reliability without Hermeticity," managed by MCC and codirected by MCC and Lehigh University.

38. J. J. Licari and L. Hughes, *Handbook of Polymer Coatings for Electronics*, Noyes, Park Ridge, NJ, 1990.

39. J. Sweet, D. Peterson, M. Tuck, and D. Renninger, "Assembly Test Chip Version 01, Description and Users Manual," Sandia National Laboratories Rep. SAND90-0755, September 1990.

40. JEDEC Standard Test Method A 101-A, "Steady-State Temperature Humidity Bias Life Test," Electronic Industries Association, Washington, D.C., July 1988.

41. JEDEC Standard Test Method A 110, "Highly Accelerated Temperature and Humidity Stress Test (HAST)," Electronic Industries Association, Washington, D.C., July 1988.

7

Three-Dimensional Packaging

Hugh L. Garvin

7.1 Drivers for Three-Dimensional Packaging

Microelectronic systems are subject to ever-increasing demands to operate at higher speeds, occupy less space, consume less power, be reliable in hostile environments, and be economically priced. Through the years, vacuum-tube electronics gave way to solid-state devices assembled on printed-wiring boards; then, in time, more of the circuitry was combined into single-chip integrated circuits (ICs). These were then further integrated into hybrid circuits and more recently into multichip modules (MCMs). The technology has advanced at an amazing pace—two-dimensional (2-D) MCM packaging satisfies most current requirements for mainframe computer business data processing, military signal processing systems, and personal computers. Modern packaging technology has advanced to the point of nearly achieving the theoretical maximum density obtainable in 2-D packaging. Any future improvements will require further integration of circuitry into the third dimension. In the ultimate integration, we can expect semiconductor manufacturing techniques—particularly the advances in semiconductor-on-insulator (SOI) and molecular beam epitaxy (MBE) technology—to be used to fabricate circuitry and active devices in a monolithic 3-D structure with the total circuitry constituting a cube of active components. Then only dielectric layers will be needed to provide the electrical isolation between the active devices and their interconnections. However, before this ambitious technology is perfected, we can expect significant advances in three-dimensional (3-D) packaging of IC chips and wafer-level integrated active circuitry.

As with multichip modules in general, the three main drivers for 3-D packaging are higher volumetric density, higher system performance, higher numbers of input/output (I/O) contacts, and lower operating power. Higher operating speed and lower power consumption can be realized because of the shorter signal (interconnect) paths between circuits that result from 3-D packaging. Figure 7-1 compares multichip modules in 2-D and 3-D configurations and the resulting volume-reduction benefits derived from the 3-D version. Even when IC chips, approximately 10 × 10 mm and 0.5 mm thick, are closely packed on an edge-connected circuit board, the highest density of active circuitry will be 1 to 2 cm^2/cm^3, and the average interconnect running to the edges and back into an IC on another board may be 10 cm long. On the other hand, if the boards (or carriers) are stacked tightly together with the vertical interconnections made internal to the assembly, densities of 10 to 15 cm^2/cm^3 can be realized by using standard (unthinned) IC chips and carriers. By thinning the chips and carriers, the density may be further increased by another order of magnitude, and the average connection length can be reduced to 1 cm

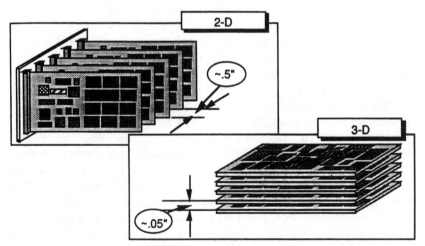

Figure 7-1. Schematic comparison of 2-D and 3-D implementations. The major direct benefit of 3-D packaging is the 10-fold reduction in system volume by decreasing module spacings from 0.5 in to 0.05 in; indirect benefits are reduced weight, reduced power, and higher clock speeds.

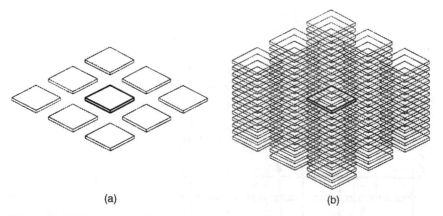

(a)   (b)

Figure 7-2. Two- versus three-dimensional architectures. Comparison of number of chips within 2-D nearest-neighbor distance for 2.5-mm z and 1.5-cm x and y pitches. (*a*) 2-D MCM nearest-neighbor chips (8); (*b*) 3-D MCM chips in same distance (116).

or less. This point is illustrated dramatically in Fig. 7-2, where the nearest-neighbor 2-D arrangement shows that eight chips can be connected to a central chip by a lead length that is equivalent to a chip dimension. When a 3-D assembly is made with the carriers closely stacked (about seven carriers in the height of one chip dimension) and with internal vertical interconnections, as many as 116 chips can be contacted by the same interconnect distance. Consequently, the delay in transferring an electric signal from one chip to another can be as short as 1 ns, thus permitting very fast circuit operations. The number and complexity of operations that can be performed in a given small volume can be very high.

This chapter presents an overview of the work in progress to develop and utilize microelectronics packaged in 3-D configurations. The presentation has been organized by circuit functions rather than in chronological order of development. The applications of 3-D packaging in digital signal processing and controls, memories, and radio-frequency (microwave or millimeter wave) circuitry are discussed. There have been several surveys of 3-D packaging reported in the literature and in government contract final reports.[1–4] Readers are referred to these reports for supplementary discussions of the technology.

7.2 Basic Categories

Before we discuss details of the various approaches to high-density packaging, it is useful to examine the cross sections of two basic cate-

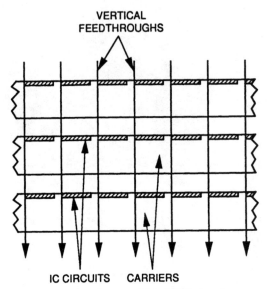

**VERTICAL
FEEDTHROUGHS**

IC CIRCUITS   CARRIERS

Figure 7-3. A 3-D assembly of WSI circuits; example of category I.

gories of 3-D interconnection features. One category (Fig. 7-3) consists of very high-density, 3-D packaging of wafer-scale integrated circuits. In this approach the individual semiconductor layers are processed with many feedthroughs before the active circuitry is fabricated adjacent to these connection lines. Each feedthrough is formed with a z-axis connection to permit many layer-to-layer interconnections which are distributed throughout the semiconductor. This configuration is referred to as category I. Clearly the multiplicity of parallel connections permits simultaneous data processing operations to be made in many paths, for example, 16,000 paths in a 128 × 128 element 3-D signal processor. The wafer levels are designed to perform predetermined functions, and between stacked levels are flexible vertical interconnections. These permit the wafer levels to be changed, allowing operational changes in design. The main feature of the category I modules is that the location (array) of vertical interconnections is fixed while the circuits are designed and built around them.

A 3-D MCM in which the IC chips are "off the shelf," selected from a variety of sizes and shapes, has a cross section shown in Fig. 7-4 and is referred to as category II. In this approach the vertical interconnections must be moved to go between the chips, since standard IC chips do not have feedthroughs to their back surfaces. By judicious arrangement, all the small chips may be arranged together on a single layer with many vertical interconnects, while another layer may carry larger chips and the vertical connections may be less dense. Thus the density

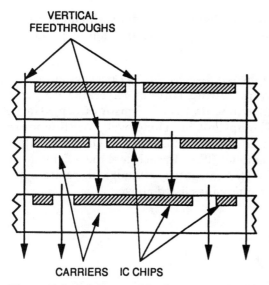

Figure 7-4. A 3-D assembly of various IC chip types; example of Category II.

achieved and the degree of 3-D interconnection depend on the similarity of the chips assembled and the flexibility in design or system function that is required. Most industrial 3-D packaging approaches are modifications of this second category.

Regardless of the category, several technical guidelines should be followed:

1. Circuit-to-circuit or chip-to-chip interconnections for 3-D packaging require the same expertise as for 2-D packaging when the circuit functions and the packaging densities increase.

2. Space-consuming wires, ribbons, or TABs must frequently be shaped to conform to the circuitry for efficient vertical stacking.

3. Vertical (or z-axis) interconnections must be made through or between the active semiconductor substrates, through the carrier substrates, and between adjacent stacked carriers.

4. Fast circuits that dissipate relatively large amounts of power require that thermal conduction paths be built into the package so that chips and circuitry are maintained at low operating temperatures.

5. Testing should be performed at various stages of assembly because of the high financial investment in circuitry and integration processes. Where possible, fault tolerance should be designed into the circuitry.

6. Low-cost, fast-turnaround, and benign procedures should be developed or made available to rework modules that fail inspection or testing.

In the following paragraphs we discuss the 3-D packaging designs of several manufacturers including Hughes Aircraft Company, General Electric, Texas Instruments, Irvine Sensors Corp., Harris Corp., Martin Marietta, Thompson-CSF, Honeywell, Westinghouse, Rockwell International, InterChip Systems, Norton Diamond Films, and Integrated System Assemblies (ISA). In each case the 3-D interconnections fall into one of the two categories, although there are many ramifications in meeting the guidelines listed above.

7.3 Digital Signal Processing Circuitry

7.3.1 Wafer-Scale Integrated 3-D Packaging

In 1982, Hughes Research Laboratories began developing a high-performance 3-D computer[5] based on the category I, 3-D integration approach. The concept (Fig. 7-5) involves stacking and assembling silicon wafers, each containing arrays of processing circuitry. Two underlying technologies were essential to this approach: thermomigrated feedthroughs, which provide the highly conductive, electrically isolated signal paths through the silicon wafers, and microbridge interconnects, which carry signals between the wafers.

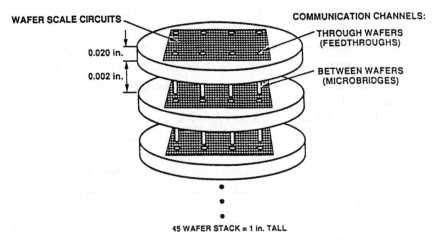

Figure 7-5. The Hughes Aircraft Company's concept of 3-D computer.

Thermomigrated Feedthroughs. The feedthroughs are produced by the migration of aluminum droplets through the silicon wafers at an elevated temperature, as illustrated in Fig. 7-6. An array of aluminum dots are evaporated on the surface of an n-type silicon wafer, which is then heated to approximately 1000°C. While this temperature is maintained, a thermal gradient of roughly 150°C/cm is applied to the wafer, so that the surface bearing the droplets is the cooler of the two. The thermal gradient produces a concentration gradient of dissolved silicon in the aluminum droplets, which in turn drives a diffusive flux of silicon from the forward face of the droplet to the back. Transported in this manner, the silicon recrystallizes at the rear of the droplets while more material is dissolved from the bulk at the forward ends. The trails of regrown silicon left by droplets incorporate aluminum, which is a p-type dopant. Thus, highly conductive p-type vias are formed which penetrate the wafer. These vias are junction isolated from each other by the wafer substrate. A

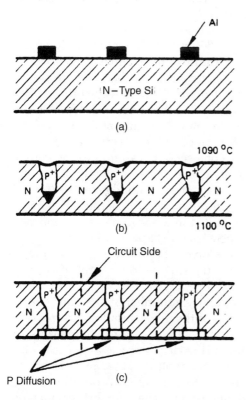

Figure 7-6. Thermomigration of aluminum dots. (*a*) Aluminum dots are deposited on silicon wafer; (*b*) At high temperature in the presence of a thermal gradient, the aluminum migrates through the wafer; (*c*) Wafer surfaces are polished after thermomigration, and a catcher diffusion is added.

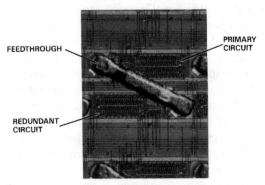

FEEDTHROUGH

PRIMARY
CIRCUIT

REDUNDANT
CIRCUIT

Figure 7-7. A 3-D circuit with a microbridge
interconnect at one node of a 32 × 32 array.

photomicrograph of a 32 × 32 array circuit illustrating the placement of a
feedthrough is shown in Fig. 7-7.

Microbridge Interconnections.   Besides the formation of thermomigrated feedthroughs, a second key feature of the Hughes approach is the
use of area arrays of microbridge interconnections between the wafers.
Air-gap microbridges were first developed by Lepselter at Bell
Laboratories to increase the density of 2-D interconnections in thin-film
circuits.[6] Arrays of microbridges are photolithographically batch-fabricated on planar substrates by first depositing a sacrificial "profiling
layer." This profiling layer consists of a set of thick lumps of photoresist
(about 2 mils tall) with gently sloping sidewalls—one lump at each site
that is to be occupied by a microbridge. The microbridge metals (typically copper or copper alloys with an outer surface coating of solder or
gold) are sequentially sputter-deposited in a vacuum system, and photolithography is used to define their lateral shape. In the final step, the
underlying photoresist lump is dissolved. Freestanding metal bridges
are thus formed which loop above the surface to a height of approximately 2 to 3 mils.

The microbridge concept incorporates a number of attributes important to the design and construction of the 3-D computer. For example,
essential to any technology chosen is the ability to compensate for
warping of the wafer that occurs during circuit processing. Also, when
a large number of interconnections are employed, they must have very
low associated parasitic impedances to minimize propagation delays
and power dissipation in the signal drivers. The microbridge intercon-

nects satisfy both requirements. Microbridges have been shown to be elastically compressible over their full height. By standing roughly 2 mils away from the surface of the wafer, microbridges are able to deform up to that amount, thereby compensating for any anticipated wafer distortion. This separation from the surface and the associated air gap also ensure a very low parasitic capacitance. Finally, it has been shown that the resistance of a mated pair of interconnects is less than 2 Ω. When the levels of the computer are stacked together, spacers (about 2 mils thick) which are inserted limit the degree to which the bridges are deformed. The microbridge connections have been operated successfully as assembled. However, for a permanent assembly, the microbridges may be solder-coated and fused together by vapor-phase heating to cause the solder to flow again.

The interconnect logic in the wafer-scale integration (WSI) 3-D assembly differs from that of normal edge-connected 2-D boards. Rather than a distribution of circuit types as shown in Fig. 7-8, the 3-D computer is constructed of levels of the same functional type, as shown in Fig. 7-9. Thus, a processing sequence can occur along a vertical column, in fact, along many thousands of columns at the same time, rather than horizontally from one board to another. In the early demonstrations of a 32 × 32 array (5 levels high), the computer performed 10^{10} operations per second in a small, handheld unit, shown in Fig. 7-10. Advanced versions of the 3-D computer having a 128 × 128 array (17 levels high) are now going into operation, and a processing capability as high as 10^{12} operations per second is expected to be realized.

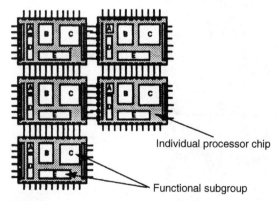

Individual processor chip

Functional subgroup

Figure 7-8. Conventional parallel processor with conventional arrangement of chip functions in a 2-D MCM.

Functional subgroup

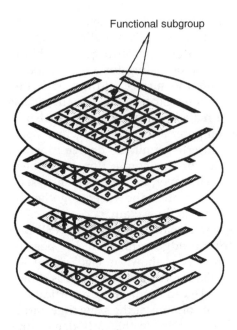

Figure 7-9. 3-D parallel processor with arrangement of IC functions for 3-D packaging.

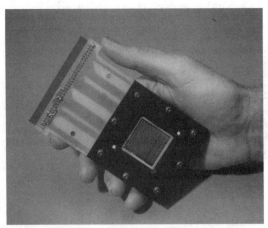

Figure 7-10. The feasibility demonstration 3-D computer: a 32×32 array of processors distributed over five vertically stacked wafers. There are 1200 communication channels through each wafer and between each pair of wafers.

7.3.2 Multichip Module Packaging

The reasons factors for using 3-D packaging are similar to those that have stimulated wide interest in 2-D WSI and MCM technologies. These technologies promise substantial reductions in size and weight, some reduction in power dissipation, and significant improvements in electrical performance. While these improvements are desirable, there is always concern about the cost and reliability of any new technology. For the military in particular, it is important that any new packaging technology maintain its integrity and performance over long mission lifetimes and in a variety of severe operating environments. In commercial applications, cost and performance are the primary considerations. Beyond these factors, it is important to consider future trends in semiconductor device capabilities and packaging techniques that could influence the attractiveness of a particular packaging approach. Thus, to be widely applicable, the 3-D MCM approach must support

- A variety and mixture of semiconductor chip technologies
- Adequate heat removal to the next level of packaging
- High electrical performance at speeds well above 100 MHz
- Extendability to an arbitrary number of layers
- High-pin-count capability to the next level of packaging
- Survivability (for military systems)
- Repairability
- Reliability

The highly regular I/O structures of memory chips have led to high volumetric efficiencies in their packaging. However, signal and data processing circuits consisting of high-performance IC chips are much more complex and difficult to integrate. If commercially available, off-the-shelf chips are to be used, their sizes will vary, the contact positions will vary with the IC layout, and few contacts can be bussed together in common. Investigators at Kirkland Air Force Base, New Mexico, have compiled a series of relevant papers on advanced microelectronics packaging. In one article[3] the needs of the Strategic Defense Initiative Office (SDIO) are described for space-based systems requiring extraordinary computational and/or data storage capabilities. Constructing electronic subsystems that support these requirements and still meet size, weight, and power constraints is particularly challenging, if not impossible, with conventional packaging approaches. While the emerging MCM technologies are a step in the right direc-

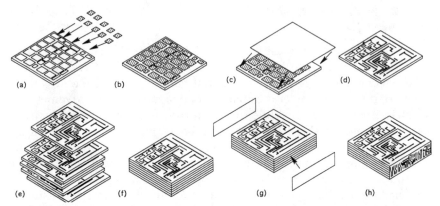

Figure 7-11. Fabrication sequence of 3-D high-density interconnect (HDI) module. (*a*) Component placement in 2-D substrate; (*b*) substrate ready for processing; (*c*) creation of patterned overlay; (*d*) completed 2-D HDI substrate; (*e*) placement of 2-D substrates into a 3-D assembly; (*f*) assembled stack; (*g*) formation of patterned overlay on edges of 3-D HDI assembly; (*h*) completed 3-D module ready for assembly in hermetic package.

tion, it is clear that additional density will be gained by exploiting the third dimension in packaging.

General Electric's approach to 3-D MCM packaging, partially funded by Phillips Laboratory and the Advanced Research Projects Agency (ARPA), is a direct extension of its 2-D high-density interconnect (HDI) process.* In the HDI process (see Chap. 3), the chips are attached inside recesses of a substrate (Fig. 7-11) then interconnected from the tops of the chips by a patterned overlay consisting of multiple layers of a polyimide dielectric and copper conductors. Several complete HDI substrates can be processed into a 3-D assembly through stacking and lamination (Figs. 7-11*a* through *e*). Fabrication of a 3-D HDI module is completed through formation of additional patterned overlays on the edges of the substrate to interconnect conductors on the edges of individual modules. A completed assembly of the RISC four-layer memory stack is shown in Fig. 7-12. The HDI module is a high-density packaging approach that differs from the die-stacking approaches that have been used for memory packaging, but still represents an edge interconnection of multichip modules rather than a true 3-D approach as represented by categories I and II. General Electric has reported using laser-drilled feedthroughs in the carrier substrates to permit the direct internal interconnection of modules for 3-D operation. Meanwhile, demonstrations of the densities and perfor-

*Rights to the HDI and MHDI processes now reside with Martin Marietta resulting from the acquisition of General Electric Aerospace in 1993.

Figure 7-12. The RISC four-layer memory stack. First demonstration of General Electric's 3-D HDI technology. Stack height is minimized by performing HDI processing on the side. This unique "on the edge" interconnect yields the lowest "through the stack" thermal impedance and highest speed.

mance achieved by the HDI approach have been made, and samples of the approach will be tested in space flights as early as 1994.

Many firms which are systems integrators of military weapons are exploring options for 3-D packaging of multichip modules. Honeywell and Texas Instruments in their approaches to the Aladdin Program— seeking to develop a "computer in a can"—have developed a cylindrical arrangement similar to that shown in Fig. 7-13. In the Honeywell approach, aluminum nitride carriers were used to integrate the IC chips in a 2-D configuration on both sides with feedthroughs interconnecting the back-to-back circuitry. The TI Aladdin structure is a single-sided MCM which uses a silicon substrate with a thin-film multilayer interconnect. In either case, the level-to-level interconnects were formed on the periphery of the carriers.

The signal processing packaging design (SPPD) program required that a several hundred MFLOP processor be packaged in a 75-g package, no larger than 250 cm^3, and that it withstand a 100°C temperature rise, adiabatically, over a long flight. The Rockwell International group[7] used the design approach shown in Fig. 7-14. The processor ICs were mounted on AlN carriers that were stacked and interconnected by flexible cable at the edges. This SPPD program achieved a maximum density of 74 g in 65 cm^3.

Hughes Research Laboratory's approach to the computer-in-a-can challenge combined the WSI 3-D computer architecture with layers of integrated 3-D MCM carriers to process input signals into the comput-

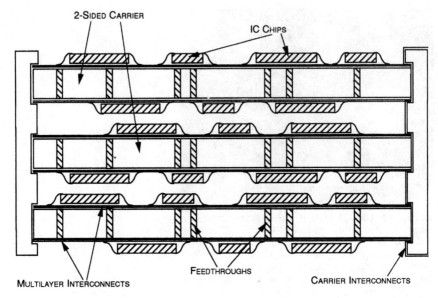

Figure 7-13. Honeywell's approach to 3-D packaging of a "computer in a can" for the Aladdin Program.

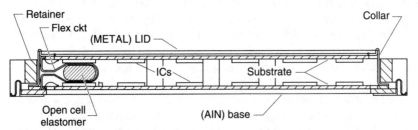

Figure 7-14. SPPD package cross section. (*Courtesy of Rockwell International.*)

er and condition the output drive signals. These levels, shown in Fig. 7-15, consist of AlN carriers with multilayer polyimide dielectric and metal interconnect traces. The chips were attached in recessed wells on the opposite side and interconnected by metal traces patterned on a vacuum-laminated Kapton overlayer film, a process similar to the General Electric/Martin Marietta HDI process. Vertical holes were laser-drilled and filled with metal through the AlN carriers. Finally, microspring bridges were used to carry signals from one level to another in the internal spaces, as depicted in category II, Fig. 7-3.

Harris Corp.[8] approached the goal of increased packaging density, particularly for mass-memory products for aerospace applications, by using low-temperature cofired ceramics (LTCCs) shaped into "carrier

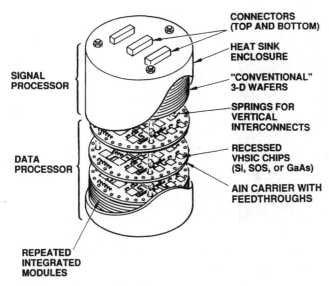

Figure 7-15. Assembly of 3-D computer.

tubs." The "green" ceramic tape is prepared with interconnect traces and recessed cavities to receive one or two memory chips or a field-programmable gate array control circuit. The interconnect traces are led to perforated holes that are later cut to form the castellation contacts around the edge of the tubs. After the ceramic is fired and the chips are integrated, the tubs are stacked together and bonded by sealing glass. The vertical interconnections are made of conductive epoxy in the castellation recesses around the edges. The blind castellations for 3-D module interconnection and hermetically sealed LTCC packages provide a useful combination of technologies to exploit for producing 3-D multichip modules. The density improvements provided by 3-D modules, particularly memory, are important in meeting aerospace and avionic size, weight, and power reduction goals.

Integrated Systems Assemblies (ISA) is investigating techniques for 2-D interconnection of IC chips onto carriers[9] by filling the spaces between chips with epoxy (see Fig. 7-16). Laser-drilled and -filled feedthroughs are made through the epoxy spacers; then the carrier substrates are ground away to make room for integration to another level of ICs so that, in fact, a "carrierless" assembly is produced.

InterChip Systems has described several 3-D MCM applications[10,11] among which are a 256-Mbyte main memory for a small high-performance computer board, a four-chip guidance package for an artillery shell, and a three-chip set for a portable antitheft device. In all three cases the Thompson-CSF process is used, wherein chips are aligned

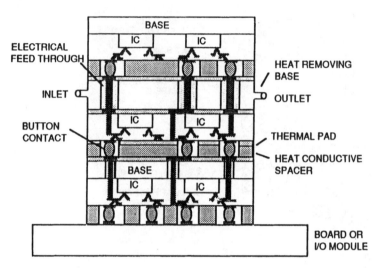

Figure 7-16. A 3-D stacked advanced MCM by ISA.

and molded together into rectangular blocks, then interconnected with traces along the side of the 3-D MCM. The Thompson-CSF process can be used with either bare or packaged dice. Prior to 3-D stacking, the bare dice are tested on carriers that are also used for alignment. The packaged dice are assumed to be known good dice based on full testing by the IC vendors.

In the memory package, 140 memory chips were bonded onto 5-mil-thick copper carriers that were stacked with spacings to permit air cooling (less than 200 linear ft/min) which limits the rise in junction temperature to less than 20°C.

The second 3-D MCM application was designed to add intelligence to munitions such as 155-mm howitzer shells, submunitions, and small rockets. The electronic components are required to function (albeit briefly) after the shells impact. Needless to say, the packaging requirements for this application are very severe. The electronic components must be able to withstand high acceleration, both positive and negative, up to 100,000 g while the electrical contacts must be maintained at all times. In addition, the package must be small, easy to install, and easy to replace. Finally, because of the high production quantities required, small artillery shells must be produced by low-cost manufacturing processes.

The 3-D MCM was designed as a cylinder that fits inside a cavity. One side of the cylinder was flattened to provide a surface for electrical connections produced by pressure contacts. The shape was therefore that of a typical soup can with one flat side. The conventional rectangular molding process was modified to create the cylindrical shape. To eliminate discrete packages, the bare dice were bonded to carriers and stacked. The carriers provided metallization that fanned out the die connections to test pads and also offered the capability of burn-in testing. The test pads were probed to test the dice at full speed prior to assembly in the 3-D configuration.

The assembled modules, chips, and their wire bonds were then embedded and made rigid by an epoxy resin, strengthening the module against the intense forces experienced in launching an artillery shell. Mechanically weak components such as printed-circuit boards, packages, and unsupported wire bonds are eliminated by using this 3-D approach. The custom-molded 3-D cylinders were designed to fit snugly into precision-machined cavities while external electrical connections were made through pressure contacts between the flat sides of the 3-D modules and the connectors mounted in the machined cavities.

A third 3-D module designed by InterChip Systems adds a security function to a variety of products such as cars, car radios, televisions, and computers. The modules, when activated remotely, will deactivate the circuitry in stolen items and report their locations to aid the retrieval of stolen property. The module design contains three sections, each to be implemented in a custom integrated circuit: a communications function (essentially a stripped-down pager or cellular phone), a security identification unit, and a unit location function. The design specifies 3-D packaging for the three-chip module. By implementing chip-to-chip connections within the module along the sides of the 3-D MCM, module I/O connections to the system are minimized.

The chips, as bare dice, are aligned in the 3-D MCM parallel with the printed-circuit boards ("pancake style"). The total thickness of the 3-D MCM is approximately 120 mils. The 3-D construction, which again encapsulates all components in epoxy, is inherently secure. Shielding is provided by metallizing all sides of the module and grounding the large areas.

In 1992, the Naval Command, Control, and Ocean Surveillance Center (NCCOSC) awarded a 3-D interconnect structures program to Honeywell and Coors Electronic Package Company (CEPC). The objective of the program is to develop cost-effective manufacturing and assembly processes for 3-D stacking of MCMs. The proposed approach is based on vertically stacking and interconnecting single- or double-sided aluminum nitride cofired ceramic substrates.[12] The program builds on a technology base developed on an earlier NCCOSC program conducted by Hughes, CEPC, and W. R. Grace, in which processes were developed for producing large-area hot-pressed aluminum nitride with a high density of tight-tolerance vias.[13] Under the current program Honeywell will design the test vehicles and demonstrate modules; develop the assembly, sealing, and stacking processes; and characterize the electrical, thermal, and mechanical performance of the 3-D structures. CEPC will develop manufacturing processes for the AlN cofired ceramic substrates and spacer bars.

The examples described are applications in which IC chips are integrated into compact 3-D assemblies to satisfy specific product requirements. They do not involve the large number of parallel interconnections and the high degree of sophisticated circuit interactions required in the compact computer assemblies. However, the ability to assemble a few chips into a compact mini-MCM will find many commercial applications in future microelectronics products.

7.4   MCM Memory Systems

High-speed imaging systems have pushed the state of the art in data storage. Today's imaging systems can have as many as 9 million pixels per frame with applications that require several seconds of data storage at about 100 frames per second. This translates into a data storage bandwidth in the gigabit-per-second range, which is beyond the capability of magnetic storage media. A potential solution is a solid-state recorder using standard static random access memory (SRAM) chips packaged by utilizing the 3-D approaches demonstrated by Texas Instruments[14] and Irvine Sensors.[15]

The first significant implementation of the approach is in the use of pretested and selected IC memory chips to form an even higher density of memory. This is logical because all the chips are chosen to be of one design, the data contacts can be made massively parallel, and the particular chip-select (addressing) lines are few. Figure 7-17 shows

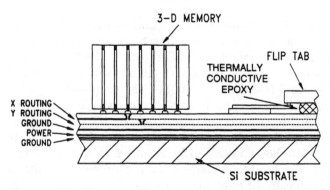

Figure 7-17. Silicon-on-silicon cross section.

schematically the manner in which the chips are stacked. These designs offer flexibility as well as a robust packaging concept that will provide mechanical and thermal survivability through aggressive data storage environments.

In 1987 Texas Instruments (TI), sponsored by ARPA, developed a generic high-density memory packaging process.[16] The primary goal of this program was to package 4 Gbits of memory in less than 100 in^3. In the TI approach, termed the *memory cube*, memory chips such as the 256K × 6-bit dynamic random access memory (DRAM) and the 1-Mbyte × 1-bit DRAM were processed, while still in wafer form, with an interconnect overlay pattern that carried the contacts to one edge (or at most two edges). After dicing and testing, tape automated bonding (TAB) leads were attached, and the chips were stack-bonded together with the TAB leads protruding from the edge. The multiple leads were then soldered onto the signal leads of a carrier board, providing close packing of memory into a small footprint of the circuit board, as shown in Fig. 7-18.

The 3-D memory technology was developed by TI to support the avionics, space, workstation, and mainframe computer markets where conventional memory packaging approaches could not meet the performance (power, size, weight, and speed) requirements of the application.

As a result of the 3-D technology development activities, TI has accrued a broad base of experience in producing 3-D memory cubes and MCM 3-D memory products. Over 2000 memory cubes, each typically consisting of 8 dice have been built to date and used in various MCM designs, including their use in combination with planar mounted logic chips. These MCMs range in density from having six memory cubes, for

Figure 7-18. Memory cube mounted on printed circuit board. (*Courtesy of Texas Instruments.*)

Figure 7-19. A 1.3-Gbit memory unit, 160 stacks of 8 dice per stack. (*Courtesy of Texas Instruments.*)

TI's Dual C40 MCM, to 160 memory cubes, for TI's 1.2-Gbit solid-state recorder (SSR), shown in Fig. 7-19. Memory cubes also offer benefits in system performance. Because of the closer spacing of memory chips, propagation delay times are reduced, and the system operating speed potential is usually increased. Further significant reductions in overall system power requirements have been demonstrated as a result of the reduced current driver requirements for 3-D memory cubes.

In the ARPA 3-D program, TI completed thermal modeling and testing of 3-D memory modules to optimize the thermal path from the junction of the memory chip to the package base. Key results of the thermal analysis indicated that

- No significant thermal enhancements to the design or materials used in the memory modules were necessary because of the relatively low operating power of the memory chips.

- Thermal vias in the multilayer polyimide interconnect are essential to minimize the junction temperature and to prevent a temperature rise in the module.

- Based on a chip dissipating 1 W with 100 percent duty cycle, there will be a temperature rise of 13.8°C above the base of the package.

In addition, elevated temperature aging of the modules at 150°C for 168 h showed no failures. These results give a good indication of the robustness of the 3-D packaging approach and its applicability to the most severe application environments.

Although it is currently expensive, TI predicts that, in volume production, its 3-D packaging technique will approach cost parity with conventional memory packaging technology. This is possible because in high volume the largest portion of the total system cost will be the cost of materials. Thus, in comparing conventional with 3-D packaging, clearly there is far less material cost in a 3-D system. In the long term, 3-D memory products are expected to revolutionize the mass-memory storage industry.

In response to a USAF need to increase the effective memory density of 2-D MCMs, a program was established with Irvine Sensors Corp. to develop a low-profile die stacking approach that resulted in components that physically resembled an integrated circuit, but were, in fact, four times denser.

In this second example of a category II approach, Irvine Sensors Corp. has pioneered a unique technology for stacking IC chips in order to address a difficult problem associated with space surveillance. The company's progress in 3-D packaging research led to the development of its "full stack" or "cubing" technology. The technology further evolved into the company's memory short stack which Irvine Sensors has successfully applied to both DRAM and SRAM memory chips. The short stack technology was first successfully demonstrated in a USAF program which integrated the low-profile stacks into the General Electric/Martin Marietta HDI process.

The Irvine Sensors Corp. process begins by procuring chips in wafer form. Since these wafers are selected "off the shelf," some modification is required before the dice can be stacked. This modification involves rerouting the IC pads by processing thin-film metal leads to one edge of the chip. The wafer is then thinned to approximately 10 mils and diced. The chips are tested and bonded together in an assembly fixture with a thin layer of adhesive between each chip. The adhesive provides good heat transfer and CTE matching. The fixture containing the stacked chips is then placed in an oven and baked to cure the adhesive and permanently bind the chips into a cube. The "face" of the stack is then etched to expose the thin-film metal leads. After etching, several layers of polyimide are deposited over the face of the stack; and vias are etched, metallized, and patterned. The pad to thin-film metal lead interconnect that is formed is referred to as a *T connect*. At this point, Irvine Sensors Corp.'s standard full-stack fabrication process is complete. The full stack can be bonded with solder bumps to a substrate for eventual connection to external circuitry (Fig. 7-20).

The memory short stack (shown in Fig. 7-21) consists of 4 to 10 memory ICs sandwiched between a bottom blank chip and ceramic cap chip. The top cap chip provides pads suitable for wire bonding, TAB, solder surface mounting, or chip-on-board attachment. Short stacks are bonded together with a thermoplastic adhesive to form full stacks. This adhesive is different from and of lower strength than that

Figure 7-20. Full 3-D stack or "cube" of 70 ICs laminated together and interconnected in a package provides 1°C/W temperature rise. (*Courtesy of Irvine Sensors Corp.*)

Figure 7-21. Memory short stack consisting of 10 memory chips occupies the same space as a single IC chip. (*Courtesy of Irvine Sensors Corp.*)

originally used to bind chips together. To separate the short stacks from the full stack, the stack is merely heated to the softening temperature of the adhesive while a shear force is applied to slide one short stack off another. The adhesive bonding the chips within the short stacks remains rigid and is unaffected by the shear force.

Irvine Sensors Corp.'s 3-D process, which has been carefully defined and patented, removes barriers to heat dissipation and provides an order-of-magnitude improvement over other packaging technologies.

Heat transfer is not a significant issue with the short stack simply because fewer chips are involved. The resulting product is a very dense cube that computer system designers can now apply to reduce size, weight, power, and other impediments to system performance. Since the technology reduces the number of interconnections between chips, this increases system reliability and reduces the potential for system failure. In addition, the memory short stack is compatible with existing memory applications, so designers can incorporate it into current products without the need for significant redesign.

Irvine Sensors and IBM, in June 1992, announced the signing of a joint agreement to develop and commercialize the full-stack (cubing) technology. These two companies are working to develop cubed products and the manufacturing technologies required to produce them in volume at reasonable cost.

Dense-Pac Microsystems, of Garden Grove, California, has advertised 3-D MCM memories consisting of prepackaged memory chips directly stacked to form a low-cost, fairly compact memory unit that can be assembled onto a printed-circuit board.

7.5 Microwave Packaging

Another application of 3-D microelectronic packaging that differs from memory and signal processing but draws on much of the same technology is the integration of solid-state microwave circuits. Whereas vacuum-tube microwave systems have used mechanically swept antennae for transmitting and receiving radar signals, the new active element, phased-array modules utilize solid-state transmit/receive (T/R) modules. These T/R modules are considered one of the most promising technologies for future ground-based, airborne, and space-based radar applications. Major benefits include performance improvements and reductions in size and weight. Feasibility and validation T/R microwave modules for many new systems have been demonstrated from small prototype quantities and limited production quantities. Costs are still quite high due to complex designs, the need for precision fabrication, the costs of parts and materials, and the general lack of adequate assembly, test, and automation equipment. Manufacturing technology programs issued by the Department of Defense[17] seek to establish and demonstrate a low-cost manufacturing capability for large quantities of complex T/R modules for inclusion in active element phased-array radar systems. Contractors, including Martin Marietta, TI, Hughes, and Westinghouse, are engaged in programs to demonstrate high-volume producibility at a projected production rate of 1000 modules per day and a target cost not to exceed

$400 per module. Materials and processing techniques for microwave hybrid circuits have been described in the book by Richard Brown.[18]

Good beam control (low side lobes) at the operating frequency of these modules (up to about 12 GHz) dictates that the emitting and receiving antennae be spaced a half wavelength apart[19]—about 0.5 in at these frequencies. With this center-to-center spacing, it is not possible to incorporate all the necessary active and passive components on one 2-D layer; therefore, 3-D packaging is required, as shown schematically in Fig. 7-22. The solid-state monolithic microwave integrated circuits (MMICs) are individually small enough that they can be arranged in a 2-D array satisfying these spacings, but the remaining circuitry that provides power, phase information, and selectivity between high-power transmission and very low-power reception requires 3-D integration. This is especially important since these arrays are planned to be distributed over an area on the surface of an aircraft or a missile. The active elements and the other circuitry must be located below the surface, and they must feed the antenna distributed on the surface.

Martin Marietta's microwave high-density interconnect (MHDI) technology brings the benefits of semiconductor-style batch processing to GaAs MMIC-based T/R modules.[20] MHDI avoids the needs for wirebond connections and allows a diverse range of chips to be integrated in a microwave multichip module (MMCM), offering typically 50 percent size reduction compared to conventional chip-and-wire or

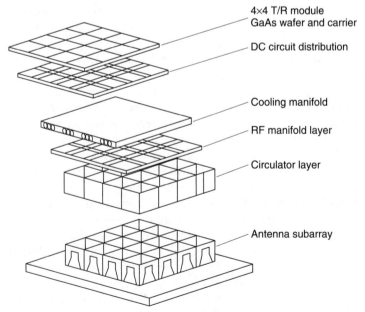

Figure 7-22. Subarray coplanar packaging concept.

flip-chip modules. Figure 7-23 shows a photograph of a 5-W C-Band MHDI T/R module measuring 1.6 × 0.93 in. This approach is similar to that used in the HDI packaging of digital signal processor MCMs (see Chap. 3). The chips are attached in preshaped wells in the substrate (made of ceramic or metal), then overlaid with Kapton polyimide film with laser-ablated vias and a copper metallization. The well depths are tailored to position the tops of all chips coplanar with the substrate surface. A resist is patterned using laser lithography to define the conductor traces. Figure 7-24 shows the cross section of a 3.5-W X-Band module made with three layers of Kapton/metal on a

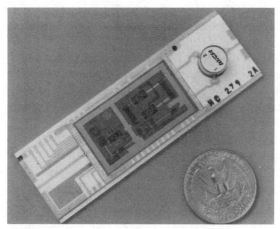

Figure 7-23. A 5-W, C-band transmit/receive multichip module based on monolithic microwave integrated circuits (MMICs). Processed by the MHDI process. (*Courtesy of Martin Marietta Corp.*)

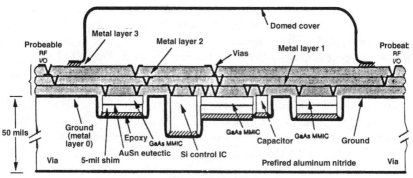

Figure 7-24. Cross section of X-band transmit/receive module. (*Courtesy of Martin Marietta Corp.*)

50-mil-thick aluminum nitride substrate. Standard 4-mil GaAs MMIC dice are eutectically mounted to 5-mil-thick copper/molybdenum/copper shims and then epoxied into the substrate chip wells. The shim thickness and a controlled epoxy bondline were determined by finite element analysis to be the optimum solution to minimizing the tensile stress in the GaAs dice arising from the CTE mismatch between AlN (4.1 ppm/°C) and GaAs (5.8 ppm/°C). The reliability of this approach was confirmed by temperature cycling 90 MMICs from − 75°C to + 250°C without any failures.

Although the GaAs MMICs were designed to be used in an air dielectric medium, the presence of the 4 mils of polyimide (dielectric constant about 3.2) over the MMICs showed little degradation in performance up to 10 GHz. For use at higher frequencies, laser ablation is used to remove portions of the dielectric.

Hughes has teamed up with Coors and W. R. Grace in an Air Force contract[21] to produce a new substrate technology for T/R module integration. This substrate is a hot-pressed AlN, a material that has the same multilayering capability of low-temperature cofired ceramic (LTCC), but approximately 100 times better thermal conductivity—which is required in microwave applications. Inexpensive thick-film screen printing is utilized which is adequate for the low interconnect densities required for microwave circuits such as the T/R modules. The AlN shrinks only 0.1 percent during hot pressing, providing excellent registration for 3-D packaging. Tungsten metallization is used for the inner-layer conductor traces, but when lower resistance is required, the outer surfaces may be gold-plated.

Hughes has integrated its flip-chip technology for MMICs into the multilayer AlN substrate. The flip-chip bonded dice obviate the need for wire bonds and provide a robust assembly, better microwave performance, and longer mean time to failure. The heat generated in the 2.5-W high-power amplifier (HPA) MMIC chip can be dissipated through thermal bumps formed on the chip that are close to the sources of heat. Increases in chip power to 4.5 W and higher are also planned to be dissipated through thermal bumps. The I/O bumps on the periphery of the chip have the added benefit of taking up the strain resulting from the slight mismatch between GaAs and the AlN substrate. The chips have survived several hundred severe thermal cycles with no discernible loss of performance. The CTE of AlN falls between that of Si and GaAs, allowing reliable flip-chip attachment of both device types on the same substrate which are required for the T/R modules.

Several multilayer substrates can be integrated by elastomeric interconnects so that the T/R module design can fit into very tight lattice spacings. Substrate-to-substrate alignment for interconnects can be ensured by punching the connections on the adjacent top and bottom green tapes laid over each other. The small 0.1 percent shrinkage will, at most, cause a misalignment of 1 mil over 1 in. Thick-film screen-printing accuracies can be maintained to better than 1 mil over 1 in. Also, the elastomeric and fuzz button interconnections provide for easy assembly and disassembly.

Earlier T/R modules were arranged perpendicular to the radiator and were integral to the lattice geometry. If the lattice design changes, the T/R module package must also change. The current tile packaging approach requires only that the unit area for a single T/R element be the same as or smaller than the unit area of the lattice; hence the lattice can change shape without causing a change in the tile geometry. This capability can provide greater standardization of T/R module packaging.

In summary, the wirebond-free technologies developed by Martin Marietta and Hughes lend themselves to efficient, economical integration of microwave circuits into 3-D configurations for active phased array antennae. The GaAs power chips pose the greatest challenge to the integration. In the Martin Marietta approach described above, the chips were thinned to the standard 4 mil thickness to facilitate heat transfer down through the eutectically bonded back surface to the heat sink. The Kapton overlay is not in the thermal path. In the Hughes approach, the custom GaAs chips were not thinned, eliminating some yield loss, and have flip-chip bonding bumps added to facilitate die bonding and thermal transfer from the active circuitry directly to a carrier and the heat sink. Both the Martin Marietta MHDI approach and Hughes flip-chip approach permit use of chips of any thickness.

The new generation of extremely dense microwave and millimeter-wave packaging and interconnect technologies are evolving into a combination of what has traditionally been separate MMIC wafer-processing and package-assembly techniques. The boundary between chip and package is becoming indistinct and necessitates a concurrent engineering approach to chip and package. In turn this approach is demanding smarter CAD tools to accurately model the 3-D electromagnetic and thermal performance. Existing tools are unable to model the full port-to-port performance of a T/R module taking into account all active and passive devices as well as the 3-D interfaces between chip, substrate, and housing. A major limitation still remains adequate CPU power to process such a calculation.

In an era of high-mix, low-volume production, there is a strong need for packaging processes compatible with flexible manufacturing lines that can be retooled with very short down-time. Automation is already well established in the assembly of microwave modules. Great

advances have been made during the past 5 years in the accurate damage free placement of GaAs dice using pick-and-place robots, precision
epoxy dispense systems, and fully automated eutectic die attach.
Microwave multichip modules now have the appearance of large
chips, manufactured and probe-tested in wafer form. Such similarities
with semiconductor processing brings an economy-of-scale to reduce
unit cost. Performance at high frequency is enhanced by eliminating
wirebonds, and thus reducing cost by increasing the number of modules that pass a given RF performance specification. Major efforts are
under way, using statistical process control data to accurately model
unit cost, as the technology moves toward applications in both military and commercial sectors.

7.6 Critical Technologies for 3-D Packaging—Vertical Interconnections

In the discussions above, it is clear that significant advances have been
made in tightly arranging individual IC chips in 2-D packages and
then further integrating them into a dense 3-D stack. A technology that
is critical to this 3-D packaging involves the achievement of vertical
interconnections. This includes feedthroughs running through individual carrier levels and interconnections between adjacent levels. These
connections may be flexible or rigid depending upon the assembly
approach and the expectations as to the need for disassembly and
reassembly of the unit. Some substrate feedthrough approaches
include

- Thermomigrated Al traces
- Laser-drilled holes, filled with silver-glass frit and fired
- Punched holes in green ceramic tape that are filled with metal and
  cofired with the multilayer interconnect carrier
- Laser-drilled holes that are through-plated
- A variety of conductive wires, pins, meshes, or epoxies that are
  inserted into drilled holes in the substrates

All these feedthrough approaches have been used with some success.
The demands have been for low and reliable resistance in front-to-
back connections—typically, less than 0.1 Ω per lead. If the leads are to
match the spacings of the present high-performance IC chips, then
feedthroughs as small as 2-mil diameter spaced on 4-mil centers may

be needed. These dimensions are achievable by the Nd-YAG laser drilling processes, but these are sequential operations that may be costly to perform. Vias in cofired ceramic assemblies (alumina or AlN with refractory metal conductors) have been produced as small as 6 mils in diameter, and this has been adequate for the packaging of the microwave assemblies. Simultaneous formation of the feedthroughs on an entire carrier makes the process more economical. The insertion of wires or meshes can be a low-cost operation for small numbers of units, but relatively large holes are required if reliable, low resistance is to be achieved.

Interconnects between levels may also include (1) "fuzz buttons"— small clusters of wire mesh that are compressed to provide the contact, (2) the lithographically prepared microspring bridges, (3) reflowed solder bumps, (4) deformable gold and indium bumps, and (5) the anisotropic, z-axis connecting elastomers—containing filamentary or metal traces oriented such that compression causes the low-resistance interconnects. The best approach for a particular assembly must be chosen on the basis of the number and size of contacts required, the parallel nature of the opposing carrier surfaces, the temperature expected in the operation, and the expected need to disassemble and reassemble the units.

The 3-D modules developed by Irvine Sensors Corp., and Thompson-CSF, described above, employed edge metallized patterns to interconnect the levels in a stacked assembly. This is thought of as a different approach from the techniques in which carrier-to-carrier interconnects are made internal to the structures so that the shortest possible signal paths can be produced from chips on one level to chips on levels above or below.

Because of the high-bandwidth signal-carrying capability of fiber optics and optical interconnections these techniques are being studied as carrier-to-carrier interconnects for future MCMs. In the most direct application, signal traces on one carrier level can be merged into an optical node which can be connected to the other carrier levels with higher bandwidth than edge connections or flexible leads now permit. 3-D carrier-to-carrier interconnection can be performed optically by including solid state light sources and detectors aligned to permit internal vertical signal transfer between the carriers. Another approach that has been pursued uses holographic lenses and mirror structures placed over the circuit that has light-emitting diode lasers. These are critically arranged so that optical signals can be transferred to receiving nodes on the same, or on adjacent carriers. Although the results to date have been feasibility demonstrations only, the huge bandwidth capability of optical signal generation and transfer makes it possible that high-performance signal-processing systems will eventually incorporate optical (photonic) interconnects as standard features.

Another issue in 3-D packaging that demands consideration is thermal control. When high-performance, high-power-dissipating chips are integrated into a 2-D carrier (such as the SEM-E interconnect boards), the challenge is to transfer the heat as directly as possible to a heat sink so that the IC circuitry is not allowed to reach temperatures that will deteriorate its expected lifetime. In 2-D packaging, this can always be accomplished by providing thermal conduction paths (vias) through multilevel interconnects (which contain the thermally resistive isolating layers) to the heat sinks. In the 3-D packages, the thermal paths are more complex because the heat-generating IC chips are buried deep in layers of stacked circuitry. However, there are several factors that tend to mediate the concerns. First, carriers within the 3-D stack can be made of high-thermal-conductivity materials. Instead of silica ceramic or alumina, silicon or aluminum nitride, having 40 to 60 percent the thermal conductivity of copper, are being employed. Therefore, these are excellent for carrying heat away from the embedded circuitry. As a further improvement, synthetic diamond substrates may be used. ARPA has supported development of diamond structures and films produced by chemical vapor deposition that have a thermal conductivity approximately 10 times greater than aluminum nitride and 100 times greater than 96 percent alumina. A 3-D stack of silicon, aluminum nitride, or diamond substrates with IC chips in good thermal contact is like building a computer, not in a can, but as a solid block. Since most of the package consists of the thermally conductive carriers, the maximum temperature rise in the operating internal chips can be held to safe limits. The work by Norton Diamond Films[22,23] of producing a 3-D diamond cube module (Fig. 7-25) and even coating the AlN carriers with diamond also advances the thermal response of the assembly.

If it can be guaranteed that the interconnect substrates (carriers) satisfactorily transfer the heat to the edge contacts and eventually to the heat sinks, then the next concern is to extract heat from the heat-generating circuitry within the chips. To do this, circuits must be analyzed not only as to their electrical function but also as to their thermal characteristics. An example is the 3-D packaging of digital memories. In both examples described above (TI and Irvine Sensors), the packaging raised no thermal concerns because the thermal demands of memory are minimal and can be treated as a cool circuit to be added to the carrier board. In contrast, the integration of high-speed multiplexing and demultiplexing (MUX and DEMUX) functions operating at gigahertz frequencies generate considerable localized heat (1 to 2 W/mm^2) in a

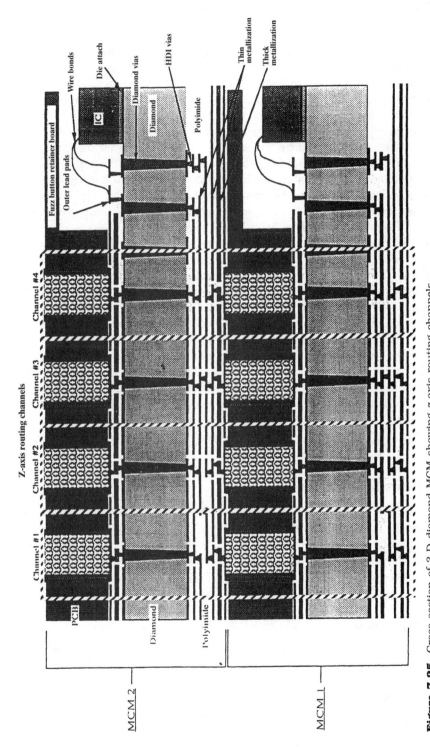

Figure 7-25. Cross section of 3-D diamond MCM showing z-axis routing channels in diamond carriers. (*Courtesy Norton Diamond Film*)

circuit, and this heat must be dissipated. Large-area computer processing chips also demand the same regard for thermal heat dissipation. However, in a 3-D assembly, with appropriate electrical speed design and thermal design, these modules are seen to be readily achieved. In the microwave packages for antenna arrays, the higher the microwave power per T/R module (typically 8 W per channel is being developed today), the greater will be the need for proper thermal design. The packaging progress described above makes use of high-performance MMIC chips that are integrated into AlN or diamond carrier assemblies; and to guarantee the heat extraction, liquid- or vapor-phase heat-exchanger levels are included to ensure the reliable operation of the radio-frequency T/R phased-array assembly.

Between the extremes of the low-power, massive digital memory units and the high-power, radio-frequency phased-array antennae modules is the broad area of signal processing circuitry. These units are being developed to demonstrate the ability to integrate ICs into 2-D and 3-D assemblies. The volumetric constraints plus speed and power of circuit operations will dictate that 3-D packaging will be the best solution to the packaging challenge. As long as 2-D packaging satisfies the system demands, then the microelectronics community is well served. But as performance demands increase, 3-D packaging will be required. And although at present the cost may be high because of the specialized developments, it may eventually be justified in terms of the significant improvement realized in terms of total system performance.

Acknowledgments

The author is pleased to acknowledge the contributions and advice of Michael J. Little, Capt. James Lyke, and Dr. Paul Cooper in preparing this overview. In addition, we appreciate the information freely shared in the many reports of progress made in 3-D MCM packaging by the referenced companies.

References

1. M. J. Little, *Signal Processing Systems Packaging—1*, Final Report for contract no. F30602-89-D-0100, Hughes Research Laboratories, Malibu, CA, 1992.
2. E. Palmer, C. Ankerson, J. Schroeder, G. Dooley, and J. Cooke, *Signal Processing Systems Packaging—2*, Final Report for contract no. F30602-89-D-0099, Harris Corp., Melbourne, FL, 1992.

3. J. C. Lyke , *Relevant Papers in Advanced Packaging: A Compilation*, Phillips Laboratory, Applied Systems Branch, Kirkland, AFB, NM, 1992.

4. J. C. Lyke, R. J. Wojnarwaski, R. Saia, P. McConnelee, G. A. Forman, B. Gorowitz, and M. C. Ciccarelli, "Development and Application of Practical Three-Dimensional Hybrid WSI Technology," *Proceedings GOMAC*, **18**, 1992.

5. M. J. Little, R. D. Etchells, J. Grinberg, S. P. Laub, J. G. Nash, and M. W. Yung, "The 3-D Computer," *Proceedings IEEE International Conference on Wafer Scale Integration*, January 1989.

6. M. P. Lepselter, "Air Insulated Beam-Lead Crossover for Integrated Circuits," *Bell System Technical Journal*, 48, February 1968.

7. R. Scannell, "SPPD Packaging Status," *Proceedings ICEMM*, 1993.

8. E. G. Palmer and C. M. Newton, "3-D Packaging Using Low-Temperature Cofired Ceramic (LTCC)," *International Journal of Microcircuits and Electronic Packaging*, **16**(4):279, 1993.

9. J. Kohl, "Stacked AMCM Technology," Integrated Systems Assemblies, Woburn, MA, private communication, 1993.

10. S. K. Ladd, "Designing 3-D Multichip Modules for High Volume Applications: Three Case Studies," *Proceedings ICEMM*, 1993, p. 417.

11. S. K. Ladd and J. E. Mandry, "3-D Memory Packaging for Workstations," *Electronic Packaging and Production*, November 1993.

12. R. J. Jensen, W. F. Jacobsen, J. M. Loy, S. L. Palmquist, C. J. Speerschneider, and R. K. Spielberger, "Three-Dimensional Interconnect Structures for AlN Multichip Modules," *Proceedings GOMAC*, **19**, 1993.

13. *Manufacturing Technology For VLSIC Packaging*, Final Report N66001-88-C-0181, Naval Command, Control, and Ocean Surveillance Center, June 1993.

14. D. Frew and G. Beene, "High Density Memory Packaging Technology for High Speed Imaging Applications," Texas Instruments, Dallas, TX, private communication, 1993.

15. K. Lian, "3-D Stacking Technology," Irvine Sensors Corp., Costa Mesa, CA, private communication, 1993.

16. L. Mowatt, D. Walter, J. Kacines, and L. Roszel, "MCM Laminate plus Overlay Technology Development," Texas Instruments, Dallas, TX, private communication, 1993.

17. *Special Technology Area Review on Microwave Packaging Technology*, Report of DoD Advisory Group on Electron Devices, Office of the Undersecretary of Defense for Acquisition, Washington, DC, February 1993.

18. R. Brown, *Materials and Processes for Microwave Hybrids*, International Society for Hybrid Microelectronics, Reston, VA, 1991.

19. L. R. Whicken and J. D. Murphy, "RF-Wafer-Scale Integration: A New Approach to Active Phased Arrays," *Proceedings International Conference on Wafer Scale Integration*, January 1992.

20. P. D. Cooper, J. D. Kennedy, R. J. Street, W. R. Kornrumpf, and W. M. Marcinkiewicz, "Microwave High Density Interconnect (MHDI)—A Methodology for Batch Manufacturing T/R Modules," *Proceedings Conference on GaAs Manufacturing Technology*, Atlanta, GA, May 1993.

21. J. J. Wooldridge, "T/R Module Packaging Technology," *Proceedings Conference on GaAs Manufacturing Technology*, Atlanta, GA, May 1993.

22. T. J. Moravec, R. C. Eden, and D. A. Schaefer, "The Use of Diamond Substrates For Implementing 3-D MCMs," *Proceedings ICEMM*, 1993.

23. R. C. Eden, "Applicability of Diamond Substrates to Multichip Modules," *Proceedings ISHM*, 1991.

8

Inspection and Testing

Howard L. Davidson

Rapid and economical electrical testing remains one of the more diffi-cult aspects in making the transition of multichip modules (MCMs) from prototypes into volume production. This is particularly true in the merchant marketplace where an MCM may contain integrated cir-cuits from several manufacturers and where no special provisions have been made to facilitate testing either at the die level or at the assembled MCM level. There are three types of testing that are applied to MCMs: (1) in-process inspection, (2) bare-die functional testing and speed sorting, and (3) final module functional and screen testing. The details of what tests are performed and how they are performed are very sensitive to how the MCM interconnect substrate is constructed, how the chips have been designed, how the chips are assembled, and where the module will be used. There is a strong dependence between the initial design and whether a given MCM can be economically test-ed. If design-for-test guidelines are not initially followed, it may be impossible to debug a module that fails at system test. This generally happens if there are insufficient register and pad visibility and control for the logic chips, inadequate access to memory devices, or chips that are obscured by a commodity chip that has no scan capability.

8.1 In-Process Inspection and Testing

There are several approaches to designing and building an MCM (see Chap. 3). The specific inspection and test procedures used during processing to ensure quality and yield are closely coupled to the fabrication technology used, i.e., whether it is MCM-D, MCM-C, MCM-L, or combinations of these. Two other elements of process flow are important in deciding on the test strategy: how the dice are mechanically and electrically attached to the substrate (wire bond, TAB, flip-chip) and whether the chips are attached first, prior to forming the interconnect structure, or last, after the multilayer interconnect has been fabricated.

8.1.1 Electrical Testing of Interconnect Substrates

Electrical testing of interconnect substrates, regardless of the type, consists in electrically testing for open and short circuits of the various nets. Historically, open- and short-circuit testing of printed-circuit boards and hybrid circuit substrates has been done with "flying-probe" testers.[1] These testers typically consist of two probes on high-speed positioners. If both ends of a net are accessible, continuity can be measured directly. However, two-probe testing does not directly detect short circuits unless each trace is measured against itself and against all other traces. Thus testing for short circuits with a two-probe tester becomes an N^2 problem.

Both open and short circuits can be detected with a single flying-probe capacitance-measuring system. This rapid, low-cost method is widely used for multichip module substrates, be they ceramic, thin film, or plastic laminate. The single-probe method measures capacitance between each net and a nearby ground plane or a metal chuck on which the part is placed during measurement. The chuck then serves as the reference ground plane. One side of a capacitance meter is connected to ground while the other side is connected to the probe. The probe is then brought into contact with each pad on the substrate. Open circuits are detected by measuring the capacitance to reference ground structures from both ends of a trace. An intact trace will measure the same from either end. An open trace will produce a lower value for at least one of the measurements. The sum of the end values will be close to the expected value. Short circuits are detected by measuring the capacitance of the trace to a reference ground. If the trace is short-circuited to another trace, the measured capacitance will be high, approximating the sum of the expected trace capacitances. If the two traces run next to each other or over each other for much of their length, the capacitance will be less than the sum of the individual values, but still higher than expected.

The expected capacitance values may be obtained by either modeling the substrate or measuring a hopefully representative sample of good substrates. Modeling is the preferred method, because the accuracy of available software for capacitance modeling is quite good and because it is possible to directly incorporate a spread of process parameters into the model. Models should be validated against a number of substrates of known properties. A plot of modeled capacitance versus measured capacitance for traces on MCM-D substrates is shown in Fig. 8-1. Short- and open-circuit nets detected during the measurements lie well outside the modeled limits.

MicroModule Systems (MMS) uses a commercial automatic wafer prober to move substrates under the probes and in so doing obtains five capacitance measurements per second. MMS reports a test escape rate of 0.5 ppm per trace.[2] A diagram of the capacitance measurement

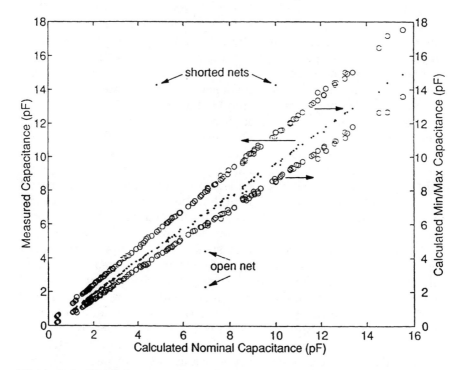

Figure 8-1. Modeled versus measured net capacitance with open- and short-circuit nets shown.

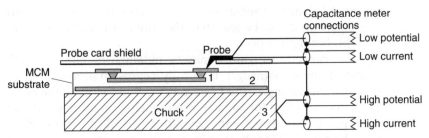

Figure 8-2. Setup for capacitance probing for open and short circuits in MCM substrate (the net under test, internal ground, and chuck are labeled 1, 2, and 3, respectively).

setup is shown in Fig. 8-2. By using high-speed positioners similar to those used to move the heads in hard-disk drives, the probes can be moved to new locations in less than 10 ms. Modern instruments can now perform 1000 capacitance measurements per second. In practice, this means that by using an optimally sorted path it is possible to test up to 100 nets per second.

IBM has developed a slightly different approach by using a controlled-force high-speed actuator as a direct z-axis drive for the probes while the substrate is moved under them by using a high-speed x/y table.[3] IBM reports 40 to 50 tests per second over each 12-mm-square die site.

Electron beam (e-beam) testing can also be used to inspect for open and short circuits in interconnect substrates.[4,5] Electron-beam testing is based on the "voltage contrast" mode of a modified scanning electron microscope. An electron beam is directed for a short time to the pad being tested so that an electric charge accumulates on the pad and on its associated trace. Next the beam is scanned over the substrate surface. All pads that are connected to the charged pad will appear different from the uncharged pads, thus providing direct open-and short-circuit data. A cross section of the e-beam process is given in Fig. 8-3. The charged nets may then be discharged by a flood electron beam, sometimes combined with a low background pressure of gas.

The electron beam tester may be subject to a false-positive form of error. For example, if two segments of a net are connected by a minute sliver of conductor, then the small amount of charge that is injected will equilibrate between the two segments and the line will be read as continuous. If the two-probe flying probe is used for continuity testing, it can detect a high-resistance net when the metal is severely necked down. Electron beam testers that have been built to date have been used for experimental purposes and have not yet been introduced in high-volume production testing.

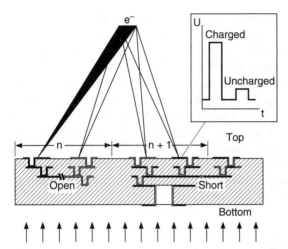

Figure 8-3. MCM substrate testing by charging and reading pads with an electron beam.[4]

None of the methods discussed so far applies any significant electrical stress to the substrate. This issue, however, has been addressed by IBM through its development of probing systems that allow simultaneous contact to 26,000 pads on the top and bottom surfaces of its MCM-C interconnect substrates.[3] The tester supplies 250-V, 20-mA signals, high enough to detect leakage between two conductors that have a latent insulation defect. The current is high enough to expose necked-down or cracked conductors. The system, however, is rather massive, there being 27 mi of wire used to connect the probe heads to the electronic components.

8.1.2 Optical Inspection

Automatic optical inspection has been used successfully by IBM for inspecting the individually processed layers of its ceramic circuit boards prior to laminating and firing. Automatic inspection is done with custom-built equipment that compares an image of the layer to a stored database which represents a good part. The machine recognizes defects such as narrowed conductor lines, bridged lines, and isolated spots of conductive paste. In principle, this equipment can be used to inspect the layers of other parallel or sequentially processed interconnect substrates. Optical inspection, for example, is well suited to in-process monitoring of the "chips first" overlay approach used by

General Electric for its high-density interconnect (HDI) process. Equipment designed for inspecting both printed-wiring boards and integrated circuits is commercially available and can be adapted for MCM inspection with minimal effort.

Visual inspection is feasible more for verifying workmanship than for exhaustively evaluating the integrity of the layers. Visual inspection is best suited to detect die attachment and wire bond placement and morphology; physical damage to the dice, interconnections, or conductor traces; and other assembly abnormalities. On the other hand, flip-chip bonded devices cannot be optically inspected since the interconnections are hidden from view. In such cases, electrical testing or ultrasonic microscopy may be used.

8.1.3   Screen Tests for Substrates

In processes where the interconnect substrates are fabricated separately, some screen tests may be performed prior to assembly, provided the tests are nondestructive, do not jeopardize the long-term reliability of the part, and are capable of detecting any latent defects. The tests that are used usually consist of a subset of the Mil-Std-883 thermal and mechanical screens that have been utilized for many years as 100 percent nondestructive tests for hybrid microcircuits. However, note that hybrid circuits are hermetically sealed prior to being subjected to the screen tests, and hence the substrate and circuitry are well protected. In screen testing MCM substrates, care must be taken in handling and in preventing contamination and condensation of moisture, especially on the cold cycle of any temperature cycling or thermal shock testing.

Liquid-to-liquid thermal shock has proved to be a good screen for assessing layer-to-layer adhesion of thin-film polyimide/copper substrates, LTCC multilayer ceramic substrates, and plastic-laminated structures. Almost any defect that weakens the bond between layers will be stressed sufficiently to cause delamination or cracking that can be visually detected. Also poorly designed or processed vias will often fracture and become open circuits during thermal shock and can be detected by the capacitance test for open and short circuits.

Sacrificial test coupons may be designed and incorporated in the area surrounding the active substrate, so that statistical quality-control data and data on the reproducibility of the materials and processes can be obtained. For example, test structures to measure the capacitance and dielectric constants for the polyimide layers have been incorporat-

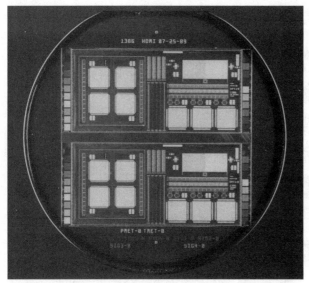

Figure 8-4. A polyimide/aluminum interconnect substrate at wafer stage showing test capacitors at each end.

ed at the wafer stage (Fig. 8-4). Daisy-chain vias and serpentine conductor lines can also be incorporated on each wafer and the continuity checked to monitor the quality and reliability of the interconnections. Figure 8-5 shows the design of a test structure consisting of pairs of serpentine conductors running in the x and y directions on adjacent layers with metal crossing over underlying metal. The structure has been used to test for both intralayer and interlayer short and open circuits. The test structure shown in Fig. 8-6 consists of a chain of vias between adjacent layers of metal which allows continuity between signal layers and via resistance measurements to be made.

Finally, a crack monitor can be incorporated. The crack monitor consists of a thin conductor line (15 to 25 μm wide) that is formed around the perimeter of the substrate culminating in two pads that can be probed for continuity. The crack monitor may be formed on the substrate (such as silicon) or between layers (such as polyimide) or both. Any fracturing or excessive bowing of the substrate or cracking of the dielectric will cause an open circuit in the crack monitor line which can be measured.

Reliability test data can be obtained by extending the number of cycles or the time at temperature until failures occur. Often, the

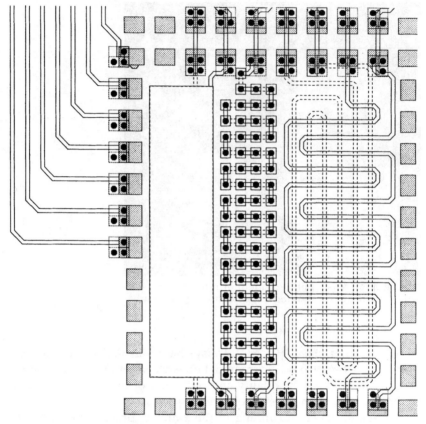

Figure 8-5. Test structure showing pairs of serpentine conductor.s (25-µm wide conductor lines on 100-µm pitch).

absence of failures after 1000 temperature cycles or 1000 h of accelerated aging is considered sufficient to ensure long-term reliability.

8.2 Integrated-Circuit Die Testing

Because of the greater functionality and density of MCMs, there are two issues for die testing that are more pronounced than for hybrid circuits. The first is obtaining and assembling "known good dice" to increase first-pass yields and reduce rework. The second issue is to "design for testability," i.e., to design dice with test features that allow

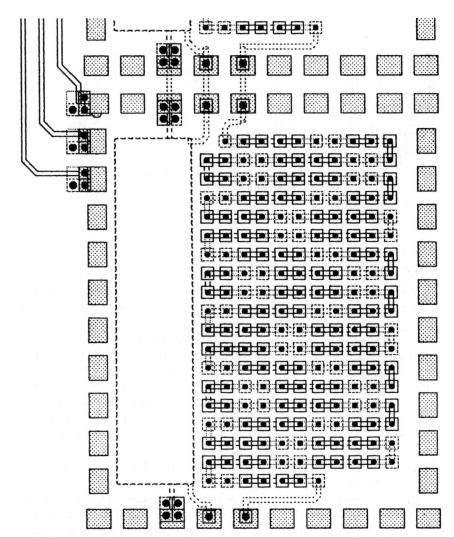

Figure 8-6. Test structure showing via chains (15-µm vias on 60-µm pitch).

easy testing and fault diagnosis. Module yield decreases exponentially with the fraction of bad dice, as shown in Fig. 8-7 for modules containing 4, 8, 20, and 45 dice. The unavailability of fully tested, sorted, and burned-in dice and dice that have been designed for ease of testing has been a major factor in delaying the wider commercial application of MCMs.

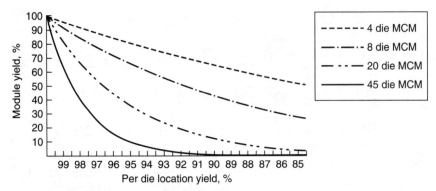

Figure 8-7. Module yield as a function of die yield.

8.2.1 Wafer Sorting and Bare-Die Testing

In the traditional process flow of IC manufacturing, wafers are probe-tested as they come off the end of the line. The wafers are tested by using a stimulus-response IC tester and wafer prober. The dice are connected to the tester through a probe card, usually composed of fine titanium or tungsten needles or metallized ceramic blades epoxied to a printed-circuit board. However, these probe cards do not have low enough parasitic reactances to test high-speed, high-pin-out chips at their full design speed. Some "conventional" probe cards designed to operate in a 50-Ω controlled-impedance environment have been used to test chips at high speed. With extreme care and the ability to tolerate more than 10 percent crosstalk, it is possible[6] to test dice up to 500 MHz. An example of this type of probe card is shown in Fig. 8-8. Note the low profile of the probe wires over the carrier board.

It is much more difficult to perform full-speed testing of CMOS chips that do not have the drive capability to operate in a 50-Ω environment. In practice, a set of "functional" test vectors is used to exercise most of the part for correct logical operation at low speed. In addition, some parametric measurements may be performed that allow the expected speed of a die to be inferred.

The demand for known good dice for MCMs has spurred the development of alternatives to the conventional epoxy ring and needle probe cards. The most promising of these are several varieties of membrane probe cards.[7] Membrane probe cards consist of a flexible circuit board having metal bumps instead of metal needles, similar to a bumped TAB frame. The electrical properties of the flexible circuit can be quite good, particularly if the membrane consists of a low-dielectric-constant film such as polyimide and supports a ground plane and

Figure 8-8. Probe card for IC testing. (*Courtesy of Cerprobe Corp.*)

vias. Impedance can be controlled down to the IC pad at speeds in the gigahertz range. Besides the improved electrical performance, testing ICs with membrane probes is much more benign than testing with needle probes. Gold-dot bumps developed by Packard-Hughes Interconnect produce little damage to IC pads after repeated probing. They also contact a very small area, leaving a large untouched area for wire bonding. Figure 8-9 shows the electrical and mechanical benefits

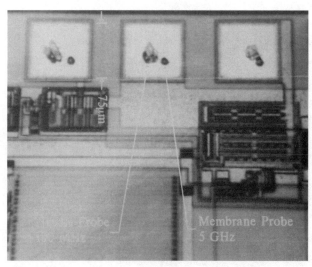

Figure 8-9. Micrograph contrasting indentations in 75-μm IC pad left by titanium needle probes after 100,000 touches and membrane probes after 300,000 touches. (*Courtesy of Packard-Hughes Interconnect.*)

derived from using a membrane probe over a needle probe. A schematic for a two-part membrane probe card manufactured by Packard-Hughes Interconnect is shown in Fig. 8-10, and a scanning electron micrograph showing the morphology and reproducibility of gold bumps, 62 μm high on 100-μm pitch, is shown in Fig. 8-11.

A barrier to wider acceptance of the current generation of membrane probe cards has been the difficulty in obtaining reproducible contacts to aluminum pads. Aluminum pads on ICs are covered with a native oxide of approximately 20 Å. Needle probes scrub across the surface and break through the oxide, whereas most membrane probes

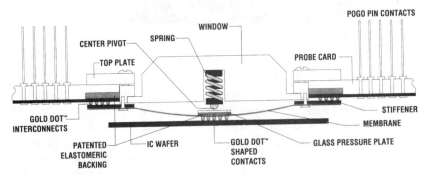

Figure 8-10. Two-part membrane probe assembly. (*Courtesy of Packard-Hughes Interconnect.*)

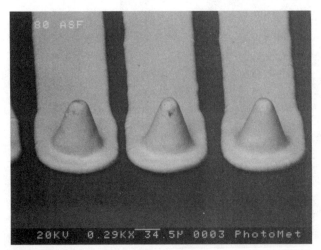

Figure 8-11. Scanning electron micrograph depicting 62-μm-high bumps on a 100-μm-pitch membrane probe. (*Courtesy of Packard-Hughes Interconnect.*)

touch down softly and cannot be scrubbed across the pads. Several techniques have been tried to facilitate penetration, among which are the embedding of small sharp particles in the bumps, sliding the membrane a small distance as it touches down, and arranging the bumps to tilt as they touch down. It is likely that more than one of these methods will prove satisfactory and that membrane probe cards will be used routinely for wafer testing of complex high-speed devices.

Some IC processes currently used or being developed have gold pads either as the default metal system or as an optional final step. Dice that are bumped for TAB are available with gold-covered bumps and work well with gold-bump membrane probes. Adding a gold top layer to the pads, with appropriate barrier metals, is a viable near-term approach while the problem of probing aluminum pads is being solved.

In wafer probing, the locations of dice that have passed are logged. After the dice are separated, they are packaged for burn-in and final testing. Final electrical testing is performed by using an automatic handler to insert and remove the dice from a socket which is connected to the same model tester that was initially used. The electrical characteristics of the socket must be close to those that the packaged part will encounter after final assembly. This permits speed sorting in addition to final functional testing.

Any process for producing or ensuring known good dice must conform to established process flows; otherwise the handling costs will be high. The dice should be handled and tested in automatic handlers for both burn-in and final testing. Fixtures for handling and testing bare dice fall into three groups: small permanent carriers, temporary packages or carriers, and bare-die sockets. TAB is a good example of a small permanent carrier; soft wire bonding into a ceramic pin grid array or in a ceramic carrier (see Testable Ribbon Bonding in Chap. 6) is an example of a temporary package; and bumped membranes are examples of bare-die sockets.

Small Permanent Test Carriers. TAB is one of the oldest methods for testing bare dice and a good example of using a test carrier that permanently stays with the die after assembly. TAB is a well-established technology in the single-metal-layer form. Most commonly the TAB frames are built on a roll of polyimide film such as Du Pont's Kapton of the same size as either 35- or 70-mm photographic film. This has allowed the easy adaptation of well-established technology for handling rolls of film in manufacturing. The inner-lead pattern of the TAB frame matches the pad pitch of the ICs. The outer-lead pattern is fanned out to accommodate the substrate to which the TAB tape will be bonded. The outer-lead pattern continues to the edge of the frame where relatively large pads are provided for the test contact. This fan-out introduces a significant inductance between the contact and the chip which can result in severe ground bounce and crosstalk when chips are tested that have a high number of pins operating at high speeds. However, TAB frames, albeit at a higher cost, can be designed for high-speed testing by fabricating them with ground planes in addition to the signal fan-out layers.

Temporary Test Packages or Carriers. Temporary packages or carriers for testing can easily be adapted to handle small numbers of engineering prototypes or small to medium quantities of production parts. This approach depends on the ability to reproducibly connect to the bond pads so that, after the bonds are removed, the pads are still suitable for final bonding and assembly. One variation consists in attaching the die to a cofired ceramic package with a thermoplastic die attachment adhesive and then wire-bonding, using a schedule that purposely produces bonds of marginal low strength yet that are electrically good. It is important that the die pads be left in a bondable condition after the temporary bonds are pulled off. The packaged parts may be burned in and electrically tested. The bonds for the parts that pass are carefully removed, and the package is heated to the softening temperature of the adhesive to detach the die. The dice are then cleaned, inspected, and committed to MCM assembly.

The testable ribbon bonding (TRB) approach, previously discussed in Chap. 6, is another variation of temporary package testing. According to this process, gold ribbon, longer than normally used, is automatically bonded to the die. The dice are then bonded to a ceramic test carrier according to the normal bonding schedule. Once the dice are burned-in and tested, the ribbon is excised by laser, leaving the permanent bonds on the carrier which are discarded along with the carrier.

Still another variation is to use an easily removed version of General Electric's overlay process.[8] First, the dice are attached to a carrier; then a polymer film is laminated over them, vias are laser-etched, and metal is deposited and photopatterned (Chap. 3). The die pads are patterned so that they fan out to larger pads for the contact. If many dice are to be tested on one carrier, the power, ground, and signals can be bussed in order to gang burn in the dice. Exactly how the metal pattern is laid out depends on the details of the die and test method. If the die has comprehensive self-testing, it may be sufficient to supply power and a clock and to monitor just one pad.

IBM has its own process for testing flip-chip bonded memory chips.[9] The solder bump chips are reflowed on a multilayer ceramic test substrate whose solder pads have been reduced in size to produce weaker bonds than normal. After burn-in and testing, the dice are mechanically sheared off. The good dice are reheated to reflow the solder balls and are then committed to assembly.

Bare-Die Sockets. Currently available bare-die sockets are closely related to the membrane probe cards already described. Typically bare-die sockets consist of a bumped multilayer flexible circuit that has a metal support ring on one side and a photolithographically patterned alignment frame on the other. The dice, which must be sawed to finer than normal tolerances, are aligned by dropping them on the membrane and gently pushing two of the sawed edges against the alignment feature. A clamp is applied to hold the die in place, and the membrane and frame are placed into a carrier socket that can be handled by ordinary burn-in and test equipment (Fig. 8-12). This technique has been developed by MicroModule Systems and Texas Instruments.[10] As with wafer probing, the natural oxide on the aluminum bonding pads can be a persistent problem. The elevated-temperature exposure during burn-in exacerbates the problem by increasing the rate of oxidation and by mechanically stressing the pressure contacts, exposing more of the contact surface to oxidation. The gold-to-gold technique can also be applied to burn in sockets. In this case, it is important to ensure that the barrier metallization has adequate integrity to prevent diffusion during burn-in.

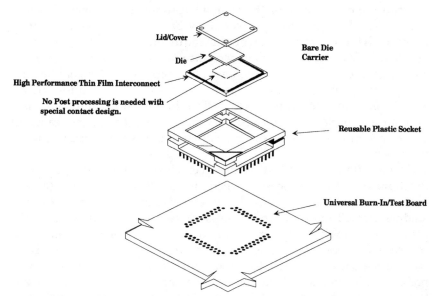

Figure 8-12. Bare-die test and burn-in socket. (*Courtesy of MicroModule Systems.*)

Anisotropic electrically conductive adhesives and elastomeric films are also being evaluated as temporary contacts between the dice and carrier pads (Chap. 6). The films consist of a synthetic polymer filled with conductive particles such as metals or metal-coated polymer microspheres. Although anisotropic films have been qualified as connectors for attaching flexible circuits to rigid printed-wiring boards, it has proved difficult to get reliable connections between the slightly recessed, oxidized IC pads and the carrier pads, at least with the early versions of the material. Active work is in progress to improve the capabilities of these films, and they are likely to become a viable alternative in pretesting dice.

8.2.2 Special Considerations for Dice Intended for MCMs

If a chip set is initially designed to be used only in MCMs, the chip testing problem can be exacerbated while module test problems may very well be reduced. There are two ways in which chip testing may become more difficult for MCMs than for conventional circuits. One way is that chips with area array pads are used in some MCMs. Currently chips may have several hundred pads, but in the near future they are expected to have several thousand pads, which can easily overrun the capabilities of the available probing techniques. The second way is that the output drivers may be scaled to have just barely enough drive current to meet the required system timing when installed in MCMs. The input capacitance of the IC tester and associated fixturing can easily be large enough to slow these drivers beyond the specified timing. If chips are to be specifically designed for MCMs, a comprehensive "design for test" discipline must be enforced. As a minimum, complete register visibility and control, boundary scan, and some form of built-in self-test (BIST) that allows speed sorting should be present.

On-module memory testing involves chips either with processor functions or without processor functions. In many cases where there is processor capability it is possible to exercise memory arrays by first loading a small loop into the processor's registers and then using the processor to exercise the memory. If the processor is defective, of course it will not be possible to test the memory chips. If there is no local processor or if access to the memory array is inconvenient, it is possible to build a small state machine adjacent to the memory arrays to exercise it. This approach becomes attractive and can significantly reduce testing time even in the presence of a local processor if there are multiple large memory arrays present. The state machines can be connected to their own test ports for convenience.

The most common form of BIST is a variation of signature analysis.[11] The generic implementation consists of configuring the boundary scan shift register cells of the input pins of the chip as a linear feedback shift register that can be clocked at least as fast as the chip as a whole. Similarly, the output pins are configured as a linear feedback shift register. The input register and the global clock are run at the first desired test frequency. The outputs from the chip accumulate in the output pins' linear feedback shift register. The value of this "signature" is read and compared to the known correct value. This test may be repeated for each speed bin if more than one speed grade of the part is required. A binary search will quickly find the actual maximum speed of a part if the information is needed for characterization or for process control purposes.

Sometimes it is useful to configure internal latch ranks as linear feedback shift registers in a fashion similar to the method just described. This allows speed verification to be performed on a section-by-section basis. This is useful for debugging and for certain chips, such as frequency counters and synthesizers, where the maximum clock rate is not the same for all sections of the chip.

For best reliability in speed sorting, the initial value of the input register must be carefully chosen to ensure that one or more of the simulated worst-case timing paths are exercised during the test. In general, a different preset will be required for each path to be exercised. This speed test is a statistical exercise of the logic. The vector set is pseudo-random and is not necessarily representative of any actual operating condition that the chip will experience in service. Nonetheless, the probability of a failing chip's producing the correct signature is vanishingly small for any reasonably long sequence. On-board memory should, in general, be exercised by using an algorithmic pattern generator (APG) cell local to the array. This can provide comprehensive test coverage for a small expenditure in area.

The common flow for using the on-board test features to test a chip is to sequentially verify more and more of the chip. The first step is to exercise the test port and prove that it is possible to scan bits through the whole scan ring or rings. The second step is to verify that all inputs can recognize a change of level and that all outputs are capable of producing a change of level. Finally, the functionality of each logic block in the chip is verified, generally by setting the contents of a latch rank, clocking the chip, and reading the contents of the next latch rank. If all combinatorial blocks show correct function, the on-board memory arrays, if any, are exercised. If everything has passed up to this point, the BIST function is used to determine how fast the chip will run.

Besides the features mentioned, other features can be incorporated into a chip to make initial module bring-up easier. Examples include bypass modes for any phase lock loop clock distribution or generation mechanisms, a control register that allows external memory on the module to be reached directly through the control or processor chip pins, and an algorithmic pattern generator for testing external memory on the module. A summary of die testability is given in Table 8-1.

8.3 Module Diagnosis and Testing

Problems encountered in MCM testing are conceptually identical to those encountered in testing a complex printed-wiring board from an edge connector. The methods developed for doing this are closely related to the methods already described for testing ICs from their pins. In both cases it is necessary to make special provisions for testing at the

Table 8-1. Suggested Integrated-Circuit Testability Features

Function block	Testability Feature
I/O cell	Boundary scan
Register	Scan read/write access
Memory array	External read-write access mode Dedicated test state-machine
Cache memory	External access to tag bits External read/write access mode Dedicated test state-machine Disable mode
Error Correction Code memory	Disable mode External read/write of check bits Dedicated test state-machine
ROM	External read access
Logic block	Scan input and output access Pseudorandom pattern generator/checker
Phase-lock loop clock generator	Bypass mode
Integrated circuit	Joint test action group Bring critical signals to pads Pseudorandom pattern generator/checker

design stage to prevent the problem from becoming intractable. As in the case of ICs, one should remember that diagnosis is not the same situation as production testing, where only one bit of information, "good or bad" (pass or fail), is produced. For diagnosis many bits of information must be produced to understand and correct the problem.

8.3.1 Diagnosis

Fault diagnosis is almost always required before the onset of volume production testing. With modern concurrent design methodology, both the module and chips are likely to be new and unproven designs when the first prototypes are built. Few companies have put in place the necessary disciplines so that the first modules built to the initial design are functional. A common scenario involves building a few prototype modules, running them on a tester or installing them in a breadboard, and then finding incomprehensible behavior or no behavior at all. At this point, fabrication errors need to be distinguished from design errors.

Being able to bypass phase-lock loop (PLL) clock circuitry is important in initial debugging. If the clock is erratic, this causes a large variety of nonreproducible problems. That variable is eliminated by forcing the clock directly from a stable test system. If the PLL cannot be disabled, it is important to provide a test point that allows observation of the regenerated clock with an oscilloscope to verify proper behavior. At this point, being able to selectively disable portions of the chips can be helpful. One approach for microprocessor-based modules is to disable the on-chip cache memory and memory management functions and to run the processor core "bare." Turning on the functional blocks in increments can then allow quick isolation of faults and associate them with major functional blocks.

If the module does not display a power-to-ground short circuit or other obvious defect, it is necessary to analyze the module systematically. If boundary scan was implemented, as described above, a pattern should be run to verify that the scan ring is intact. If it is not, the problem is likely to be intractable. At this point it may be appropriate to remove the silicon dice from the module and retest them through the IC test process. Although this may be expensive, it will be even more expensive to proceed with production quantities without a good understanding of the likely failure modes and mechanisms of the module or chip set.

If the scan ring is intact, it is possible to verify the connectivity of the module. This can be done with a small set of vectors. These vectors should be coded so that the results can be interpreted. It is possible to convert the vector output to something like "net 0317 is open" or "nets 1159 and 4483 are short circuited together." If the boundary scan finds a defect in the wiring, the module must be reworked. If the boundary scan indicates that all the wires are correct, it is appropriate to exercise each chip individually through the joint test action group (JTAG) port and other test features such as BIST and on-board APG for memory array testing.[11,12] As with the boundary scan, it is necessary at this stage to provide results that can be interpreted by designers.

If no errors are found during chip testing, verify that the chips can communicate with each other. This can be done by using boundary scan directly or by using the BIST on one chip to drive pseudorandom vectors between chips at full clock speed. If the direct test verifies communications and the BIST produces correct signatures at full speed, the path between the chips tested can be assumed to be good.

If all the above tests prove unsuccessful in diagnosing the problem, the most likely remaining fault lies with the memory chips. Memory chips are often commodity devices that lack any of the test features incorporated in custom-designed dice. Often memory chips can be tested only by using the rest of the module as a tester.

For a module that contains a processor, it is possible to exercise the memory by executing a small program in the processor. Other useful approaches include having special hardware on one of the chips that can directly execute a memory test or having a bypass mode that allows the internal memory chips to be buffered out through the signal pins of one of the chips.

After completion of the procedure described above, it is possible to assemble a module that has good interconnections and good dice and that functions correctly. Even so, the module may still display subtle logic or timing errors that are manifested only under some specific conditions. The problem, then, is no longer with the module, but reverts to one of design verification.

8.3.2 Production Testing

Once the module and chip designs have been verified to operate correctly, production testing must be addressed. Unlike the initial debugging and diagnostic testing, the constraint in production testing is to ensure that a module that passes final test does, in fact, function correctly. This should be done as quickly as possible with the minimum investment of capital equipment and test development effort.

Many tests that were designed for debugging can be reused in production testing. The strategy is much the same except that the vectors no longer require tuning for human interpretation. The usual sequence is as follows:

- Power the module.
- Verify correct supply current draw.
- Exercise the interconnect with boundary scan.
- Exercise each chip with its own BIST and scan.
- Exercise any memory present.
- Perform a final set of full module vectors that are equivalent to running system diagnostics.

For small runs, a general-purpose IC tester or a board tester (for final test) may be used. For large production runs, e.g., over 1000 pieces per month, special-purpose testers are recommended. A particularly interesting case occurs when the module is a processor core, an accelerator, or an I/O device for a computer. In such cases it is quite easy to modify the target system into a tester. In the most straightforward scenario, the module is loaded into the system, and the system runs its normal diagnostic suite. In a more specialized scenario, the system is equipped with an adapter card that allows JTAG to be run on several modules in parallel. Actually running the module in its target system is the most thorough test that can be performed. For example, with workstations, merely booting the operating system will run several billion vectors through a processor core. If the system boots successfully, it can then exercise the module even further by running applications that are known to heavily stress system resources. A sequence for testing commercial multichip modules is shown in Fig. 8-13.

8.3.3 Special Situations

Several special test situations are likely to be common for the foreseeable future. Two examples are the testing of low-cost, "few chip" modules and the testing of modules assembled with chips that have no special test provisions. In both cases somewhat different debugging and test methods than those already described must be used.

Multichip modules having only two chips are representative of the low-cost products likely to be produced and marketed by semiconductor manufacturers. An example is a processor chip packaged with its coprocessor chip in a plastic package. This module will appear to the user as a single packaged chip and will probably be offered and treated as such. These modules must be manufactured by a process closely related to that used to manufacture ICs, including chip attachment,

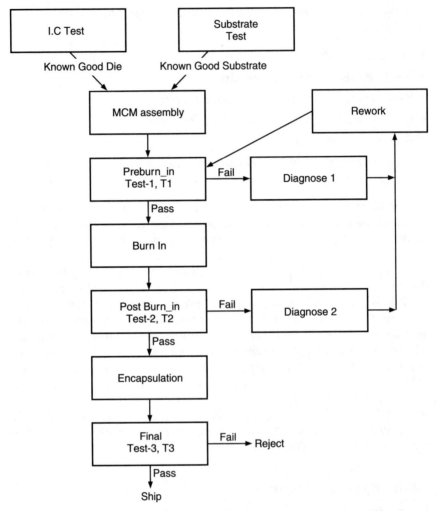

Figure 8-13. Test flow for commercial MCMs.

wire bonding, and plastic encapsulation. To fit into the existing process flow, testing must be made to resemble chip testing. The chip manufacturer has an advantage in already having a complete logic and timing model for the chips and an established test setup. These can be used to generate good coverage ATPG (automatic test pattern generation) patterns that can run on a conventional IC tester. High yields can be realized because the initial functional testing performed at wafer probe is more comprehensive than that usually performed on single chips.

Initial debugging can be performed by assembling several modules without encapsulating them so that their chip pads can be probed. A socketed module is placed in the probe station and exercised either by a benchtop IC tester or by an example of the target system. As described in the section on diagnosis, the module is incrementally exercised. Any unexpected behavior is examined in detail from the module pins and from internal points. This testing approach is equivalent to having both chips in a single-chip package on a printed-wiring board (PWB). For test purposes, once the module netlist and layout have been validated, the module may be treated as a single large chip.

A more difficult situation involves the testability of large MCMs where no special provisions for testing have been made at the chip level. Unfortunately, such modules compose a large fraction of current designs. In these cases, it is helpful to first build a "pseudomodule," i.e., a printed-wiring assembly built from prepackaged dice that is the functional equivalent of the MCM. While this board may not run as fast as the final module, it should display the same functional characteristics so that this board can be debugged by using conventional PWB debugging techniques that depend on having access to all the IC pins. The debugged board version can then be used as the basis for developing the module netlist and layout. Computer checking should be used to verify that the two designs are in fact equivalent.

If the board version were to go into production, it would most likely be tested on a "bed of nails" board tester. These testers have one probe per IC pin and directly verify the integrity of the interconnections. These testers are also capable of isolating each chip by overdriving any inputs that emanate from other chips. This combination of full access and chip isolation has proved effective in manufacturing high-volume products. Converting the board design to an MCM may make many of, or all, the internal connections inaccessible to the tester. In that case, the only feasible test may be to operate the module in the target system.

If a module is built in a "chips first" overlay process (see Chap. 3), small test points can be brought to the surface from all the IC pads. This allows at least continuity testing by using a two-probe, flying-probe tester. When an array probe such as that developed by IBM is available, the equivalent of a bed-of-nails tester can be used. In some cases, an area array pressure connector can be used to access the test pads.

Still another option entails adding extra ICs to the module to facilitate testing. Integrated-circuit chips, for example, may be used to add boundary scan capability to modules that do not have it designed in. A partial boundary scan can thus be performed through a test port. In using this approach, it is important to model accurately the impact on

Table 8-2. Suggested Module Testability Features

Problem	Testability feature
Inaccessible signal	Add test points to module.
	JTAG for module.
	JTAG for each major chip.
	Add scan chips to signal paths.
Memory chips behind logic chips	Add bypass mode to logic chips.
	Add self-test to memory chips.
	Add memory test control logic.

yield of using the extra chips. Maly et al. have suggested an interesting alternative where the scan logic is built into an "active substrate."[14]

In summary, in all cases where test access is difficult after the module has been assembled, it is important that the dice used be tested and known to be good prior to assembly. A combination of known good dice and tight process controls can produce MCMs in high yields and can meet the final test of operating successfully in the final product. Some suggested features that make module testing and diagnosis easier are given in Table 8-2.

8.3.4 Final Screen Testing

A series of screen tests are often applied on a 100 percent basis to all modules after they have passed final electrical tests and are required for high-reliability military and life support systems. The tests are considered nondestructive and are intended to provoke infant mortality failures and to detect workmanship defects as may occur in adhesive attachment, wire bonding, or hermetic sealing. Most of the screen test procedures are defined in Mil-Std-883 or are derived from them. Burn-in under bias is generally performed for both military and commercial modules, but the duration and temperature conditions vary considerably—the more severe conditions are used for the military and space applications. Other screen tests include constant acceleration, temperature cycling, thermal shock, mechanical shock, fine and gross leak testing, particle impact noise detection (PIND), and x-ray inspection. These tests and test specifications were first developed for integrated circuits and then were modified for hybrid microcircuits and are now being used, again with some modifications, for MCMs. Figure 8-14 is an example of a sequence of acceptance screen tests that can be used for assembled multichip modules.

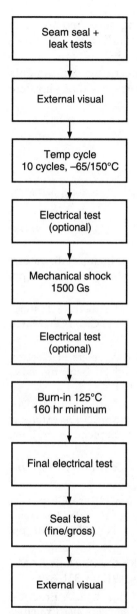

Figure 8-14. Acceptance screen test sequence for multichip modules.

High-reliability commercial modules are being screened by tests specifically designed to reflect actual operating conditions. The automotive industry, for example, has developed screening tests that simulate conditions encountered by under-the-hood electronic components while the medical electronics industry uses screen tests that simulate body conditions.

References

1. R. R. Tummala, "Multichip Packaging—A Tutorial," *Proceedings of the IEEE,* **80**(12); December 1992, pp. 1924–1941.

2. F. C. Chong, MicroModule Systems, Cupertino, CA, personal communication, November 1993.

3. F. Crnic and T. H. Morrison, "Electrical Test of Multichip Substrates," *Proceedings International Conference on Multichip Modules,* Denver, 1993, pp. 422–428.

4. M. Brunner and R. Schmidt, "Electron-Beam MCM Substrate Tester," *Proceedings of the IEEE Multichip Module Conference,* Santa Cruz, CA, March 1993, pp. 62–68.

5. M. Brunner, D. Winkler, and B. Lischke, "Circuit Parameters in Electron Beam Short/Open Testing," in A. Heuberger and H. Benking, eds., *Microcircuit Engineering 84,* Academic Press, London, 1985.

6. E. Subramanian and R. Nelson, "Enhanced Probe Card Facilitates At-Speed Wafer Probing in Very High Density Applications," *Proceedings of the IEEE International Test Conference,* Baltimore, MD, September 1992, pp. 936–939.

7. J. Bond, "Membrane Technology Advances Wafer Probes," *Test and Measurement World,* August 1991, p. 74.

8. R. A. Fillion, R. J. Wojnarowski, and W. Daum, "Bare Chip Test Techniques for Multichip Modules," *Proceedings of the 40th Electronic Components and Technology Conference,* Las Vegas, NV, May 1990, pp. 554–558.

9. J. L. Chu, H. R. Torbai, and F. J. Towler, "A 128 Kb CMOS Static RAM," *IBM Journal of Research and Development,* **35**(3), May 1991, pp. 321–329.

10. R. Roebuck, F. Agahdel, and C. Ho, "Known Good Die: A Practical Solution," *Proceedings of the Second International Conference and Exhibition on Multichip Modules,* Denver, April 1993.

11. P. H. Bardell et al., *Built-in Test for VLSI Pseudo Random Techniques,* Wiley, New York, 1987.

12. IEEE Standard 1149.1, *Standard Test Access Port and Boundary Scan Architecture*, IEEE Standards Board, New York, May 1990.

13. T. C. Russell and Y. Wen, "Electrical Testing of Multichip Modules," chap. 13 in D. A. Doane and P. D. Franzon, eds., *Multichip Module Technologies and Alternatives*, Van Nostrand Reinhold, New York, 1993.

14. W. Maly, D. Feltham, A. Gattiker, M. Hobaugh, K. Bakus, and M. Thomas, *Multichip Module Smart Substrate System*, CMU-SRC Report CMUCAD-93-61, Carnegie Mellon University, Pittsburgh, PA, September 1993.

9
Rework and Repair

It is, of course, desirable not to have to repair or rework any module. With the maturity of any process, quality workmanship, high production, and automation, rework can be virtually eliminated. For the low-end consumer multichip modules (MCMs), rework is usually considered not cost-effective, and any nonfunctioning parts are treated as throwaways. However, because of the complexity, size, and high cost of very large-scale integrated-circuit (VLSI) dice, high-density MCMs, especially the larger MCM-D types, cannot be considered throwaways. Even with the assurance of using known good pretested dice, failures may occur during assembly, handling, and testing; and reworking the modules, even several times, is considered cost-effective.

Reworking high-density MCMs is more difficult and riskier than reworking hybrid circuits for the following reasons:

1. MCMs contain more active and more expensive dice per unit area than hybrids do. The ratio of silicon die area to substrate area for MCMs can be 80 percent or greater, thus increasing the probability of one or more faulty dice having to be replaced. In MCMs there is also a greater propensity for wires and wire bonds to be damaged during handling and testing. A 2×4 in MCM, for example, may contain over 40 large integrated circuits (ICs) interconnected by 2000 to 3000 wire bonds.

2. The much larger dice used in MCMs require greater force to remove because of their larger surface areas. The application of a larger force to dislodge the dice risks damaging both the adjacent good dice and the substrate. Custom dice are now approaching 1-in-square, which, not too long ago, was the size of an entire substrate.

3. Most high-performance MCMs, such as the MCM-Ds, are fabricated by using polyimide dielectric. Polyimides or other organic

dielectrics are softer and more easily damaged than their ceramic counterparts used in multilayer thick-film circuits.

4. The conventional process of removing a die by applying a lateral shear or twist force to the edge of the die while heating the entire module is riskier in MCMs because the dice are closely spaced, leaving little room for the shear tool.

9.1 Rework of Adhesive-Attached Dice and Substrates

A general procedure for removing faulty dice that are attached with adhesive involves first excising the leads (wire, TAB, or ribbon) and then reducing the strength of the adhesive so that the die can be mechanically separated. The use of specially designed tools to grip the dice by the edges and then dislodge the dice by applying a torque or shear force or both while simultaneously heating the module is a generic method that has been used successfully for many years to detach epoxy-attached ICs. Localized heating can be applied by forcing a jet of hot air or nitrogen gas to the tops of the dice. Even so, the entire module should be heated to between 80 and 150°C to accelerate the process. Rework procedures are generally manual and require skill to avoid damaging other parts of the module. Heating should be controlled to avoid thermal degradation of wire bonds, devices, and materials.

This technique, although successful for reworking hybrid microcircuits and some MCMs, cannot be used for the very high-density MCM-D where there is little space between the dice for inserting a tool and where the hot gas can affect the integrity of neighboring closely spaced chips. In such cases, attaching a thermode to the top of the die with an adhesive having a higher strength and higher softening temperature than that used to originally attach the die, then heating and simultaneously applying a tensile and torque force through the thermode, has been demonstrated to be a benign procedure[1,2] (Fig. 9-1). Devices as large as 0.580 in square can be removed and replaced up to five times on the same mounting pad.

Although chips attached by using the conventional thermosetting epoxy adhesives can be removed by the thermode method, many companies are converting from thermosetting adhesives to thermoplastic adhesives to facilitate removal. Unlike thermosetting adhesives, which soften only gradually after they reach their glass transition temperature and still retain much of their strength (Fig. 9-2) thermoplastic adhesives melt over a narrow temperature range and lose all their

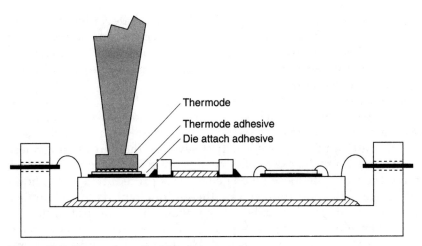

Figure 9.1 Thermode method for die removal.

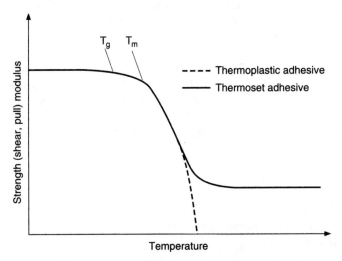

Figure 9.2 Relative strengths of thermosetting and thermo-plastic adhesives. T_m = melt temperature of thermoplastic adhesive; T_g = glass transition temperature of thermosetting adhesive.

strength. Thermoplastic adhesives are also suited to fast-turnaround production processing since they liquefy quickly at their melting temperatures and solidify rapidly on cooling, avoiding hours of curing in controlled-atmosphere ovens.[3]

After a die has been removed, the bonding site must be cleaned to remove adhesive residues. Again, this is more difficult for thermosetting adhesives than for thermoplastics. In fact, whatever thermoplastic residues remain will remelt and fuse with newly applied adhesive. For thermosetting adhesives, solvent cleaning alone is not sufficient to remove residues—some mechanical scrubbing with a nylon brush, e.g., may be necessary.

There are occasions when it is also desirable to salvage a substrate already assembled in a package, e.g., if the package is a fine or gross leaker and the costs of the substrate and dice are high. In such cases, the substrate can be removed by first severing the wires that connect the substrate to the package pins, then heating the entire module, and prying the substrate loose by one edge. This procedure is risky, however, since if the substrate cracks, the entire circuit is lost. Alumina substrates, being mechanically stronger than silicon substrates, stand a better chance of surviving this procedure. Alternative methods have been reported, but none are widely used by the industry. One approach involves sliding a heated wire, through which current is passed, between the substrate and the package, softening the adhesive, and slicing through the adhesive. Another approach is to provide a resistance heater on the underside of the substrate to generate additional localized heat at the adhesive interface.

Adhesive selection is a key factor in the ease and success of reworking a substrate. However, there is a tradeoff—an adhesive that is more easily reworked may not provide the strength required to pass certain constant-acceleration and thermal-cycling tests. As with die attachment, thermoplastic adhesives, especially in film form, may provide the best compromise.

9.2 Rework of Metallurgically Attached Dice and Substrates

In some high-power modules, both dice and substrates must be metallurgically attached with either low- or high-temperature melting solders to maximize thermal transfer. High-temperature solders such as gold-germanium, gold-tin, or gold-silicon melt at temperatures higher than 320°C, while soft solders are low-melting-temperature alloys (<300°C) including most lead-tin, indium-tin, and lead-tin-silver alloys. These solder alloys are generally applied as preforms. However, mechanical scrubbing of silicon die-to-gold surfaces forms a silicon-gold eutectic bond without requiring an alloy preform. Silicon-gold eutectic attachment is widely used to attach small dice in single-chip packages. Devices attached with high-temperature melting solders are generally more difficult to rework due to the high temperatures involved, not only in detaching the die but also in reapplying and reprocessing the new solder. On the other hand, chips attached with low-temperature melting solders, with certain precautions, can be easily reworked.

Rework of chips attached with silver-glass is not practical because of the high softening temperature of the glass (>400°C) and the potential thermal damage to the circuit. Silver-glass has been limited to the attachment of dice in single-chip, throwaway packages.

In assembling a module, a hierarchy of solders with progressively lower melting temperatures should be used; e.g., a substrate should be attached with a higher-temperature melting solder than the die, otherwise the temperature required to detach the die would also detach the substrate. Solder-attached dice are not as easily removed as epoxy-attached dice, since rework temperatures for solders are much higher than those for epoxies. Cleanup of reworked chip sites is more difficult, since it requires removal of old solder alloy and reapplication of fresh solder. In most cases, flux or a reducing atmosphere are necessary to ensure good wetting. Repeated reworking at high temperatures also risks extended exposures of bimetallic wire bonds (Au-Al) with consequent degradation of strength and electrical conductivity.[4]

9.3 Rework of Flip-Chip, Solder-Attached Dice

IBM has been using flip-chip attached dice on a production scale in its mainframe computer modules since 1964, when IBM first introduced the C(4) process (see Chap. 6). IBM has developed rework procedures whereby the solder-bump dice are easily removed either mechanically or by localized heating and melting of the solder balls. After chip removal, residual solder on the substrate pads is removed by a process called *dressing*. Solder accumulation at sites which experience multiple replacements can cause short circuits between adjacent pads. Solder dressing is also necessary to ensure joint integrity by maintaining a solder volume that is optimum for thermal-fatigue life. Excess solder may be removed by wicking onto a material such as porous copper having a high wettability for solder. It is reported that the reliability of reworked solder-attached flip-chip dice is as good as that of the originally attached dice.[5]

9.3.1 Mechanical Methods

Mechanical methods for removing flip-chip attached dice involve grip-ping the die at the edges with a tool coated with Teflon or other poly-meric material to avoid damaging the chip (Fig. 9-3). Applying a torque force both clockwise and counterclockwise will separate the solder in a random fashion.[6] Although mechanical removal is the simplest and most direct method, there is a limit to how large a chip can be removed without damaging adjacent dice or the sublying substrate pads. Chips having 300 or more solder ball connections tend to fracture at corners due to the higher force required for their removal. Purely mechanical methods also require sufficient space around the die which is not always the case with densely packed dice. Thus for high-density MCMs, thermal methods or mechanical methods assisted by thermal or ultrasonic energy are more widely used and are less risky.

9.3.2 Thermal and Mechanical Methods

Combining thermal and mechanical methods facilitates removal of large flip-chip bonded dice, as it does for wire-bonded adhesive-attached dice. The heat must be localized so that only the solder balls of the die to be removed are melted. Heat may be applied by infrared radiation from a light source or by a focused jet of hot inert gas. In general, heat is focused on the back of the die. A careful thermal bal-ance must be maintained to prevent melting the solder balls of adja-cent dice. This is accomplished by preheating the entire module to a temperature below the melting temperature of the solder balls and then focusing heat on the die to be removed. When the solder balls melt, the chip is removed by using a vacuum probe. Preheating the entire module is also a good practice since this reduces the thermal

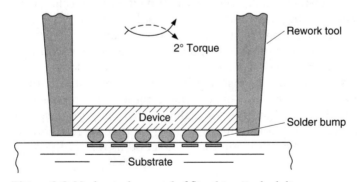

Figure 9.3 Mechanical removal of flip-chip attached die.

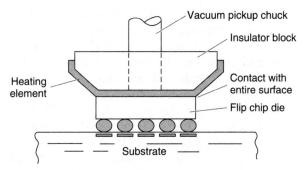

Figure 9.4 Thermal conduction method for flip-chip die removal.

gradients among the various materials of construction of the MCM, thus minimizing stresses during rework.

Another thermal method involves contacting a heating element to the back side of the die, thus transferring heat to the solder balls by conduction. Again, once the solder melts, a vacuum chuck is activated to pick up the die (Fig. 9-4). Approximately half the solder of each solder ball remains on the substrate pads and must be removed. To dress the solder, a porous copper block similar in size to the chip is pressed onto the chip site and heated. As the solder melts, it wicks into the copper, leaving behind a planar thin layer of solder.[7]

9.4 Rework of TAB Attached Dice

The removal and replacement of TAB attached dice are somewhat different from those of flip-chip or wire-bonded dice and can be more difficult. In removing TAB dice, one must contend with not only the removal of an adhesive (as with wire-bonded dice) but also the greater strength of the wider TAB leads and the limited freedom in outer-lead rebonding to another site. TAB leads may be bonded by solder reflow or by thermocompression or thermosonic bonding. Whether the die is bonded face-up (normal TAB) or face-down (flip-TAB), a thermally conductive adhesive is used to attach the die to the substrate prior to making the TAB connections. Therefore, die removal involves two steps: removal of the TAB leads from the substrate and detachment of the die. In removing the outer-lead bonds, different techniques are used depending on whether the leads are solder-attached or thermocompression or thermosonically bonded.

Removal of thermosonically or thermocompression-bonded TAB leads without damaging the substrate bond pads is difficult. Of major concern in detaching these leads is their high strengths (30 to 60 g), the high shear forces required, and consequently the risk in damaging the bonding sites on the interconnect substrate and even the substrate itself. One procedure involves shearing the leads with precisely controlled miniature shearing tools. The process consists of two steps: (1) The TAB leads are cut or sheared at the die, and the die is removed, leaving the outer TAB lead stubs attached to the substrate. (2) The stubs are sheared off the substrate by an automatic wire bonder with a chisel-shaped shearing tool. The process requires precise control of the height and movement of the tool pushing and peeling each TAB lead from the die site without touching or damaging the substrate. Best results are obtained by tacking down the TAB outer-lead bonds at initial bonding, using lower-than-optimum bonder power. After the MCMs are electrically tested, faulty chips can be readily removed and the TAB lead stubs sheared without damaging the substrate bonding sites. After rework is completed and the circuit is functioning properly, the TAB outer leads are rebonded according to the optimum bonding schedule.

Several alternate approaches have been described. One consists in cutting the leads of the new die shorter and rebonding in-board. This, however, requires precision shortening of the leads of the replacement die by using special excising tools. A second approach consists in leaving the remnant of the outer lead, bonding it more securely, then using this as an overbonding site for the new die. The replacement film adhesive must be thicker (approximately 1.5 mils) to compensate for the extra thickness due to the remnant lead (Fig. 9-5).[8]

In removing solder-attached, fine-pitch (< 8 mils), outer-lead bonds (OLBs), tools have been designed that will conduct heat to the bonds. Heat should be delivered quickly and locally to the die to avoid inadvertent separation of adjacent dice. After the solder melts, the leads are detached and excess solder remaining on the pads is removed by wicking onto a copper mesh. Next an adhesive-attached thermode is used to soften the die-attachment adhesive and to mechanically detach the die from the substrate. The procedure and equipment used are similar to those previously discussed and represented in Fig. 9-1. As with other rework methods already described, selective application of heat to the die being removed is critical to success, and thermoplastic preform adhesives are more easily removed than thermosetting types. Adhesives may be filled with aluminum nitride, diamond, or beryllia to improve their thermal conductivities.

Cleanup of residual adhesive and solder is necessary before a new

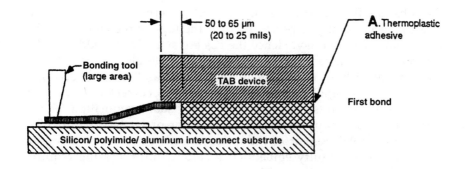

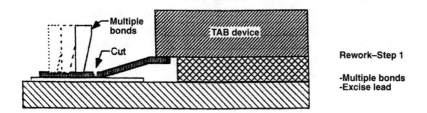

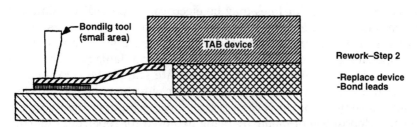

Figure 9.5 Rework of TAB leads using the initial lead, overbonded to the substrate bond pad, as the new bond pad[8].

die is reattached. Excess solder can be removed by wicking onto a copper mesh that is part of the OLB detachment tool.[9]

9.5 Rework of HDI Overlay Modules

The General Electric high-density interconnect (HDI) process presents a unique situation for repair and rework since the chips are attached to the substrate first, before the interconnect patterns are fabricated. The interconnect circuitry is processed over the tops of the chips on polyimide film as an overlay. Hence the ICs are actually embedded below the interconnect layers, making the chips inaccessible. General Electric has developed both in-line repair procedures and repair procedures after full processing. In-line repair is used to correct metallization open or short circuits which are found by visual examination. This is done either by reapplying the metal layer or by laser-controlled removal of metal. Repair of the interconnection overlay structure or replacement of a faulty embedded IC after full processing requires complete removal of the overlay structure. Removal of the overlay is accomplished by first heating the structure to 260°C to soften the thermoplastic adhesive and then peeling away the overlay film. A selective etching of the copper and solvent cleaning of the adhesive restore the substrate to its original condition. A defective chip is removed by again heating the substrate to soften the chip-attachment adhesive, also a thermoplastic. After the chip is replaced, the entire overlay interconnect structure must be recreated.[10]

9.6 Repairing Conductor Lines

Occasionally conductor lines may be electrically open (discontinuous) due to faulty photoetching (poor photoresist coverage, defects in the photoresist due to particle contaminants, or excessive undercutting during etching) or due to inadvertent mechanical damage. If the damage is extensive, the interconnect substrate should be scrapped. If minor, and with customer approval, jumper wires may be used to electrically close the gap. However, several novel techniques involving the selective deposition of metal are being explored for repairing thin conductor traces. A unique method involves coating the faulty area or the entire surface with a solution of a metalloorganic compound, drying to evaporate the solvents, and then exposing to a focused laser or to an electron beam to decompose the metalloorganic coating to the pure metal, thus bridging the gap. The unexposed coating is then washed away. Commercially available metalloorganic solutions such as the organic complexes or salts of gold, copper, nickel, and silver may be used (see Chap. 3). Copper, for example, can be deposited from aqueous or from aqueous glycerol solutions of copper formate by exposure to the Nd YAG laser beam.[11] This technique can be used to rework conductor traces of both thin- and thick-film circuits, either before or

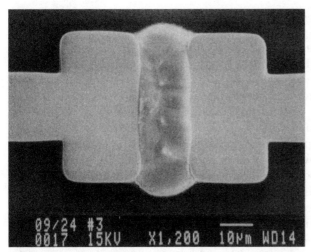

Figure 9.6 Open circuit in a gold conductor bridged with copper laser-deposited from copper formate. (*Courtesy of R. Miracky, MCC.*)

after assembly. The resulting single-film thicknesses range from 500 to 2500 Å but, if necessary, may be increased by repeating the process[12] or by selective electroless plating. Electrical sheet resistances of 40 mΩ/square[13] and resistivities[14] below 10 μΩ•cm have been reported. In gold conductor lines, 10- to 20-μm-wide gaps have been successfully bridged by laser-depositing copper from a copper formate solution (Fig. 9-6). The thickness of the copper deposits (4 to 6 μm) provided sufficient conductivity without subsequent plating.

A variation of this process involves first patterning a palladium catalytic seed layer, then electroless plating with copper or gold. A palladium-doped coating is deposited on the surface and irradiated by a programmed laser beam in selected areas, and the unirradiated material is washed away. This process has also been reported successful in fabricating fast-turnaround interconnect substrates[15] (see also Chap. 3).

Still another method, called *laser chemical vapor deposition* (LCVD), employs a focused laser beam to decompose and deposit a metal film from one or more gaseous compounds. In one example, metallic tungsten was deposited by chemically reducing gaseous tungsten hexafluoride in the presence of hydrogen adsorbed on a surface,[16] as follows:

$$WF_6 + 3H_2 \rightarrow W + 6HF$$

The laser was focused through a window of a vacuum chamber containing the part and the gases. The reduction reaction occurred only in the laser-exposed regions of the substrate. As with the metalloorganic process, very thin films are produced and often require thickening to get the required electrical conductivity. Two-micron-thick films have been produced by double processing.[17]

Electrical open circuits in thick-film conductors may be repaired by touching up with conductor paste and refiring, but this procedure can be performed only prior to assembling the chips due to the high firing temperatures involved. Applying conductor paste by direct writing (see Chap. 3) is a precise method for bridging open conductor traces or constructing a missing trace.

Adjacent conductor lines may also be bridged with metal, resulting in electrical short circuits. This may again be due to faulty photoresist processing or to metal migration. Bridged metal conductors are easier to repair than open conductors. The bridged metal can be removed by either mechanical scribing or laser ablation. In either case, care must be taken to not damage the underlying dielectric or conductor traces. Figure 9-7 shows a 15-μm-wide gold conductor line on a polyimide substrate that was laser-cut without damaging the underlying polyimide.

9.7 Delidding and Relidding Hermetically Sealed Packages

Hermetically sealed hybrid or multichip module packages sometimes fail during burn-in or thermal and mechanical screen tests. Two of the most common failure modes are bad dice and lifted or damaged wire bonds. Depending on the value of the module, it may be scrapped or reworked. If the circuit is to be reworked, first the seam-welded metal lid must be removed without damaging the circuit. The sealing surface also must not be damaged so that a new lid can be subsequently welded. Although lids can be removed manually by carefully grinding them off on a grinding machine, lids are best removed by using commercially available precision cover removers. The delidding machine consists of a circular silicon carbide saw blade having several hundred teeth. As the saw cuts into the metal at about 75 r/min, the cuttings are vacuumed and collected, preventing them from depositing onto the circuit. A slot is cut at a preset depth wider than the weld area but not as wide as the package wall. Generally, if the wall width is 40 mils, the width of the cut is 30 mils inward from the edge. This procedure also prevents metal particles from falling into the package.

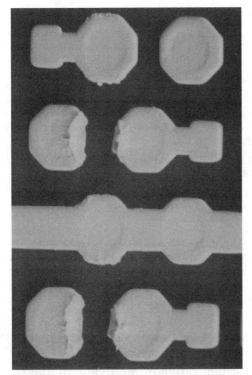

Figure 9.7 Scanning electron micrograph of laser-cut conductor lines, 15 µm wide. (*Courtesy of R. Miracky, MCC.*)

Packages up to 3-in-square may be delidded with the standard equipment, but up to 5-in-square packages may be delidded with special tooling. After delidding, diagnosing the failure, and correcting the problem, a new lid is welded, following all the procedures used in the initial sealing. Should the module fail again, it may be delidded and resealed again, but of course there is a point of diminishing return where it will be cheaper to scrap the module. Some space and high-reliability programs will not allow any delidding, although most programs allow up to three deliddings.

References

1. C. S. Iwami, J. J. Licari, and C. T. Nakagawa, "Adhesive and Device Removal Requirements for Multichip Modules," *Proceedings Fifth SAMPE Conference*, Los Angeles, June 1991.

2. U.S. Patent 5164037, Apparatus for removing semiconductor devices from high-density multichip modules, November 17, 1992.

3. A. Shores, "Adhesive Bonding Hybrid Microcircuit Substrates with a Thermoplastic Film," *SAMPE Quarterly*, **19**, April 1988.

4. G. Harmon, *Wirebond Reliability and Yield*, International Society of Hybrid Microelectronics, Reston, VA.

5. K. J. Puttlitz, "An Overview of Flip-Chip Replacement Technology on MLC Multichip Modules," *International Journal of Microcircuits and Electronic Packaging*, **5**(3), 1992.

6. R. D. McNutt and K. Schink, "Chip Removal and Chip Site Dressing," *IBM Tech. Disc. Bull.* **24**(2):1288–1289, July 1981.

7. S. K. Ray, K. Beckham, and R. Master, "Flip-Chip Interconnection Technology for Advanced Thermal Conduction Modules," *Proceedings IEEE*, 1991.

8. L. A. Hughes et al., "Improved Die Attach and Rework Techniques for MCM-D Circuits Using Thermoplastic Adhesives and Flip-TAB Devices," *Proceedings International Conference and Exhibition on Multichip Modules*, Denver, April 1993.

9. M. Bertram, "Repair Method for Solder Reflow TAB OLB," *International Tape Automated Bonding Symposium*, 1990.

10. R. A. Fillion and C. A. Neugebauer, "High Density Multichip Packaging," *Solid State Technology, Hybrid Supplement*, September 1989.

11. R. Miracky et al., "Laser Customization of Multichip Modules," *Proceedings GOMAC*, November 1991.

12. J. A. Pacinelli, "Metallo-Organics: Prior Technology Becomes State-of-the Art," *Circuits Manufacturing*, March 1987.

13. C. Howlett et al., "A Thick Film Solution to High-Speed, High-Density Multilayer Interconnection," *Hybrid Circuit Technology*, April 1990.

14. H. G. Muller, "Photothermal Formation of Copper Conductors on Multichip Module Substrates Using a Nd:YAG Laser," *IEEE Transactions on Components, Hybrids, and Manufacturing Technology*, **16**(5), August 1993.

15. T. Hirsch, R. Miracky, and S. Sommerfeldt, "Laser Customization of a Multichip Module Substrate," *Transactions of the IEEE*, **1**(3), September 1993.

16. R. F. Miracky, "Selective Area Laser Assisted Processing for Microelectronic Multichip Interconnect Applications," *Proceedings Materials Research Society Symposium*, **129**: 547–558, 1989.

17. R. Miracky, Microelectronics and Computer Technology Corp., Austin, TX, personal communication, January 5, 1994.

Conversion Factors

Thermal Conductivity Units and Equivalents

Cal/(s•cm•°C)	BTU/(h•ft•°F)	W/(m•K)	W/(cm•°C)	W/(in•°C)
1	241.9	418.6	4.186	10.63
4.13×10^{-3}	1	1.73	0.0173	0.044
2.39×10^{-3}	0.578	1	0.01	0.0254
0.239	57.8	100	1	2.54
0.094	22.74	39.4	0.394	1

Pressure and Force Conversion Factors

Multiply	By	To obtain
Pascals	1.45×10^{-4}	Pounds/in^2
Pounds	4.44	Newtons
Dynes/cm^2	1.45×10^{-5}	Pounds/in^2
Newtons/mm^2	1×10^{-3}	Gigapascals

Dimension Units and Equivalents

Mil	Inch	Centimeter	Millimeter	Angstrom	Micrometer	Microinch
1	0.001	0.00254	0.0254	254,000	25.4	1000
1000	1	2.54	25.4	2.54×10^8	25,400	1.0×10^6
394	0.394	1	10	1.0×10^8	1.0×10^4	3.94×10^5
39.4	0.0394	0.1	1	1.0×10^7	1.0×10^3	3.94×10^4
3.94×10^{-6}	3.94×10^{-9}	1.0×10^{-8}	1.0×10^{-7}	1	1.0×10^{-4}	3.94×10^{-3}
0.0394	3.94×10^{-5}	1.0×10^{-4}	0.001	10,000	1	39.4
0.001	1.0×10^{-6}	2.54×10^{-6}	2.54×10^{-5}	254	0.025	1

Abbreviations, Acronyms, and Symbols

Å	Angstrom ($= 1 \times 10^{-8}$ cm)
Ag	Silver
Al	Aluminum
ALU	Arithmetic logic unit
APG	Algorithmic pattern generator
ARPA	Advanced Research Projects Agency
ASCM	Advanced Spaceborne Computer Module
ASEM	Application-specific electronic module
ASIC	Application-specific integrated circuit
ASTM	American Society for Testing and Materials
ATPG	Automatic test pattern generation
Au	Gold
BCB	Benzocyclobutene
BIP	Bonded interconnect pin
BIST	Built-in self-test
BT	Bismaleimide triazine
C	Capacitance
C_m	Mutual capacitance per unit length of coupled transmission line
C_0	Capacitance per unit length of transmission line
C(4)	Controlled collapse chip connection

CAD	Computer-aided design
CAE	Computer-aided engineering
CAM	Computer-aided manufacturing
CCT	Cooper & Chyan Technology
CIM	Computer-integrated manufacturing
CLA	Centerline average
CMOS	Complementary metal oxide semiconductor
COB	Chip on board
cP	Centipoise
CPU	Central processing unit
CTE	Coefficient of thermal expansion
CVD	Chemical vapor deposition
CVI	Chemical vapor infiltration
dB	Decibel
DEMUX	Demultiplexer
DIP	Dual in-line package
DFT	Design for test
DRAM	Dynamic random access memory
DRC	Design rule checking
DSC	Differential scanning calorimetry
ECL	Emitter coupled logic
EMI	Electromagnetic interference
F_p	Footprint dimension of chip
FEM	Finite element modeling
FEP	Fluorinated ethylene propylene
FR-4	Fire retardant (epoxy glass-reinforced laminate)
g	Acceleration due to gravity
GHz	Gigahertz
GPa	Gigapascal
HAST	Highly accelerated stress test
HDI	High-density interconnect
HDL	Hardware description language
HDMI	High-density multichip interconnect
Hi-TEA	Hierarchical Test and Economics Advisor
HPA	High-power amplifier
HTCC	High-temperature cofired ceramic

I_{ground}	Current flowing in a ground lead
I_{signal}	Current flowing in a signal lead
IC	Integrated circuit
ILB	Inner-lead bonding
I/O	Input/output
IPC	Institute of Interconnecting and Packaging Electronic Circuits
ITO	Indium tin oxide
J	Joule
JTAG	Joint test action group
k	Dielectric constant
KGD	Known good die (dice)
Klb/in²	1000 lb/in²
L	Inductance
L_{eff}	Effective inductance of a ground lead
L_m, L_{mutual}	Mutual inductance per unit length of coupled transmission line
L_0, L_{self}	Self-inductance per unit length of transmission line
LSI	Large-scale integration
LTCC	Low-temperature cofired ceramic
mA	Milliampere
MBE	Molecular beam epitaxy
MCM	Multichip module
MFLOP	Million floating operations
MHz	Megahertz
MMIC	Monolithic microwave integrated circuit
MIPS	Million instructions per second
MOD	Metalloorganic decomposition
MPa	Megapascal
MSFC	Marshall Space Flight Center
MTBF	Mean time between failure
MTTF	Mean time to failure
MUX	Multiplexer
Nc	Number of chips in a module
Nd	Neodymium
nH	nanohenry
nm	nanometer
NMP	N-methyl pyrrolidone

ns	nanosecond
ODA	*ortho*-dianiline
OLB	Outer-lead bonding
pA	picoampere
PCB	Printed-circuit board
Pd	Palladium
PDSE	Packaging design support environment
PECVD	Plasma-enhanced chemical vapor deposition
pF	picofarad
PI	polyimide
PLL	Phase-lock loop
PMDA	Pyromellitic dianhydride
POLYHIC	Polymer hybrid interconnect (trade name of AT&T)
ppm	parts per million
Pt	Platinum
PTF	Polymer thick film
PTFE	Polytetrafluoroethylene
QFP	Quad flat pack
QTAI	Quick-turnaround interconnect (trade name of MCC)
R	Resistance
$R_c(t)$	Probability of correct component operation at time t
R_{on}	On resistance of an output stage
$R_s(t)$	Probability of correct system operation at time t
RAM	Random access memory
RC	Resistance-capacitance
RH	Relative humidity
RIE	Reactive ion etching
RISC	Reduced instruction set computer
RT	Room temperature
RTL	Register travel level
RwoH	Reliability without hermeticity
SAM	Standard avionics module
SCOB	Sealed chip on board
SDIO	Strategic Defense Initiative Office
SEM	Standard electronics module; also scanning electronic microscope

SLA	Scanning laser ablation
SLC	Surface laminar circuit (trade name of IBM)
SMT	Surface mount technology
SOI	Semiconductor on insulator
SRAM	Static random access memory
SURF	Santa Cruz ULSI routing framework
t	Time
$t_{charging}$	RC charging delay
$T(g)$	Glass transition temperature
t_{tof}	Time-of-flight delay
T_m	Melting temperature
TAB	Tape automated bonding
TCM	Thermal control module
Ti	Titanium
TFE	Tetrafluoroethylene (Teflon)
TGA	Thermogravimetric analysis
TMA	Thermomechanical analysis
T/R	Transmit/receive
TRB	Testable ribbon bonding
μ_0	Permeability of free space
μm	Micrometer (micron)
ULSI	Ultra-large-scale integrated
V_{noise}	Switching noise magnitude
VHSIC	Very high-speed integrated circuit
VLSIC	Very large-scale integrated circuit
W	Tungsten
WSI	Wafer-scale integration
YAG	Yttrium aluminum garnet
Z_0	Characteristic impedance of transmission line
Z_{0e}	Even-mode impedance of coupled transmission line system
Z_{0o}	Odd-mode impedance of coupled transmission line system

Index

About the Author

James J. Licari, an internationally recognized authority on materials and processes for electronics, is a consultant and Chief Scientist at AvanTeco Corp. in Whittier, California. He was previously Program Manager and Chief Scientist at Hughes Aircraft and Laboratory Manager at Rockwell International, Autonetics Division. Dr. Licari has more than 30 years of materials research and production engineering experience, including new product development projects at American Cyanamid, American Potash and Chemical Company, and Remington-Rand Laboratories of Advanced Research. He holds seven patents, and is the author of more than 100 publications, including three other books: *Plastic Coatings for Electronics, Hybrid Microcircuit Technology Handbook,* and *Handbook of Polymer Coatings for Electronics.*